T0301979

Energy Project Financing:
Resources and Strategies for Success

Energy Project Financing: Resources and Strategies for Success

Albert Thumann, P.E., C.E.M.
Eric A. Woodroof, Ph.D., C.E.M., CRM

Published 2020 by River Publishers
River Publishers
Alsbjergvej 10, 9260 Gistrup, Denmark
www.riverpublishers.com

Distributed exclusively by Routledge
4 Park Square, Milton Park, Abingdon, Oxon OX14 4RN
605 Third Avenue, New York, NY 10017, USA

Library of Congress Cataloging-in-Publication Data

Thumann, Albert.
Energy project financing : resources and strategies for success / Albert Thumann, Eric A. Woodroof.
p. cm.
Includes index.
ISBN 0-88173-597-3 (alk. paper) -- ISBN 978-8-7702-2273-0 (electronic) -- ISBN 978-8-7702-2393-5 (distribution by taylor & fancis : alk. paper)
1. Energy conservation--Finance. 2. Industries--Energy conservation--Finance. I. Woodroof, Eric A. II. Title.

HD9502.A2T5187 2008
658.2'6--dc22

2008017540

Energy Project Financing : Resources and Strategies for Success / Albert Thumann, Eric A. Woodroof.
First published by Fairmont Press in 2009.

Routledge is an imprint of the Taylor & Francis Group, an informa business

0-88173-597-3 (The Fairmont Press, Inc.)
978-8-7702-2393-5 (print)
978-8-7702-2273-0 (online)
978-1-0031-6932-1 (ebook master)

Foreword

The landscape for implementing energy efficient projects is rapidly changing. The need for energy project financing has never been greater. The factors influencing energy project financing have been brought about by legislation, oil prices surging past $120 a barrel, and the growing concern for global warming.

In December of 2007, the Energy Independence and Security Act was passed into law. This act promotes energy savings performance contracting in the federal government, and provides flexible financing and training of federal contract officers. The Energy Policy Act of 2005 reauthorizes energy service performance contracting through September 30, 2016.

The purpose of this book is to provide the key success factors for structuring a finance energy project and getting it approved by top management. The goals of the authors are threefold: First, we want to explore as many financing options as possible. Second, we want to provide the tools to make a comprehensive financial analysis. Third, we want to broaden the readers' horizons with new trends in the industry.

There are many correct ways to assemble and finance an energy management project. The number of possibilities is only limited to one's creativity. So be flexible and keep searching until you find the "win-win" deal for everyone.

Albert Thumann, PE, CEM

Table of Contents

Contributors

CHAPTER 1, 2, 6, 8, & 9

Eric A. Woodroof, Ph.D., CEM., CRM, shows clients how to make more money and simultaneously help the environment. During the past 15 years, he has helped over 250 organizations improve profits with energy-environmental solutions. He has written over 25 professional journal publications and his work has appeared in hundreds of articles. Dr. Woodroof is the chairman of the board for the Certified Carbon Reduction Manager program and he has been a board member of the Certified Energy Manager Program since 1999. Dr. Woodroof has advised clients such as the U.S. Public Health Service, IBM, Pepsi, Ford, GM, Verizon, Hertz, Visteon, JPMorgan-Chase, universities, airports, utilities, cities and foreign governments. He is friends with many of the top minds in energy, environment, finance, and marketing. He is also a columnist for several industry magazines, a corporate trainer, and a keynote speaker. Eric is the founder of ProfitableGreenSolutions.com and can be reached at 888-563-7221.

CHAPTER 3

Neil Zobler, President of Catalyst Financial Group, Inc., has been designing energy finance programs and arranged project-specific financing for demand side management (DSM) and renewable energy projects since 1985. Catalyst, a specialist in energy and water conservation projects, has arranged financings for over $1 billion. Neil's clients include U.S. EPA ENERGY STAR, the Inter-American Development Bank, over 20 electric and gas utilities (including Con Edison Co. of NY, PG&E, TVA), engineering companies and vendors, and hundreds of individual companies and organizations. He speaks regularly for organizations including the Government Finance Officers Association, the Association of School Business Officials, National Association of State Energy Officers, Association of Government Leasing & Finance, and the Council of

Great City Schools and is on the task force of The American College & University Presidents Climate Commitment/Clinton Climate Initiative program. He has been published widely in finance and energy periodicals. Neil is fluent in Spanish and helped design financing programs for energy projects in Mexico , Peru and El Salvador. Neil has a BA in Finance from Long Island University (LIU) and has completed post-graduate studies in marketing at the Arthur T. Roth Graduate School at LIU. His email address is nzobler@catalyst-financial.com.

Caterina (Katy) Hatcher is the US EPA ENERGY STAR National Manager for the Public Sector. She works with education, government, water and wastewater utility partners to help them improve their energy performance through the use of ENERGY STAR tools and resources. Katy has been working for the US Environmental Protection Agency for about 11 years. She holds a degree from the University of Virginia 's School of Architecture in City Planning.

EPA offers ENERGY STAR to organizations as a straightforward way to adopt superior energy management and realize the cost savings and environmental benefits that can result. EPA's guidelines for energy management promote a strategy for superior energy management that starts with the top leadership, engages the appropriate employees throughout the organization, uses standardized measurement tools, and helps an organization prioritize and get the most from its efficiency investments.

EPA's ENERGY STAR Challenge is a national call-to-action to improve the energy efficiency of America 's commercial and industrial facilities by 10 percent or more. EPA estimates that if the energy efficiency of commercial and industrial buildings and plants improved by 10 percent, Americans would save about $20 billion and reduce greenhouse gas emissions equal to the emissions from about 30 million vehicles.

CHAPTER 5

Ryan Park is one of the most influential individuals in the downstream integration market of the solar electricity industry. He was one of the founding members of REC Solar Inc., which now installs more solar electricity systems every year than any other company in the United States.

Ryan is currently responsible for developing the commercial sales team, structuring large solar projects, and establishing new strategic partnerships. Ryan graduated with honors from California Polytechnic (Cal Poly), San Luis Obispo and is committed to improving the world through renewable energy technologies, energy efficiency, and empowering people.

His email address is: rpark@recsolar.com

CHAPTER 12

Barney L. Capehart, Ph.D., C.E.M., is a professor emeritus of industrial and systems engineering at the University of Florida, Gainesville. He has broad experience in the commercial/industrial sector, having served as director of the University of Florida Industrial Assessment Center from 1990 to 1999. He has personally conducted over 100 audits of industrial facilities and has assisted students in conducting audits of hundreds of office buildings and other nonindustrial facilities. He has taught a wide variety of courses and seminars on systems analysis, simulation, and energy-related topics.

APPENDIX A

David B. Pratt, Ph.D., P.E., is an associate professor and the undergraduate program director in the School of Industrial Engineering and Management at Oklahoma State University. He holds B.S., M.S., and Ph.D. degrees in industrial engineering. Prior to joining academia, he held technical and managerial positions in the petroleum, aerospace, and pulp and paper industries for over 12 years. He has served on the industrial engineering faculty at his *alma mater*, Oklahoma State University, for over 16 years. His research, teaching, and consulting interests include production planning and control, economic analysis, and manufacturing systems design. He is a registered Professional Engineer, an APICS Certified Fellow in production and inventory management, and an ASQ Certified Quality Engineer. He is a member of IIE, NSPE, APICS, INFORMS, and ASQ.

Chapter 1

Background on the Need for Financing Energy Projects

Eric A. Woodroof, Ph.D., CEM, CRM

INTRODUCTION

Most facility managers agree that energy management projects (EMPs) are good investments. Generally, EMPs reduce operational costs, have a low risk/reward ratio, usually improve productivity, and even have been shown to improve a firm's stock price.[1] Despite these benefits, many cost-effective EMPs are not implemented due to financial constraints. A study of manufacturing facilities revealed that first-cost and capital constraints represented over 35% of the reasons cost-effective EMPs were not implemented.[2] Often, the facility manager does not have enough cash to allocate funding, or cannot get budget approval to cover initial costs. Financial arrangements can mitigate a facility's funding constraints,[3] allowing additional energy savings to be reaped.

Alternative finance arrangements can overcome the "initial cost" obstacle, allowing firms to implement more EMPs. However, many facility managers are either unaware or have difficulty understanding the variety of financial arrangements available to them. Most facility managers use simple payback analyses to evaluate projects, which do not reveal the added value of after-tax benefits.[4] Sometimes facility managers do not implement an EMP because financial terminology and contractual details intimidate them.[5]

To meet the growing demand, there has been a dramatic increase in the number of finance companies specializing in EMPs. At a recent energy management conference, finance companies represented the most common exhibitor type. These financiers are introducing new payment arrangements to implement EMPs. Often, the financier's innovation will satisfy the unique customer needs of a large facility. This is a great service; however, most financiers are not attracted to small facilities with EMPs requiring less than $100,000. Thus, many facility managers remain unaware or confused about the common financial arrangements that could help them implement EMPs.

The authors hope that by reading this book you will have new opportunities open for you and be able to get more projects implemented!

References

1. Wingender, J. and Woodroof, E., (1997) "When Firms Publicize Energy Management Projects Their Stock Prices Go Up: How High?—As Much as 21.33% within 150 days of an Announcement," *Strategic Planning for Energy and the Environment,* Vol. 17(1), pp. 38-51.
2. U.S. Department of Energy, (1996) "Analysis of Energy-Efficiency Investment Decisions by Small and Medium-Sized Manufacturers," U.S. DOE, Office of Policy and Office of Energy Efficiency and Renewable Energy, pp. 37-38.
3. Woodroof, E. and Turner, W. (1998), "Financial Arrangements for Energy Management Projects," *Energy Engineering* 95(3) pp. 23-71.
4. Sullivan, A. and Smith, K. (1993) "Investment Justification for U.S. Factory Automation Projects," *Journal of the Midwest Finance Association,* Vol. 22, p. 24.
5. Fretty, J. (1996), "Financing Energy-Efficient Upgraded Equipment," Proceedings of the 1996 International Energy and Environmental Congress, Chapter 10, Association of Energy Engineers.

Chapter 2
Financing Energy Management Projects

Eric A. Woodroof, Ph.D., CEM, CEP, CLEP

INTRODUCTION

Financing can be a key success factor for projects. This chapter's purpose is to help facility managers understand and apply the financial arrangements available to them. Hopefully, this approach will increase the implementation rate of good energy management projects, which would have otherwise been cancelled or postponed due to lack of funds.

Most facility managers agree that energy management projects (EMPs) are good investments. Generally, EMPs reduce operational costs, have a low risk/reward ratio, usually improve productivity, and even have been shown to improve a firm's stock price.[1] Despite these benefits, many cost-effective EMPs are not implemented due to financial constraints. A study of manufacturing facilities revealed that first-cost and capital constraints represented over 35% of the reasons cost-effective EMPs were not implemented.[2] Often, the facility manager does not have enough cash to allocate funding or cannot get budget approval to cover initial costs. Financial arrangements can mitigate a facility's funding constraints,[3] allowing additional energy savings to be reaped.

Alternative finance arrangements can overcome the initial cost obstacle, allowing firms to implement more EMPs. However, many facility managers are either unaware or have difficulty understanding the variety of financial arrangements available to them. Most facility managers use simple payback analyses to evaluate projects, which do not reveal the added value of after-tax benefits.[4] Sometimes facility managers do not implement an EMP because financial terminology and contractual details intimidate them.[5]

To meet the growing demand, there has been a dramatic increase in the number of finance companies specializing in EMPs. At a recent energy management conference, finance companies represented the most common exhibitor type. These financiers are introducing new payment arrangements to implement EMPs. Often, the financier's innovation will satisfy the unique customer needs of a large facility. This is a great service; however, most financiers are not attracted to small facilities with EMPs requiring less than $100,000. Thus, many facility managers remain unaware or confused about the common financial arrangements that could help them implement EMPs.

Numerous papers and government programs have been developed to show facility managers how to use quantitative (economic) analysis to evaluate financial arrangements.[4,5,6] *Quantitative analysis includes computing the simple payback, net present value (NPV), internal rate of return (IRR), and life-cycle cost of a project with or without financing*. Although these books and programs show how to evaluate the economic aspects of projects, they do not incorporate qualitative factors like strategic company objectives, which can impact the financial arrangement selection. Without incorporating a facility manager's qualitative objectives, it is hard to select an arrangement that meets all of the facility's needs. A recent paper showed that qualitative objectives can be at least as important as quantitative objectives.[9]

This chapter hopes to provide some valuable information that can be used to overcome the previously mentioned issues. The chapter is divided into several sections to accomplish three objectives. These sections will *introduce the basic financial arrangements* via a simple example and *define financial terminology*. Each arrangement is explained in greater detail while applied to a case study. The remaining sections show *how to match financial arrangements to different projects and facilities*. For those who need a more detailed description of rate of return analysis and basic financial evaluations, refer to Appendix A.

FINANCIAL ARRANGEMENTS: A SIMPLE EXAMPLE

Consider a small company, "PizzaCo," that makes frozen pizzas and distributes them regionally. PizzaCo uses an old delivery truck that breaks down frequently and is inefficient. Assume the old truck has no salvage value and is fully depreciated. PizzaCo's management would like to obtain a new and more efficient truck to reduce expenses and improve reliability. However, they do not have the cash on hand to purchase the truck. Thus, they consider their financing options.

Purchase the Truck with a Loan or Bond

Just like most car purchases, PizzaCo borrows money from a lender (a bank) and agrees to a monthly re-payment plan. Figure 2-1 shows PizzaCo's annual cash flows for a loan. The solid arrows represent the financing cash flows between PizzaCo and the bank. Each year, PizzaCo makes payments on the principal, plus interest based on the unpaid balance, until the balance owed is zero. The payments are the negative cash flows. Thus, at time zero when PizzaCo borrows the money, it receives a large sum of money from the bank, which is a positive cash flow that will be used to purchase the truck.

The *dashed* arrows represent the truck purchase as well as savings cash flows. Thus, at time zero, PizzaCo purchases the truck (a negative cash flow) with the money from the bank. Due to the new truck's greater efficiency, PizzaCo's annual expenses are reduced, which is a savings. The annual savings are the positive cash flows. The remaining cash flow diagrams in this chapter utilize the same format.

PizzaCo could also purchase the truck by selling a bond. This arrangement is similar to a loan, except investors (not a bank) give PizzaCo a large sum of money (called the bond's "par value"). Periodically, PizzaCo would pay the investors only the interest accumulated. As Figure 2-2 shows, when the bond reaches maturity, PizzaCo returns the par value to the investors. The equipment purchase and savings

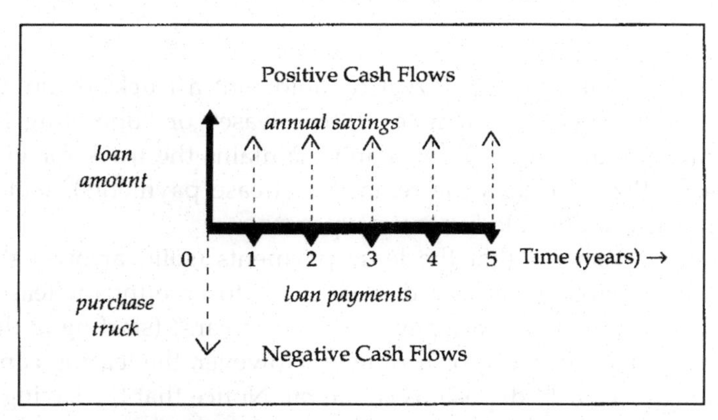

Figure 2-1. PizzaCo's Cash Flows for a Loan.

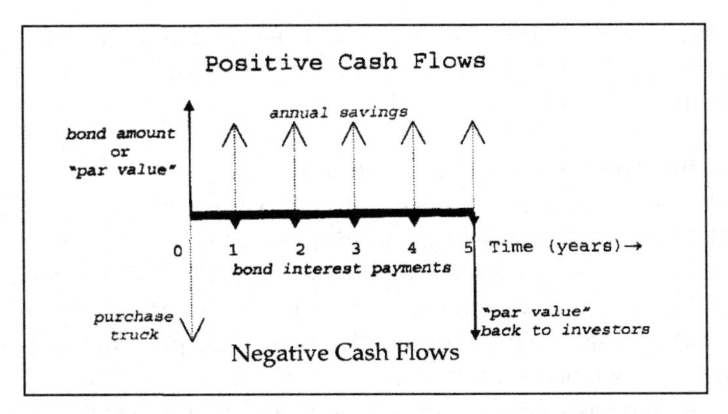

Figure 2-2. PizzaCo's Cash Flows for a Bond.

cash flows are the same as with the loan.

Sell Stock to Purchase the Truck

In this arrangement, PizzaCo sells its stock to raise money to purchase the truck. In return, PizzaCo is expected to pay dividends back to shareholders. Selling stock has a similar cash flow pattern as a bond, with a few subtle differences. Instead of interest payments to bondholders, PizzaCo would pay dividends to shareholders until some future date when PizzaCo could buy the stock back. However, these dividend payments are not mandatory, and if PizzaCo is experiencing financial strain, it is not required to distribute dividends. On the other hand, if PizzaCo's profits increase, this wealth will be shared with the new stockholders, because they now own a part of the company.

Rent the Truck

Just like renting a car, PizzaCo could rent a truck for an annual fee. This would be equivalent to a "true lease" or "operating lease." The rental company (lessor) owns and maintains the truck for PizzaCo (the lessee). PizzaCo pays the rental fees (lease payments), which are considered tax-deductible business expenses.

Figure 2-3 shows that the lease payments (solid arrows) start as soon as the equipment is leased (year zero) to account for lease payments paid in advance. Lease payments "in arrears" (starting at the end of the first year) could also be arranged. However, the leasing company may require a security deposit as collateral. Notice that the savings cash flows are essentially the same as the previous arrangements, except

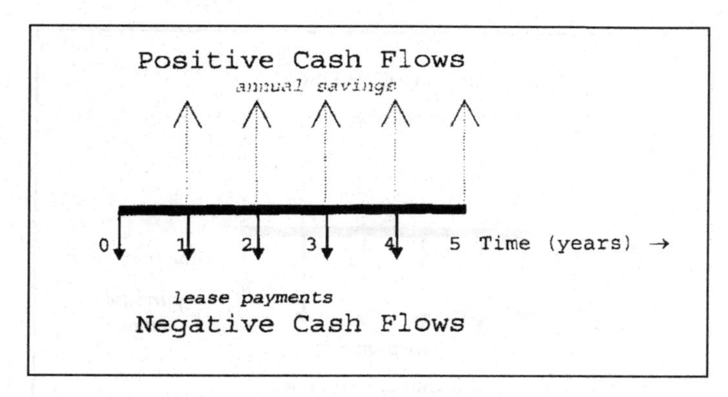

Figure 2-3. PizzaCo's Cash Flows for a True Lease.

there is no equipment purchase, which is a large negative cash flow at year zero.

In a true lease, the contract period should be shorter than the equipment's useful life. The lease is cancelable because the truck can be leased easily to someone else. At the end of the lease, PizzaCo can either return the truck or renew the lease. In a separate transaction, PizzaCo could also negotiate to buy the truck at the fair market value.

If PizzaCo wanted to secure the option to buy the truck (for a bargain price) at the end of the lease, then they would use a capital lease. A capital lease can be structured like an installment loan, however ownership is not transferred until the end of the lease. The lessor retains ownership as security in case the lessee (PizzaCo) defaults on payments. Because the entire cost of the truck is eventually paid, the lease payments are larger than the payments in a true lease, (assuming similar lease periods). Figure 2-4 shows the cash flows for a capital lease with advance payments and a bargain purchase option at the end of year five.

There are some additional scenarios for lease arrangements. A "vendor-financed" agreement is when the lessor (or lender) is the equipment manufacturer. Alternatively, a third party could serve as a financing source. With "third party financing," a finance company would purchase a new truck and lease it to PizzaCo. In either case, there are two primary ways to repay the lessor:

1. With a "fixed payment plan," where payments are due whether

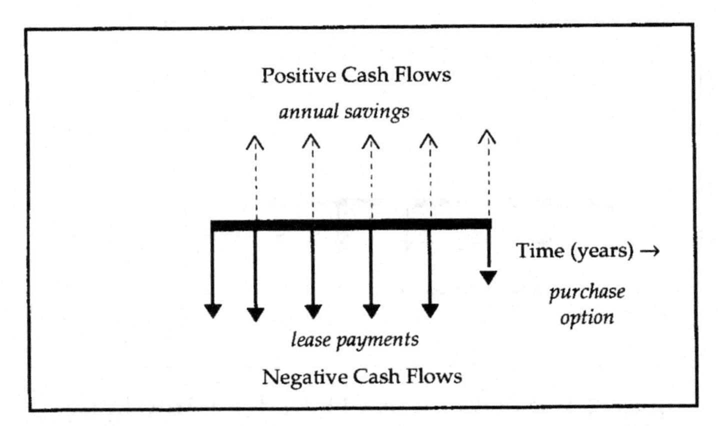

Figure 2-4. PizzaCo's Cash Flows for a Capital Lease.

or not the new truck actually saves money.

2. With a "flexible payment plan," where the savings from the new truck are shared with the third party until the truck's purchase cost is recouped with interest. This is basically a "shared savings" arrangement.

Subcontract Pizza Delivery to a Third Party

Since PizzaCo's primary business is not delivery, it could subcontract that responsibility to another company. Let's say that a delivery service company would provide a truck and deliver the pizzas at a reduced cost. Each month, PizzaCo would pay the delivery service company a fee. However, this fee is guaranteed to be less than what PizzaCo would have spent on delivery. Thus, PizzaCo would obtain savings without investing any money or risk in a new truck. This arrangement is analogous to a performance contract. A performance contract can take many forms; however, the "performance" aspect is usually backed by a guarantee on operational performance from the contractor. In some performance contracts, the host can own the equipment and the guarantee assures that the operational benefits are greater than the finance payments. Alternatively, some performance contracts can be viewed as "outsourcing," where the contractor owns the equipment and provides a "service" to the host.

This arrangement is very similar to a third-party lease. However, with a performance contract, the contractor assumes most of the risk, and the contractor is also responsible for ensuring that the delivery fee is less than what PizzaCo would have spent. For the PizzaCo example, the arrangement would be designed under the conditions below:

- The delivery company is responsible for all operations related to delivering the pizzas.
- The monthly fee is related to the number of pizzas delivered. This is the performance aspect of the contract; if PizzaCo doesn't sell many pizzas, the fee is reduced. *A minimum amount of pizzas may be required by the delivery company (performance contractor) to cover costs.* Thus, the delivery company assumes these risks:
    1. PizzaCo will remain solvent, and
    2. PizzaCo will sell enough pizzas to cover costs, and
    3. The new truck will operate as expected and will actually reduce expenses per pizza, and
    4. The external financial risk, such as inflation and interest rate changes, are acceptable.
- The delivery company is an expert in delivery; it has specially skilled personnel and uses efficient equipment. Thus, the delivery company can deliver the pizzas at a lower cost (even after adding a profit) than PizzaCo.

Figure 2-5 shows the net cash flows according to PizzaCo. Since the delivery company simply reduces PizzaCo's operational expenses, there is only a net savings. There are no negative financing cash flows.

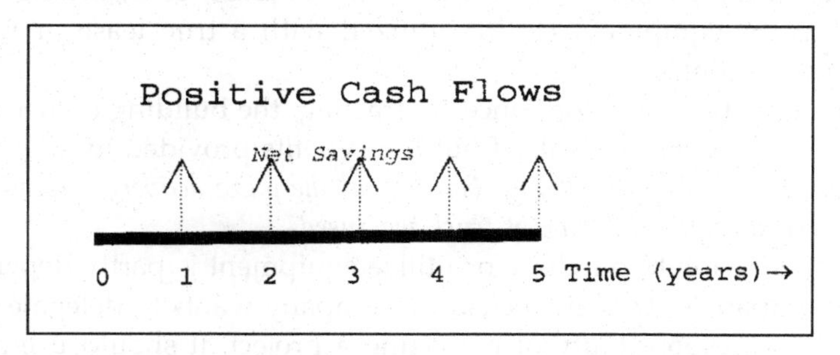

Figure 2-5. PizzaCo's Cash Flows for a Performance Contract.

Unlike the other arrangements, the delivery company's fee is a less expensive substitute for PizzaCo's in-house delivery expenses. With the other arrangements, PizzaCo had to pay a specific financing cost (loan, bond or lease payments, or dividends) associated with the truck, whether or not the truck actually saved money. In addition, PizzaCo would have to spend time maintaining the truck, which would detract from its core focus—making pizzas. With a performance contract, the delivery company is paid from the operational savings it generates. Because the savings are greater than the fee, there is a net savings. Often, the contractor guarantees the savings.

Supplementary note: Combinations of the basic finance arrangements are possible. For example, a guaranteed arrangement can be structured within a performance contract. Also, performance contracts are often designed so that the facility owner (PizzaCo) would own the asset at the end of the contract.

FINANCIAL ARRANGEMENTS: DETAILS AND TERMINOLOGY

To explain the basic financial arrangements in more detail, each one is applied to an energy management-related case study. To understand the economics behind each arrangement, some finance terminology is presented below.

Finance Terminology

Equipment can be purchased with cash on-hand (officially labeled "retained earnings"), a loan, a bond, a capital lease, or by selling stock. Alternatively, equipment can be utilized with a true lease or with a performance contract.

Note that with performance contracting, the building owner is not paying for the equipment itself but the benefits provided by the equipment. *In the Simple Example, the benefit was the pizza delivery. PizzaCo was not concerned with what type of truck was used.*

The decision to purchase or utilize equipment is partly dependent on the company's strategic focus. If a company wants to delegate some or all of the responsibility of managing a project, it should use a true lease, or a performance contact.[10] However, if the company wants to be intricately involved with the EMP, purchasing and self-managing the equipment could yield the greatest profits. When the building owner purchases equipment, he/she usually maintains the equipment and lists it as an asset on the balance sheet so it can be depreciated.

Financing for purchases has two categories:

1. *Debt Financing,* which is borrowing money from someone else or another firm (using loans, bonds and capital leases).

2. *Equity Financing,* which is using money from your company or your stockholders (using retained earnings, or issuing common stock).

In all cases, the borrower will pay an interest charge to borrow money. The interest rate is called the "cost of capital." The cost of capital is essentially dependent on three factors: (1) the borrower's credit rating, (2) project risk and (3) external risk. External risk can include energy price volatility and industry-specific economic performance, as well as global economic conditions and trends. The cost of capital (or "cost of borrowing") influences the return on investment. If the cost of capital increases, then the return on investment decreases.

The "minimum attractive rate of return" (MARR) is a company's "hurdle rate" for projects. *Because many organizations have numerous projects competing for funding, the MARR can be much higher than interest earned from a bank or other risk-free investment.* Only projects with a return on investment greater than the MARR should be accepted. The MARR is also used as the discount rate to determine the "net present value" (NPV).

Explanation of Figures and Tables

Throughout this chapter's case study, figures are presented to illustrate the transactions of each arrangement. Tables are also presented to show how to perform the economic analyses of the different arrangements. The NPV is calculated for each arrangement.

It is important to note that the NPV of a particular arrangement can change significantly if the cost of capital, MARR, equipment residual value, or project life is adjusted. Thus, the examples within this chapter are provided only to illustrate how to perform the analyses. The cash flows and interest rates are estimates, which can vary from project to project. To keep the calculations simple, end-of-year cash flows are used throughout this chapter.

Within the tables, the following abbreviations and equations are used:

EOY = End of Year
Savings = Pre-tax Cash Flow
Depr. = Depreciation
Taxable Income = Savings – Depreciation – Interest Payment
Tax = (Taxable Income)*(Tax Rate)
ATCF = After Tax Cash Flow = Savings – Total Payments – Taxes

Table 2-1 shows the basic equations that are used to calculate the values under each column heading within the economic analysis tables.

Regarding depreciation, the "modified accelerated cost recovery system" (MACRS) is used in the economic analyses. This system indicates the percent depreciation claimable year-by-year, after the equipment is purchased. Table 2-2 shows the MACRS percentages for seven-year property. *For example, after the first year, an owner could depreciate 14.29% of an equipment's value. The equipment's "book value" equals the remaining unrecovered depreciation. Thus, after the first year, the book value would be 100%-14.29%, which equals 85.71% of the original value. If the owner sells the property before it has been fully depreciated, he/she can claim the book value as a tax-deduction.**

APPLYING FINANCIAL ARRANGEMENTS: A CASE STUDY

Suppose PizzaCo (*the "host" facility*) needs a new chilled water system for a specific process in its manufacturing plant. The installed

To be precise, the IRS uses a "half-year convention" for equipment that is sold before it has been completely depreciated. In the tax year that the equipment is sold, (say year "x") the owner claims only Ω of the MACRS depreciation percent for that year. (This is because the owner has only used the equipment for a fraction of the final year.) Then on a separate line entry, (in the year "x"), the remaining unclaimed depreciation is claimed as "book value." The x* year is presented as a separate line item to show the book value treatment, however x* entries occur in the same tax year as "x."

Table 2-1. Table of Sample Equations used in Economic Analyses.

A	B	C	D	E	F	G	H	I	J
				Payments		*Principal*	*Taxable*		
EOY	*Savings*	*Depreciation*	*Principal*	*Interest*	*Total*	*Outstanding*	*Income*	*Tax*	*ATCF*
n									
n+1		= (MACRS %)*			=(D) +(E)	=(G at year n)	=(B)–(C)–(E)	=(H)*(tax rate)	=(B)–(F)–(I)
n+2		(Purchase Price)				–(D at year n+1)			

Table 2-2. MACRS Depreciation Percentages.

EOY	*MACRS Depreciation Percentages for 7-Year Property*
0	0
1	14.29%
2	24.49%
3	17.49%
4	12.49%
5	8.93%
6	8.92%
7	8.93%
8	4.46%

cost of the new system is $2.5 million. The expected equipment life is 15 years, however the process will only be needed for 5 years, after which the chilled water system will be sold at an estimated market value of $1,200,000 (book value at year five = $669,375). The chilled water system should save PizzaCo about $1 million/year in energy savings. PizzaCo's tax rate is 34%. The equipment's annual maintenance and insurance cost is $50,000. PizzaCo's MARR is 18%. Since at the end of year 5, PizzaCo expects to sell the asset for an amount greater than its book value, the additional revenues are called a "capital gain" (equals the market value – book value) and are taxed. If PizzaCo sells the asset for less than its book value, PizzaCo incurs a "capital loss."

PizzaCo does not have $2.5 million to pay for the new system, thus it considers its finance options. PizzaCo is a small company with an average credit rating, which means that it will pay a higher cost of capital than a larger company with an excellent credit rating. (As with any borrowing arrangement, if investors believe that an investment is risky, they will demand a higher interest rate.)

Purchase Equipment with Retained Earnings (Cash)

If PizzaCo did have enough retained earnings (cash on-hand) available, it could purchase the equipment without external financing. Although external finance expenses would be zero, the benefit of tax-deductions from interest expenses is also zero. Also, any cash used to purchase the equipment would carry an "opportunity cost," because that cash could have been used to earn a return somewhere else. This opportunity cost rate is usually set equal to the MARR. In other words, the company lost the opportunity to invest the cash and gain at least the MARR from another investment.

Of all the arrangements described in this chapter, purchasing equipment with retained earnings is probably the simplest to understand. For this reason, it will serve as a brief example and introduction to the economic analysis tables that are used throughout this chapter.

Application to the Case Study

Figure 2-6 illustrates the resource flows between the parties. In this arrangement, PizzaCo purchases the chilled water system directly from the equipment manufacturer.

Once the equipment is installed, PizzaCo recovers the full $1 million/year in savings for the entire five years, but it must spend $50,000/year on maintenance and insurance. At the end of the five-year project, PizzaCo expects to sell the equipment for its market value of $1,200,000. Assume MARR is 18% and the equipment is classified as 7-year property for MACRS depreciation. Table 2-3 shows the economic analysis for purchasing the equipment with retained earnings.

Reading Table 2-3 from left to right, and top to bottom, at EOY 0, the single payment is entered into the table. Each year thereafter, the savings as well as the depreciation (which equals the equipment purchase price multiplied by the appropriate MACRS % for each year)

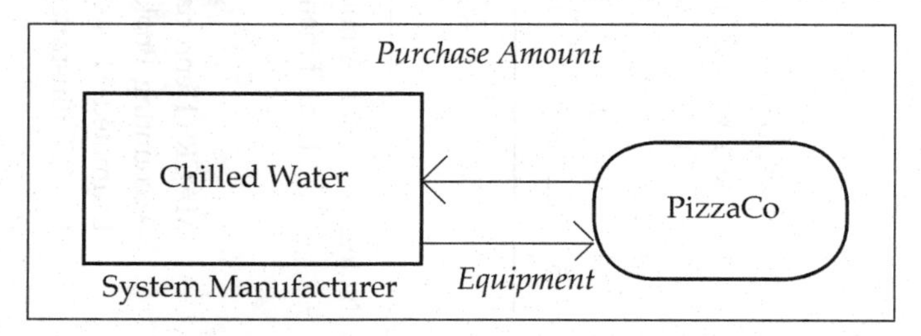

Figure 2-6. Resource Flows for Using Retained Earnings

Table 2-3. Economic Analysis for Using Retained Earnings.

EOY	*Savings*	*Depr.*	*Payments* *Principal*	*Payments* *Interest*	*Payments* *Total*	*Principal Outstanding*	*Taxable Income*	*Tax*	*ATCF*
0					2,500,000				-2,500,000
1	950,000	357,250					592,750	201,535	748,465
2	950,000	612,250					337,750	114,835	835,165
3	950,000	437,250					512,750	174,335	775,665
4	950,000	312,250					637,750	216,835	733,165
5	950,000	111,625					838,375	285,048	664,953
5*	1,200,000	669,375					530,625	180,413	1.019.588
		2,500,000							
			Net Present Value at 18%:						$320,675

Notes: Loan Amount: 0
Loan Finance Rate: 0% MARR 18%
Tax Rate 34%

MACRS Depreciation for 7-Year Property, with half-year convention at EOY 5
Accounting Book Value at end of year 5: 669,375
Estimated Market Value at end of year 5: 1,200,000
EOY 5* illustrates the Equipment Sale and *Book* Value
Taxable Income: =(Market Value - Book Value)
=(1,200,000 - 669,375) = $530,625

are entered into the table. Year by year, the taxable income = savings – depreciation. The taxable income is then taxed at 34% to obtain the tax for each year. The after-tax cash flow = savings – tax for each year.

At EOY 5, the equipment is sold before the entire value was depreciated. EOY 5* shows how the equipment sale and book value are claimed. In summary, the NPV of all the ATCFs would be $320,675.

Loans

Loans have been the traditional financial arrangement for many types of equipment purchases. A bank's willingness to loan depends on the borrower's financial health, experience in energy management, and number of years in business. Obtaining a bank loan can be difficult if the loan officer is unfamiliar with EMPs. Loan officers and financiers may not understand energy-related terminology (demand charges, kVAR, etc.). In addition, facility managers may not be comfortable with the financier's language. Thus, to save time, a bank that can understand EMPs should be chosen.

Most banks will require a down payment and collateral to secure a loan. However, securing assets can be difficult with EMPs, because the equipment often becomes part of the real estate of the plant. *For example, it would be very difficult for a bank to repossess lighting fixtures from a retrofit.* In these scenarios, lenders may be willing to secure other assets as collateral.

Application to the Case Study

Figure 2-7 illustrates the resource flows between the parties. In this arrangement, PizzaCo purchases the chilled water system with a loan from a bank. PizzaCo makes equal payments (principal + interest) to

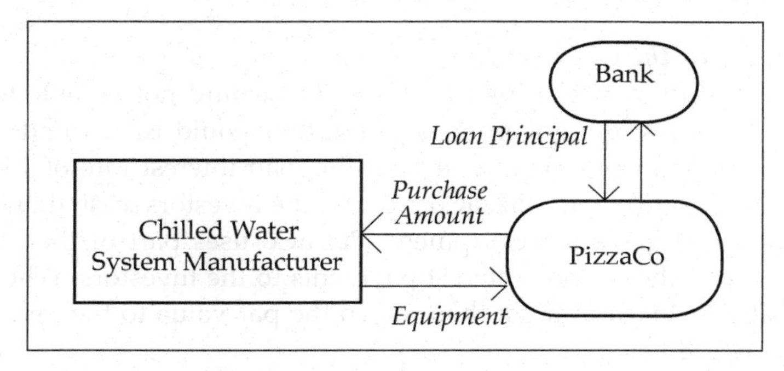

2-7. Resource Flow Diagram for a Loan.

the bank for five years to retire the debt. Due to PizzaCo's small size, credibility, and inexperience in managing chilled water systems, PizzaCo is likely to pay a relatively high cost of capital. For example, let's assume 15%.

PizzaCo recovers the full $1 million/year in savings for the entire five years, but it must spend $50,000/year on maintenance and insurance. At the end of the five-year project, PizzaCo expects to sell the equipment for its market value of $1,200,000. Tables 2-4 and 2-5 show the economic analysis for loans with a zero down payment and a 20% down payment, respectively. Assume that the bank reduces the interest rate to 14% for the loan with the 20% down payment. Since the asset is listed on PizzaCo's balance sheet, PizzaCo can use depreciation benefits to reduce the after-tax cost. In addition, all loan interest expenses are tax-deductible.

Bonds

Bonds are very similar to loans; a sum of money is borrowed and repaid with interest over a period of time. The primary difference is that with a bond, the issuer (PizzaCo) periodically pays the investors only the interest earned. This periodic payment is called the "coupon interest payment." *For example, a $1,000 bond with a 10% coupon will pay $100 per year. When the bond matures, the issuer returns the face value ($1,000) to the investors.*

Bonds are issued by corporations and government entities. Government bonds generate tax-free income for investors, thus these bonds can be issued at lower rates than corporate bonds. This benefit provides government facilities an economic advantage to use bonds to finance projects.

Application to the Case Study

Although PizzaCo (a private company) would not be able to obtain the low rates of a government bond, they could issue bonds with coupon interest rates competitive with the loan interest rate of 15%.

In this arrangement, PizzaCo receives the investors' cash (bond par value) and purchases the equipment. PizzaCo uses part of the energy savings to pay the coupon interest payments to the investors. When the bond matures, PizzaCo must then return the par value to the investors. (See Figure 2-8.)

As with a loan, PizzaCo owns, maintains and depreciates the

Table 2-4. Economic Analysis for a Loan with No Down Payment.

EOY	*Savings*	*Depr.*	*Payments* *Principal*	*Interest*	*Total*	*Principal Outstanding*	*Taxable Income*	*Tax*	*ATCF*
0				2,500,000					
1	950,000	357,250	370,789	375,000	745,789	2,129,211	217,750	74,035	130,176
2	950,000	612,250	426,407	319,382	745,789	1,702,804	18,368	6,245	197,966
3	950,000	4,372	490,368	255,421	745,789	1,212,435	257,329	187,492	116,719
4	950,000	312,200	563,924	181,865	745,789	648,511	455,885	55,001	49,210
5	950,000	111,625	648,511	97,277	745,789	0	741,098	251,973	-47,761
5*	1,200,000	669,375			530,625	180,413	1,019,588		
		2,500,000							
					Net Present Value at 18%:				$757,121

Notes: Loan Amount: 2,500,000 (used to purchase equipment at year 0)

Loan Finance Rate: 15% MARR 18%

Tax Rate 34%

MACRS Depreciation for 7-Year Property, with half-year convention at EOY 5

Accounting Book Value at end of year 5: 669,375

Estimated Market Value at end of year 5: 1,200,000

EOY 5* illustrates the Equipment Sale and *Book* Value

Taxable Income: =(Market Value - Book Value)

=(1,200,000 - 669,375) = $530,625

Table 2-5. Economic Analysis for a Loan with a 20% Down Payment,

EOY	*Savings*	*Depr.*		*Payments*		*Principal*	*Taxable*	*Tax*	*ATCF*
			Principal	*Interest*	*Total*	*Outstanding*	*Income*		
0					500,000	2,000,000			–500,000
1	950,000	357,250	302,567	280,000	582,567	1,697,433	312,750	106,335	261,098
2	950,000	612,250	344,926	237,641	582,567	1,352,507	100,109	34,037	333,396
3	950,000	437,250	393,216	189,351	582,567	959,291	323,399	109,956	257,477
4	950,000	312,250	448,266	134,301	582,567	511,024	503,449	171,173	196,260
5	950,000	111,625	511,024	71,543	582,567	0	766,832	260,723	106,710
5*	1,200,000	669,375					530,625	180,413	1,019,588
		2,500,000							
					Net Present Value at 18%:				$710,962

Notes: Loan Amount: 2,000,000 (used to purchase equipment at year 0)
Loan Finance Rate: 14% MARR 18%
500,000 Tax Rate 34%
MACRS Depreciation for 7-Year Property, with half-year convention at EOY 5
Accounting Book Value at end of year 5: 669,375
Estimated Market Value at end of year 5: 1,200,000
EOY 5* illustrates the Equipment Sale and *Book* Value
Taxable Income: =(Market Value - Book Value)
=(1,200,000 - 669,375) = $530,625

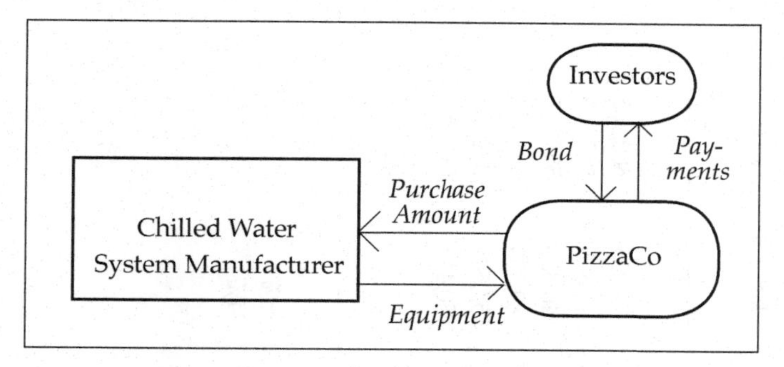

Figure 2-8. Resource Flow Diagram for a Bond.

equipment throughout the project's life. All coupon interest payments are tax-deductible. At the end of the five-year project, PizzaCo expects to sell the equipment for its market value of $1,200,000. Table 2-6 shows the economic analysis of this finance arrangement.

Selling Stock

Although less popular, selling company stock is an equity financing option which can raise capital for projects. For the host, selling stock offers a flexible repayment schedule, because dividend payments to shareholders aren't absolutely mandatory. Selling stock is also often used to help a company attain its desired capital structure. However, selling new shares of stock dilutes the power of existing shares and may send an inaccurate "signal" to investors about the company's financial strength. If the company is selling stock, investors may think that it is desperate for cash and in a poor financial condition. Under this belief, the company's stock price could decrease. However, recent research indicates that when a firm announces an EMP, investors react favorably.[11] On average, stock prices were shown to increase abnormally by 21.33%.

By definition, the cost of capital (rate) for selling stock is:

$$\text{cost of capital}_{\text{selling stock}} = D/P$$

where D = *annual dividend payment*

P = *company stock price*

However, in most cases, the after-tax cost of capital for selling stock is higher than the after-tax cost of debt financing (using loans, bonds and capital leases). This is because interest expenses (on debt)

Table 2-6. Economic Analysis for a Bond.

EOY	Savings	Depr.	Payments Principal	Payments Interest	Payments Total	Principal Outstanding	Taxable Income	Tax	ATCF
0						2,500,000			
1	950,000	357,250		375,000	375,000	2,500,000	217,750	74,035	500,965
2	950,000	612,250		375,000	375,000	2,500,000	-37,250	-12,665	587,665
3	950,000	437,250		375,000	375,000	2,500,000	137,750	46,835	528,165
4	950,000	312,250		375,000	375,000	2,500,000	262,750	89,335	485,665
5	950,000	111,625	2,500,000	375,000	2,875,000	0	463,375	157,548	-2,082,548
5*	1,200,000	669,375					530,625	180,413	1,019,588
		2,500,000							
					Net Present Value at 18%:				953,927

Notes: Loan Amount: 2,500,000 (used to purchase equipment at year 0)
Loan Finance Rate: 0% MARR 18%
Tax Rate 34%

MACRS Depreciation for 7-Year Property, with half-year convention at EOY 5
Accounting Book Value at end of year 5: 669,375
Estimated Market Value at end of year 5: 1,200,000
EOY 5* illustrates the Equipment Sale and *Book* Value
Taxable Income: =(Market Value - Book Value)
=(1,200,000 - 669,375) = $530,625

are tax deductible, but dividend payments to shareholders are not.

In addition to tax considerations, there are other reasons why the cost of debt financing is less than the financing cost of selling stock. Lenders and bond buyers (creditors) will accept a lower rate of return because they are in a less risky position due to the reasons below.

- Creditors have a contract to receive money at a certain time and future value. (Stockholders have no such guarantee with dividends.)
- Creditors have first claim on earnings. (Interest is paid before shareholder dividends are allocated.)
- Creditors usually have secured assets as collateral and have first claim on assets in the event of bankruptcy.

Despite the high cost of capital, selling stock does have some advantages. This arrangement does not bind the host to a rigid payment plan (like debt financing agreements), because dividend payments are not mandatory. The host has control over when it will pay dividends. Thus, when selling stock, the host receives greater payment flexibility, but at a higher cost of capital.

Application to the Case Study

As Figure 2-9 shows, the financial arrangement is very similar to a bond. At year zero the firm receives $2.5 million, except the funds come from the sale of stock. Instead of coupon interest payments, the firm distributes dividends. At the end of year five, PizzaCo repurchases the

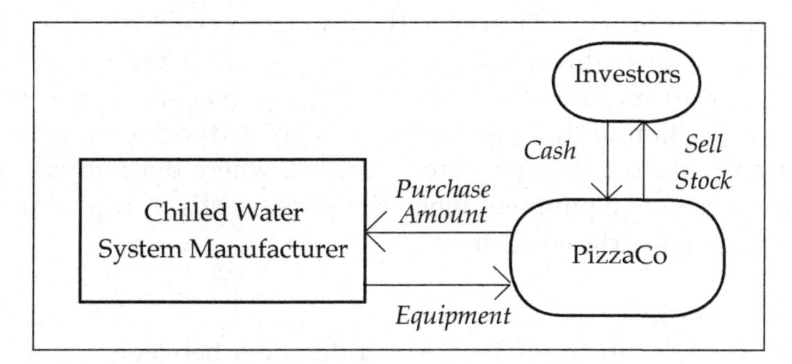

Figure 2-9. Resource Flow Diagram for Selling Stock.

stock. Alternatively, PizzaCo could capitalize the dividend payments, which means setting aside enough money so the dividends could be paid with the interest generated.

Table 2-7 shows the economic analysis for issuing stock at a 16% cost of equity capital, then repurchasing the stock at the end of year five. (For consistency of comparison to the other arrangements, the stock price does not change during the contract.) Like a loan or bond, PizzaCo owns and maintains the asset. Thus, the annual savings are only $950,000. PizzaCo pays annual dividends worth $400,000. At the end of year 5, PizzaCo expects to sell the asset for $1,200,000.

Note that Table 2-7 is slightly different from the other tables in this chapter:

Taxable Income = Savings – Depreciation, and

ATCF = Savings – Stock Repurchases - Dividends - Tax

Leases

Firms generally own assets, however it is the use of these assets that is important, not the ownership. Leasing is another way of obtaining the use of assets. There are numerous types of leasing arrangements, ranging from basic rental agreements to extended payment plans for purchases. Leasing is used for nearly one-third of all equipment utilization.[12] Leases can be structured and approved very quickly, even within 48 hours. Table 2-8 lists some additional reasons why leasing can be an attractive arrangement for the lessee.

Basically, there are two types of leases: the "true lease" (a.k.a. "operating" or "guideline lease") and the "capital lease." One of the primary differences between a true lease and a capital lease is the tax treatment. In a true lease, the lessor owns the equipment and receives the depreciation benefits. However, the lessee can claim the entire lease payment as a tax-deductible business expense. In a capital lease, the lessee (PizzaCo) owns and depreciates the equipment. However, only the interest portion of the lease payment is tax-deductible. In general, a true lease is effective for a short-term project, where the company does not plan to use the equipment when the project ends. A capital lease is effective for long-term equipment.

The True Lease

Figure 2-10 illustrates the legal differences between a true lease and a capital lease.[13] A true lease (or operating lease) is strictly a rental

Table 2-7. Economic Analysis of Selling Stock.

EOY	Savings	Depr.	Stock Transactions			Taxable	Tax	ATCF
			Sale of Stock	Repurchase	Dividend Payments	Income		
0	$2,500,000 from Stock Sale is used to purchase equipment, thur ATCF = 0							
1	950,000	357,250			400,000	592,750	201,535	348,465
2	950,000	612,250			400,000	337,750	114,835	435,165
3	950,000	437,250			400,000	512,750	174,335	375,665
4	950,000	312,250			400,000	637,750	216,835	333,165
5	950,000	111,625		2,500,000	400,000	838,375	285,048	-2,235,048
5*	1,200,000	669,375				530,625	180,413	1,019,588
		2,500,000						
					Net Present Value at 18%:			477,033

Notes: Value of Stock Sold (which is repurchased after year five 2,500,000 (used to purchase equipment at year zero)
Cost of Capital = Annual Dividend Rate: 16% MARR = 18%
Tax Rate = 34%

MACRS Depreciation for 7-Year Property, with half-year convention at EOY 5
Accounting Book Value at end of year 5: 669,375
Estimated Market Value at end of year 5: 1,200,000
EOY 5 illustrates the Equipment Sale and Book Value*
Taxable Income: = (Market Value - Book Value)
= (1,200,000 - 669,375) = $530,625

Table 2-8. Good Reasons to Lease.

Financial Reasons
- With some leases, the entire lease payment is tax-deductible.
- Some leases allow "off-balance sheet" financing, preserving credit lines

Risk Sharing
- Leasing is good for short-term asset use, and reduces the risk of getting stuck with obsolete equipment
- Leasing offers less risk and responsibility

agreement. The word "strictly" is appropriate because the Internal Revenue Service will only recognize a true lease if it satisfies the following criteria:

1. The lease period must be less than 80% of the equipment's life.
2. The equipment's estimated residual value must be 20% of its value at the beginning of the lease.
3. There is no "bargain purchase option."
4. There is no planned transfer of ownership.
5. The equipment must not be custom-made nor useful only in a particular facility.

Application to the Case Study

It is unlikely that PizzaCo could find a lessor that would be willing to lease a sophisticated chilled water system and, after five years, move the system to another facility. Thus, obtaining a true lease would be unlikely. Nevertheless, Figure 2-11 shows the basic relationship between the lessor and lessee in a true lease. A third-party leasing company could also be involved by purchasing the equipment and leasing to PizzaCo. Such a resource flow diagram is shown for the capital lease.

Table 2-9 shows the economic analysis for a true lease. Notice that the lessor pays the maintenance and insurance costs, so PizzaCo saves the full $1 million per year. PizzaCo can deduct the entire lease payment of $400,000 as a business expense. However, PizzaCo does not obtain ownership, so it can't depreciate the asset.

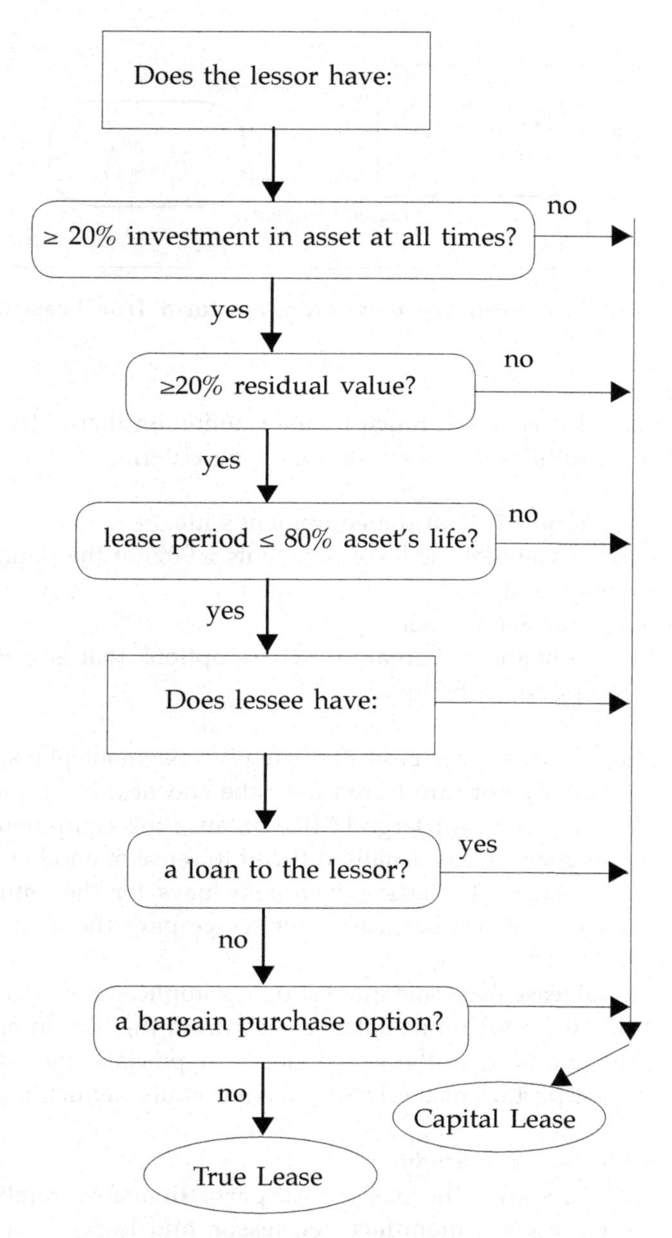

Figure 2-10. Classification for a True Lease.

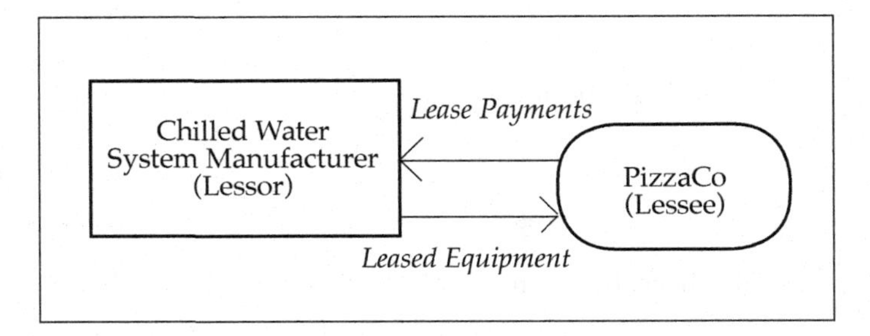

Figure 2-11. Resource Flow Diagram for a True Lease.

The Capital Lease

The capital lease has a much broader definition than a true lease. A capital lease fulfills *any one* of the following criteria:

1. The lease term ≥75% of the equipment's life.
2. The present value of the lease payments ≥ 90% of the initial value of the equipment.
3. The lease transfers ownership.
4. The lease contains a "bargain purchase option" that is negotiated at the inception of the lease.

Most capital leases are basically extended payment plans, except ownership is usually not transferred until the end of the contract. This arrangement is common for large EMPs, because the equipment (such as a chilled water system) is usually difficult to reuse at another facility. With this arrangement, the lessee eventually pays for the entire asset (plus interest). In most capital leases, the lessee pays the maintenance and insurance costs.

The capital lease has some interesting tax implications, because the lessee must list the asset on its balance sheet from the beginning of the contract. Thus, like a loan, the lessee gets to depreciate the asset, and only the interest portion of the lease payment is tax deductible.

Application to the Case Study

Figure 2-12 shows the basic third-party financing relationship between the equipment manufacturer, lessor, and lessee in a capital lease. The finance company (lessor) is shown as a third party, although it also could be a division of the equipment manufacturer.

Table 2-9. Economic Analysis for a True Lease

EOY	Savings	Depr.	Payments Principal	Payments Interest	Payments Total	Principal Outstanding	Taxable Income	Tax	ATCF
0					400,000		-400,000		-400,000
1	1,000,000				400,000		600,000	204,000	396,000
2	1,000,000				400,000		600,000	204,000	396,000
3	1,000,000				400,000		600,000	204,000	396,000
4	1,000,000				400,000		600,000	204,000	396,000
5	1,000,000						1,000,000	340,000	660,000
					Net Present Value at 18%:				$953,757

Notes: Annual Lease Payment: 400,000
MARR = 18%
Tax Rate 34%

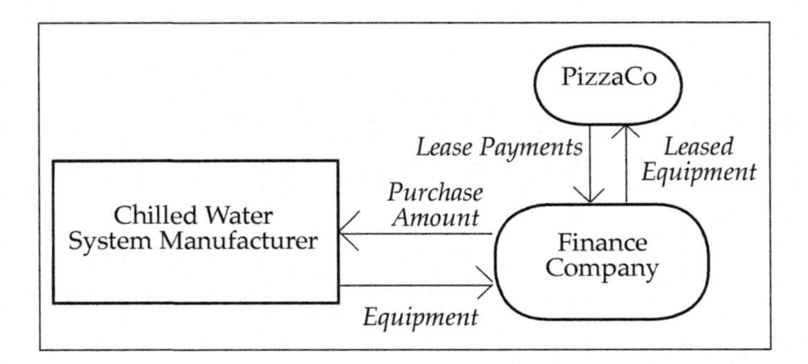

Figure 2-12. Resource Flow Diagram for a Capital Lease.

Because the finance company (with excellent credit) is involved, a lower cost of capital (12%) is possible, due to reduced risk of payment default.

Like an installment loan, PizzaCo's lease payments cover the entire equipment cost. However, the lease payments are made in advance. Because PizzaCo is considered the owner, it pays the $50,000 annual maintenance expenses, which reduces the annual savings to $950,000. PizzaCo receives the benefits of depreciation and tax-deductible interest payments. To be consistent with the analyses of the other arrangements, PizzaCo would sell the equipment at the end of the lease for its market value. Table 2-10 shows the economic analysis for a capital lease.

The Synthetic Lease

A synthetic lease is a "hybrid" lease that combines aspects of a true lease and a capital lease. Through careful structuring and planning, the synthetic lease appears as an operating lease for accounting purposes (enables the Host to have off-balance sheet financing), yet also appears as a capital lease for tax purposes (to obtain depreciation for tax benefits). Consult your local financing expert to learn more about synthetic leases; they must be carefully structured to maintain compliance with the associated tax laws.

With most types of leases, loans and bonds the monthly payments are fixed, regardless of the equipment's utilization or performance. However, shared savings agreements can be incorporated into certain types of leases.

Table 2-10. Economic Analysis for a Capital Lease.

EOY	Savings	Depr.	Payments in Advance			Principal	Taxable	Tax	ATCF
			Principal	Interest	Total	Outstanding	Income		
0			619,218	0	619,218	1,880,782		-619,218	
1	950,000	357,250	393,524	225,694	619,218	1,487,258	367,056	124,799	205,983
2	950,000	612,250	440,747	178,471	619,218	1,046,511	159,279	54,155	276,627
3	950,000	437,250	493,637	125,581	619,218	552,874	387,169	131,637	199,145
4	950,000	312,250	552,874	66,345	619,218	0	571,405	194,278	136,503
5	950,000	111,625					838,375	285,048	664,953
5*	1,200,000	669,375					530,625	180,413	1,019,588
		2,500,000							
					Net Present Value at 18%:				$681,953

Notes: Total Lease Amount: 2,500,000

However, Since the payments are in advance, the first payment is analogous to a Down-Payment

Thus the actual amount borrowed is only = $500,000 - 619,218 = 1,880,782

Lease Finance Rate: 12% MARR 18%

Tax Rate 34%

MACRS Depreciation for 7-Year Property, with half-year convention at EOY 5

Accounting Book Value at end of year 5: 669,375

Estimated Market Value at end of year 5: 1,200,000

EOY 5* illustrates the Equipment Sale and Book Value

Taxable Income: =(Market Value - Book Value)

=(1,200,000 - 669,375) = $530,625

Performance Contracting

Performance contracting is a unique arrangement that allows the building owner to make necessary improvements while investing very little money upfront. The contractor usually assumes responsibility for purchasing and installing the equipment, as well as maintenance throughout the contract. But the unique aspect of performance contracting is that the contractor is paid based on the performance of the installed equipment. Only after the installed equipment actually reduces expenses does the contractor get paid. Energy service companies (ESCOs) typically serve as contractors within this line of business.

Again, unlike most loans, leases and other fixed payment arrangements, the ESCO is paid based on the performance of the equipment. In other words, if the finished product doesn't save energy or operational costs, the host doesn't pay. This aspect removes the incentive to "cut corners" on construction or other phases of the project, as with bid/spec contracting. In fact, often there is an incentive to exceed savings estimates. For this reason, performance contracting usually entails a more "facility-wide" scope of work (to find extra energy savings) than loans or leases on particular pieces of equipment.

With a facility-wide scope, many improvements can occur at the same time. For example, lighting and air conditioning systems can be upgraded at the same time. In addition, the indoor air quality can be improved. With a comprehensive facility management approach, a "domino-effect" on cost reduction is possible. For example, if facility improvements create a safer and higher quality environment for workers, productivity could increase. As a result of decreased employee absenteeism, the workman's compensation cost could also be reduced. These are additional benefits to the facility.

Depending on the host's capability to manage the risks (equipment performance, financing, etc.) the host will delegate some of these responsibilities to the ESCO. In general, the amount of risk assigned to the ESCO is directly related to the percent savings that must be shared with the ESCO.

For facilities that are not in a good position to manage the risks of an energy project, performance contracting may be the only economically feasible implementation method. *For example, the US Federal Government used performance contracting to upgrade facilities when budgets were being dramatically cut. In essence, they "sold" some of their future energy savings to an ESCO, in return for receiving new equipment and efficiency benefits.*

In general, performance contracting may be the best option for facilities that:

- are severely constrained by their cash flows;
- have a high cost of capital;
- don't have sufficient resources, such as a lack of in-house energy management expertise or an inadequate maintenance capacity*;
- are seeking to reduce in-house responsibilities and focus more on their core business objectives; or
- are attempting a complex project with uncertain reliability, or if the host is not fully capable of managing the project. *For example, a lighting retrofit has a high probability of producing the expected cash flows, whereas a completely new process does not have the same "time-tested" reliability. If the in-house energy management team cannot manage this risk, performance contracting may be an attractive alternative.*

Performance contracting does have some drawbacks. In addition to sharing the savings with an ESCO, the tax benefits of depreciation and other economic benefits must be negotiated. Whenever large contracts are involved, there is reason for concern. One study found that 11% of customers who were considering EMPs felt that dealing with an ESCO was too confusing or complicated.[14] Another reference claims, "With complex contracts, there may be more options and more room for error."[15] Therefore, it is critical to choose an ESCO with a good reputation and experience within the types of facilities that are involved.

There are a few common types of contracts. The ESCO will usually offer the following options:

- guaranteed fixed dollar savings;
- guaranteed fixed energy (MMBtu) savings;

*Maintenance capacity represents the ability of the maintenance personnel to maintain the new system. It has been shown that systems fail and are replaced when maintenance concerns are not incorporated into the planning process. See Woodroof, E. (1997) "Lighting Retrofits: Don't Forget About Maintenance," Energy Engineering, 94(1) pp. 59-68.

- a percent of energy savings; or
- a combination of the above.

Obviously, facility managers would prefer the options with "guaranteed savings." However this extra security (and risk to the ESCO) usually costs more. The primary difference between the two guaranteed options is that guaranteed fixed dollar savings contracts ensure dollar savings, even if energy prices fall. *For example, if energy prices drop and the equipment does not save as much money as predicted, the ESCO must pay (out of its own pocket) the contracted savings to the host.*

Percent energy savings contracts are agreements that basically share energy savings between the host and the ESCO. The more energy saved, the higher the revenues to both parties. However, the host has less predictable savings and must also periodically negotiate with the ESCO to determine "who saved what" when sharing savings. There are numerous hybrid contracts available that combine the positive aspects of the above options.

Application to the Case Study

PizzaCo would enter into a hybrid contract: *percent energy savings/guaranteed arrangement.* The ESCO would purchase, install and operate a highly efficient chilled water system. The ESCO would guarantee that PizzaCo would save the $1,000,000 per year, but PizzaCo would pay the ESCO 80% of the savings. In this way, PizzaCo would not need to invest any money and would simply collect the net savings of $200,000 each year. To avoid periodic negotiations associated with shared savings agreements, the contract could be worded such that the ESCO will provide guaranteed energy savings worth $200,000 each year.

With this arrangement, there are no depreciation, interest payments or tax-benefits for PizzaCo. However, PizzaCo receives a positive cash flow with no investment and little risk. At the end of the contract, the ESCO removes the equipment. At the end of most performance contracts, the host usually acquires or purchases the equipment for fair market value; however, for this case study, the equipment was removed to make a consistent comparison with the other financial arrangements.

Figure 2-13 illustrates the transactions between the parties. Table 2-11 presents the economic analysis for performance contracting.

Note that Table 2-11 is slightly different from the other tables in

Table 2-11. Economic Analysis of a Performance Contract.

EOY	*Savings*	*Depr.*	*ESCO Payments*		*Taxable*	*Tax*	*ATCF*
			Total	*Principal Outstanding*	*Income*		
0							
1	1,000,000		800,000		200,000	68,000	132,000
2	1,000,000		800,000		200,000	68,000	132,000
3	1,000,000		800,000		200,000	68,000	132,000
4	1,000,000		800,000		200,000	68,000	132,000
5	1,000,000		800,000		200,000	68,000	132,000
				Net Present Value at 18%:			$412,787

Notes: ESCO purchases/operates equipment. Host pays ESCO 80% of the savings = $800,000.
The contract could also be designed so that PizzaCo can buy the equipment at the end of year 5.

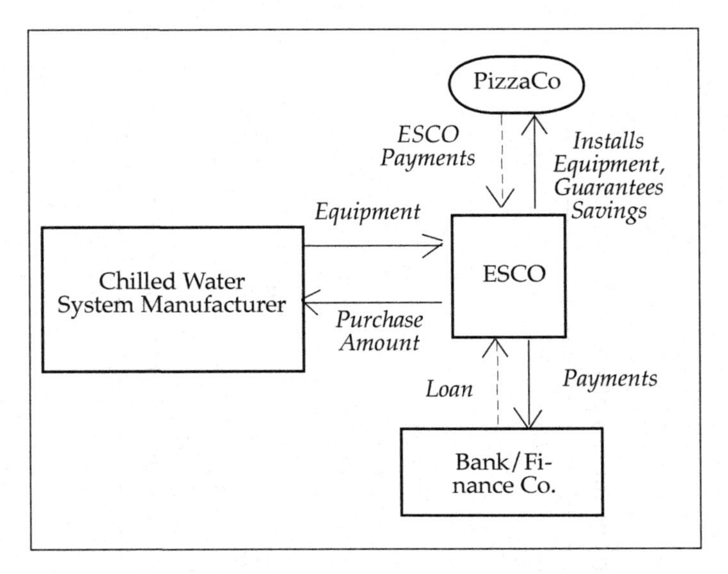

Figure 2-13. Transactions for a Performance Contract.

this chapter: Taxable Income = Savings – Depreciation – ESCO Payments.

Summary of Tax Benefits

Table 2-12 summarizes the tax benefits of each financial arrangement presented in this chapter.

Additional Options

Combinations of the basic financial arrangements can be created to enhance the value of a project. A sample of the possible combinations are described below.

* In some Performance Contracts, the Host can own the equipment and the guarantee assures that the operational benefits are greater than the finance payments. Alternatively, some performance contracts can be viewed as "outsourcing," where the contractor owns the equipment and provides a "service" to the Host.

- Third party financiers often cooperate with performance contracting firms to implement EMPs.

Table 2-12. Host's Tax Benefits for each Arrangement.

ARRANGEMENT	*Depreciation Benefits*	*Interest Payments are Tax-Deductible*	*Total Payments are Tax-Deductible*
Retained Earnings	X		
Loan	X	X	
Bond	X	X	
Sell Stock	X		
Capital Lease	X	X	
True Lease			X
Performance Contract			X

- Utility rebates and government programs may provide additional benefits for particular projects.
- Tax-exempt leases are available to government facilities.
- Insurance can be purchased to protect against risks relating to equipment performance, energy savings, etc.
- Some financial arrangements can be structured as non-recourse to the host. Thus, the ESCO or lessor would assume the risks of payment default. However, as mentioned before, profit sharing increases with risk sharing.

Attempting to identify the absolute best financial arrangement is a rewarding goal, unless it takes too long. As every minute passes, potential dollar savings are lost forever. When considering special grant funds, rebate programs, or other unique opportunities, it is important to consider the lost savings due to delay.

"PROS" & "CONS" OF EACH FINANCIAL ARRANGEMENT

This section presents a brief summary of the "Pros" and "Cons" of each financial arrangement from the host's perspective.

Loan

"Pros":

- Host keeps all savings.

- Depreciation and interest payments are tax-deductible.
- Host owns the equipment.
- The arrangement is good for long-term use of equipment.

"Cons":

- Host takes all the risk and must install and manage the project.

Bond

Has the same Pros/Cons as loan, plus:

"Pro":

- Good for government facilities, because they can offer a tax-free rate that is lower (but considered favorable by investors).

Sell Stock

Has the same Pros/Cons as loan, plus:

"Pro":

- Selling stock could help the host achieve its target capital structure.

"Cons":

- Dividend payments (unlike interest payments) are not tax-deductible.
- Dilutes company control.

Use Retained Earnings

Has the same Pros/Cons as loan, plus:

"Pro":

- Host pays no external interest charges. However, retained earnings do carry an opportunity cost, because such funds could be invested somewhere at the MARR.

"Con":

- Host loses tax-deductible benefits of interest charges

Capital Lease

Has the same Pros/Cons as loan, pus:

"Pro":

- Has greater flexibility in financing and possible lower cost of capital with third-party participation.

True Lease

"Pros":

- Allows use of equipment, without ownership risks.
- Has reduced risk of poor performance, service, equipment obsolescence, etc.
- Is good for short-term use of equipment.
- Entire lease payment is tax-deductible.

"Cons":

- There is no ownership at the end of lease contract.
- There are no depreciation tax benefits.

Performance Contract

"Pros":

- Allows use of equipment, with reduced installment/operational risks.
- Reduced risk of poor performance or service, equipment obsolescence, etc.
- Allows host to focus on its core business objectives.

"Cons":

- Involves potentially binding contracts, legal expenses, and increased administrative costs.
- Host must share project savings.

Rules of Thumb

When investigating financing options, consider the following generalities:

> Loans, bonds and other host-managed arrangements should be used when a customer has the resources (experience, financial support, time) to handle the risks. Performance contracting (ESCO assumes most of the risk) is usually best when a customer doesn't have the resources to properly manage the project. Remember that with any arrangement where the host delegates risk to another firm, the host must also share the savings.
>
> Leases are the "middle ground" between owning and delegating risks, and they are very popular due to their tax benefits.
>
> True leases tend to be preferred when:

- the equipment is needed on a short-term basis;
- the equipment has unusual service problems that cannot be handled by the host;
- technological advances cause equipment to become obsolete quickly; or
- depreciation benefits are not useful to the lessee.

Capital Leases are preferred when:

- the installation and removal of equipment is costly;
- the equipment is needed for a long time; or
- the equipment user desires to secure a "bargain purchase option."

CHARACTERISTICS THAT INFLUENCE WHICH FINANCIAL ARRANGEMENT IS BEST

There are at least three types of characteristics that can influence which financial arrangement should be used for a particular EMP. These include facility characteristics, project characteristics and financial arrangement characteristics. In this section, quantitative characteristics are bulleted with this symbol, "$." The qualitative characteristics are bulleted with the symbol, "☺." Note that qualitative characteristics are generally "strategic" and are not associated with an exact dollar value.

A few of the Facility Characteristics include:

☺ The long-term plans of facility. For example, is the facility trying to focus on core business objectives and outsourcing other tasks, such as EMPs?

$ The facility's current financial condition. Credit ratings and ability to obtain loans can determine whether certain financial arrangements are feasible.

☺ The experience and technical capabilities of in-house personnel. Will additional resources (personnel, consultants, technologies, etc.) be needed to successfully implement the project?

☺ The facility's ability to obtain rebates from the government, utilities, or other organizations. For example, there are Dept. of Energy subsidies available for DOE facilities.

$ The facility's ability to obtain tax benefits. For example, government facilities can offer tax-exempt interest rates on bonds.

A few of the Project Characteristics include:

$ The project's economic benefits: Net Present Value, Internal Rate of Return and Simple Payback.

☺ The project's complexity and overall risk. For example, a complex project that has never been done before has a different level of risk than a standard lighting retrofit.

☺ The project's alignment with the facility's long-term objectives. Will this project's equipment be needed for long-term goals?

☺ The project's cash flow schedule and the variance between cash flows. For example, there may be significant differences in the acceptability of a project based on when revenues are received.

A few of the Financial Arrangement Characteristics include:

$ The economic benefit of a project using a particular financial arrangement. The Net Present Value and Internal Rate of Return can be influenced by the financial arrangement selected.

☺ The impact on the corporate capital structure. For example, will additional debt be required to finance the project? Will additional liabilities appear on the firm's balance sheet and impact the image of the company to investors?

☺ The flexibility of the financial arrangement. For example, can the facility manager alter the contract and payment terms in the event of revenue shortfall or changes in operational hours?

INCORPORATING STRATEGIC ISSUES WHEN SELECTING FINANCIAL ARRANGEMENTS

Because strategic issues can be important when selecting financial arrangements, the facility manager should include them in the selection process. The following questions can help assess a facility manager's needs:

- Does the facility manager want to manage projects or outsource?
- Are net positive cash flows required?
- Will the equipment be needed for long-term needs?
- Is the facility government or private?
- If private, does the facility manager want the project's assets on or off the balance sheet?
- Will operations be changing?

From the research experience, a Strategic Issues Financing Decision Tree was developed to guide facility managers to the financial arrangement that is most likely optimal. Figure 2-14 illustrates the decision tree, which is by no means a rule, but it embodies some general observations from the industry.

Working the tree from the top to bottom, the facility manager should assess the project and facility characteristics to decide whether it is strategic to manage the project or outsource. If outsourced, the "performance contract" would be the logical choice.* If the facility manager wants to manage the project, the next step (moving down the tree) is to evaluate whether the project's equipment will be needed for long or short-term purposes. If short-term, the "true lease" is logical. If it is a long-term project, in a government facility, the "bond" is likely to be the best option. If the facility is in the private sector, the facility manager should decide whether the project should be on or off the balance sheet. An off-balance sheet preference would lead back to the "true lease." If the facility manager wants the project's assets on the balance sheet, the Net Present Value (or other economic benefit indicator) can help determine which "host-managed" arrangement (loan, capital lease or cash) would be most lucrative.

Although the decision tree can be used as a guide, it is most im-

*It should be noted that a performance contract could be structured using leases and bonds.

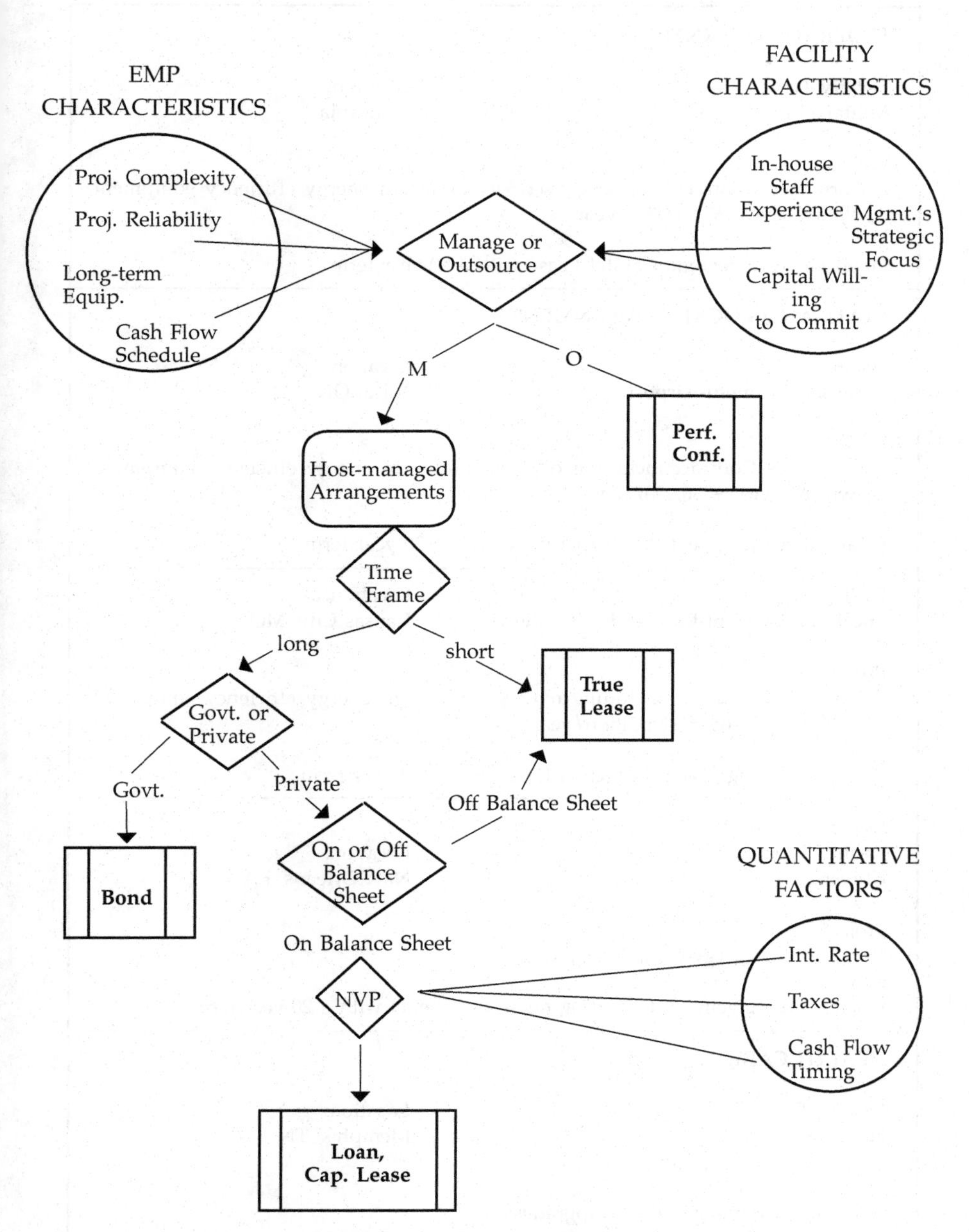

Figure 2-14. Strategic Issues Financing Decision Tree.

FEDERAL GOVERNMENT

Facility Type: Military Installation

Location: California

Project:
Performance Contract including over $40 million in energy efficiency equipment, which saves $7,000,000/year

Financial Arrangement: Capital Lease with a 20-year term

STATE AND LOCAL GOVERNMENT

Facility Type: **Local Government-Airport**

Location: Tulsa, OK

Project:
Performance Contract including over $4 million in energy efficiency equipment, which saves $380,000/year

Financial Arrangement: Municipal Bond with a 20-year term

Facility Type: **Local Government-Convention Center**

Location: Kansas City, MO

Project:
Performance Contract including over $8 million in energy efficiency equipment, which saves over $1 million/year

Financial Arrangement: Municipal Lease with a 10-year term

Education

Facility Type: University

Location: New Orleans, LA

Project:
$8 million in energy-related upgrades

Financial Arrangement: Operating Lease (Synthetic) with a 20-year term

HEALTH CARE

Facility Type: Hospital

Location: Memphis, TN

Project:
$15 million in energy-related upgrades

Financial Arrangement: Operating Lease (Synthetic) with a 20-year term

portant to use the financial arrangement that best meets the needs of the organization. The examples on the adjacent page demonstrate that any organization can be creative with its financial arrangement selection. All of these examples are for Performance Contracting projects; however, similar financial arrangements can be structured without using a performance contract.

CHAPTER SUMMARY

It is clear that knowing the strategic needs of the facility manager is critical to selecting the best arrangement. There are practically an infinite number of financial alternatives to consider. This chapter has provided some information on the basic financial arrangements. Combining these arrangements to construct the best contract for your facility is only limited by your creativity.

GLOSSARY

Capitalize

To convert a schedule of cash flows into a principal amount, called capitalized value, by dividing by a rate of interest. In other words, to set aside an amount large enough to generate (via interest) the desired cash flows forever.

Capital or Financial Lease

Lease that, under Statement 13 of the Financial Accounting Standards Board, must be reflected on a company's balance sheet as an asset and corresponding liability. Generally, this applies to leases where the lessee acquires essentially all of the economic benefits and risks of the leased property.

Depreciation

The amortization of fixed assets, such as plant and equipment, so as to allocate the cost over their depreciable life. Depreciation reduces taxable income but is not an actual cash flow.

Energy Service Company (ESCO)

Company that provides energy services (and possibly financial services) to an energy consumer.

Host

The building owner or facility that uses the equipment.

Lender

Individual or firm that extends money to a borrower with the expectation of being repaid, usually with interest. Lenders create debt in the form of loans or bonds. If the borrower is liquidated, the lender is paid off before stockholders receive distributions.

Lessee

The renter. The party that buys the right to use equipment by making lease payments to the lessor.

Lessor

The owner of the leased equipment.

Line of Credit

An informal agreement between a bank and a borrower indicating the maximum credit the bank will extend. A line of credit is popular because it allows numerous borrowing transactions to be approved without re-application paperwork.

Liquidity

Ability of a company to convert assets into cash or cash equivalents without significant loss. For example, investments in money market funds are much more liquid than investments in real estate.

Leveraged Lease

Lease that involves a lender in addition to the lessor and lessee. The lender, usually a bank or insurance company, puts up a percentage of the cash required to purchase the asset, usually more than half. The balance is put up by the lessor, who is both the equity participant and the borrower. With the cash the lessor acquires the asset, giving the lender (1) a mortgage on the asset and (2) an assignment of the lease and lease payments. The lessee then makes periodic payments to the lessor, who in turn pays the lender. As owner of the asset, the lessor is entitled to tax deductions for depreciation on the asset and interest on the loan.

MARR (Minimum Attractive Rate of Return)

MARR is the "hurdle rate" for projects within a company. MARR is used to determine the NPV. The annual after-tax cash flow is discounted at MARR (which represents the rate the company could have received with a different project).

Net Present Value (NPV)

As the saying goes, "A dollar received next year is not worth as much as a dollar today." The NPV converts the worth of that future dollar into what it is worth today. NPV converts future cash flows by using a given discount rate. For example, at 10%, $1,000 dollars received one year from now is worth only $909.09 dollars today. In other words, if you invested $909.09 dollars today at 10%, in one year it would be worth $1,000.

NPV is useful because you can convert future savings cash flows back to "time zero" (the present), and then compare to the cost of a project. If the NPV is positive, the investment is acceptable. In capital budgeting, the discount rate used is called the "hurdle rate" and is usually equal to the incremental cost of capital.

"Off-Balance Sheet" Financing

Typically refers to a True Lease, because the assets are not listed on the balance sheet. Because the liability is not on the balance sheet, the Host appears to be financially stronger. However, most large leases must be listed in the footnotes of financial statements, which reveals the "hidden" assets.

Par Value or Face Value

Equals the value of the bond at maturity. For example, a bond with a $1,000 dollar par value will pay $1,000 to the issuer at the maturity date.

Preferred Stock

A hybrid type of stock that pays dividends at a specified rate (like a bond), and has preference over common stock in the payment of dividends and liquidation of assets. However, if the firm is financially strained, it can avoid paying the preferred dividend as it would the common stock dividends. Preferred stock doesn't ordinarily carry voting rights.

Project Financing

A type of arrangement typically meaning that a Single Purpose

Entity (SPE) is constructed. The SPE serves as a special bank account. All funds are sent to the SPE, from which all construction costs are paid. Then all savings cash flows are also distributed from the SPE. The SPE is essentially a mini-company, with the sole purpose of funding a project.

Secured Loan

Loan that pledges assets as collateral. Thus, in the event that the borrower defaults on payments, the lender has the legal right to seize the collateral and sell it to pay off the loan.

True Lease or Operating Lease or Tax-Oriented Lease

Type of lease, normally involving equipment, whereby the contract is written for considerably less time than the equipment's life, and the lessor handles all maintenance and servicing; also called service lease. Operating leases are the opposite of capital leases, where the lessee acquires essentially all the economic benefits and risks of ownership. Common examples of equipment financed with operating leases are office copiers, computers, automobiles and trucks. Most operating leases are cancelable.

WACC (Weighted Average Cost of Capital)

The firm's average cost of capital, as a function of the proportion of different sources of capital: Equity, Debt, Preferred Stock, etc. *For example, a firm's target capital structure is:*

Capital Source	Weight (w_i)
Debt	30%
Common Equity	60%
Preferred Stock	10%

and the firm's costs of capital are:

before tax cost of debt = k_d = 10%
cost of common equity = k_s = 15%
cost of preferred stock = k_{ps} = 12%

Then the weighted average cost of capital will be:

$$WACC = w_d k_d (1-T) + w_s k_s + w_{ps} k_{ps}$$

where w_i = *weight of Capital Source*$_i$

$$T = tax\ rate = 34\%$$
$$After\text{-}tax\ cost\ of\ debt = k_d(1\text{-}T)$$

Thus,

WACC= (.3)(.1)(1-.34) +(.6)(.15) + (.1)(.12)
WACC= 12.18%

References

1. Wingender, J. and Woodroof, E., (1997) "When Firms Publicize Energy Management Projects Their Stock Prices Go Up: How High?—As Much as 21.33% within 150 days of an Announcement," *Strategic Planning for Energy and the Environment*, Vol. 17(1), pp. 38-51.
2. U.S. Department of Energy, (1996) "Analysis of Energy-Efficiency Investment Decisions by Small and Medium-Sized Manufacturers," U.S. DOE, Office of Policy and Office of Energy Efficiency and Renewable Energy, pp. 37-38.
3. Woodroof, E. and Turner, W. (1998), "Financial Arrangements for Energy Management Projects," *Energy Engineering* 95(3) pp. 23-71.
4. Sullivan, A. and Smith, K. (1993) "Investment Justification for U.S. Factory Automation Projects," *Journal of the Midwest Finance Association*, Vol. 22, p. 24.
5. Fretty, J. (1996), "Financing Energy-Efficient Upgraded Equipment," Proceedings of the 1996 International Energy and Environmental Congress, Chapter 10, Association of Energy Engineers.
6. Pennsylvania Energy Office, (1987) The Pennsylvania Life Cycle Costing Manual.
7. United States Environmental Protection Agency (1994). ProjectKalc, Green Lights Program, Washington DC.
8. Tellus Institute, (1996), P2/Finance version 3.0 for Microsoft Excel Version 5, Boston MA.
9. Woodroof, E. And Turner, W. (1999) "Best Ways to Finance Your Energy Management Projects," *Strategic Planning for Energy and the Environment*, Summer 1999, Vol. 19(1) pp. 65-79.
10. Cooke, G.W., and Bomeli, E.C., (1967), *Business Financial Management*, Houghton Mifflin Co., New York.
11. Wingender, J. and Woodroof, E., (1997) "When Firms Publicize Energy Management Projects: Their Stock Prices Go Up," *Strategic Planning for Energy and the Environment*, 17 (1) pp. 38-51.
12. Sharpe, S. and Nguyen, H. (1995) "Capital Market Imperfections and the Incentive to Lease," *Journal of Financial Economics*, 39(2), p. 271-294.
13. Schallheim, J. (1994), *Lease or Buy?*, Harvard Business School Press, Boston, p. 45.
14. Hines, V. (1996), "EUN Survey: 32% of Users Have Signed ESCO Contracts," *Energy User News* 21(11), p.26.
15. Coates, D.F. and DelPonti, J.D. (1996), "Performance Contracting: a Financial Perspective" *Energy Business and Technology Sourcebook*, Proceedings of the 1996 World Energy Engineering Congress, Atlanta. p.539-543.

Chapter 3

Choosing the Right Financing For Your Energy Efficiency And Green Projects With ENERGY STAR®

Neil Zobler
Catalyst Financial Group, Inc.
Katy Hatcher
U.S. Environmental Protection Agency,
ENERGY STAR Program

FINANCING EFFICIENCY PROJECTS TODAY WITH FUTURE ENERGY SAVINGS

Administrators or managers often think they must postpone the implementation of energy efficiency upgrades because they do not have the funds in their current budgets. Other barriers may include the lack of time, personnel, or expertise. As you will see later in this chapter, most organizations have access to more financial resources than they may think, and resolving the financial barrier frequently provides solutions to all the other barriers.

Postponing energy efficiency upgrades for as little as one year can prove to be an expensive decision. The U.S. Environmental Protection Agency (EPA) estimates that as much as 30% of the energy consumed in buildings may be used unnecessarily or inefficiently. The money lost due to these inefficiencies in just one year frequently totals more than all the costs of financing energy upgrades over the course of the entire financing period!

Consider this business logic: The energy efficiency project you'd like to do is not in your current capital budget; but if the costs of financing the project are less than the operating budget dollars saved from reduced utility bills, why not finance the project? The benefits of doing the project sooner rather than later are numerous, starting with improved cash flow, better facilities, using the existing capital budget for other projects, helping make your facility "green," and more. Using this logic, financing energy efficiency projects is a good business decision. Delaying the project is, however, a conscious decision to continue paying the utility companies for the energy waste rather than investing these dollars in improving your facilities.

Energy efficiency projects are unlike most other projects. With properly structured financing, you may be able to implement energy efficiency projects without exceeding your existing operating or capital budgets. Thousands of companies that participate in ENERGY STAR® know from experience that today's energy efficiency technologies and practices have saved them operating budget dollars. In fact, some of the more conservative lenders like Bank of America and CitiBank have been providing funds for energy efficiency projects for over a decade. And implementing energy efficiency projects will have a positive impact on your organization's overall financial performance as well as the environment. So why wait? This chapter will help you understand how to leverage your savings opportunities, which financing vehicles to consider, and where to find the money.

ENERGY EFFICIENCY AND GREEN BUILDINGS

Energy efficiency and indoor air quality are critical components of any green building project and are required by the U.S. Green Building Council (USGBC) to obtain Leadership in Energy and Environmental Design (LEED) certification. For example, the LEED for Existing Buildings rating system requires that certain prerequisites be met before a certification at any level can be achieved. An ENERGY STAR rating of at least 60 is one of those prerequisites. (See the FINANCING TOOLS AND RESOURCES from ENERGY STAR Section below for more information on EPA's energy performance rating system.) USGBC is also requiring that at least two points be earned under the Optimize Energy Efficiency Performance credit, which essentially requires an even higher ENERGY STAR rating.

Energy efficiency related building improvements are synergistic with other categories in the LEED process (sustainable sites, water efficiency, energy and atmosphere, materials and resources, and indoor environmental quality). For instance, points can be earned for controllability of systems, day lighting, water usage, etc.

Using the operating budget savings realized from implementing energy efficiency projects is a good way to offset the costs of making your building greener and obtaining LEED certification.

OPERATING VERSUS CAPITAL BUDGETS

Before addressing different financing options and vehicles, let's review some "accounting 101" fundamentals. Organizations make purchases by spending their own cash or by borrowing the needed funds. The impact on the balance sheet is either exchanging one asset for another when spending cash, or adding to the assets *and* liabilities when incurring debt.

To argue the advantages of one financing option versus another, it is important to be conversant with the roles of the operating expense budget and the capital expense budget in your organization. ***Capital expenses*** are those that pay for long-term debt and fixed assets (such as buildings, furniture, and school buses) and whose repayment typically extends **beyond** one operating period (one operating period usually being 12 months). In contrast, ***operating expenses*** are those general and operating expenses (such as salaries or supply bills) incurred **during** one operating period (again, typically 12 months).* For example, repayment of a bond issue is considered a capital expense, whereas paying monthly utility bills is considered an operating expense.

The disadvantages associated with trying to use capital expense budget dollars for your energy efficiency projects are as follows: (1) current fiscal year capital dollars are usually already committed to other projects; (2) capital dollars are often scarce, so your efficiency projects are competing with other priorities; and (3) the approval process for requesting new capital dollars is time consuming, expensive, and often cumbersome.

When arranging financing for energy efficiency projects in the pri-

---

*According to Barron's Dictionary of Accounting Terms, capital expenditures are "outlays charged to a long-term asset account. A capital expenditure either adds a fixed asset unit or increases the value of an existing fixed asset." Operating expenditures are costs "associated with the ... administrative activities of the [organization]."

vate sector, one of the most frequently asked questions is, "How do we keep this financing off our balance sheet?" and thereby not reflect the transaction on the company's financial statements as a liability or debt. The reasons for this request vary by organization and include: (a) treating the repayment of the obligation as an operating expense, thereby avoiding the entire capital budget process, and, (b) avoiding the need for compliance with restrictive covenants that are frequently imposed by existing lenders, which may be viewed as cumbersome to a point of interfering with the ongoing management of the company. Restrictive covenants start by requiring the borrower to periodically provide financial statements that enable the lender to track the performance of the company by calculating key financial ratios measuring liquidity (i.e., the current ratio, which is current assets versus current liabilities), leverage (debt-to-equity), and profitability margins. Covenants include maintaining financial ratios at agreed standards. If the ratios are not in compliance with these targets, the lender can call in all loans, creating serious cash flow problems for the borrower. Many of these financial ratios are improved by keeping debt off the balance sheet. Other typical covenants include limitations on issuing new debt, paying dividends to stockholders, and selling assets of the company.

While organizations in the public sector may not have to deal with the profitability and equity issues of the private sector, they do face their own challenges when incurring debt through the capital budget process, which is established by statute, constitution, or charter, and usually requires voter approval. Public sector organizations may find that the political consequences of incurring new debt may be more of a deterrent than the financial ones, particularly when raising taxes is involved.

Treating repayment of the financing for energy efficiency projects as an operating expense can keep the financing "off the balance sheet." And, because the immediate benefit of installing energy efficiency projects is reducing the operating expense budget earmarked for paying the energy and water bills, off balance sheet treatment makes sense. Post-ENRON, however, having your auditors treat financing as "off balance sheet" is becoming increasingly difficult, especially in light of The Sarbanes-Oxley Act of 2002, which established new or enhanced standards for all U.S. public company boards, management, and public accounting firms.

---

*Note, however, off balance sheet transactions will be referenced in the footnotes to the organization's financial statements.

Nevertheless, several financing vehicles do allow financing payments for energy efficiency upgrades to be treated as operating expenses. (These include operating leases, power purchase agreements, and tax-exempt lease-purchase agreements; see the FINANCIAL INSTRUMENTS section below for more information.) Regardless of the type of financing vehicle used, implementing energy efficiency projects is in the best financial interests of your organization and helps protect the environment.

FINANCING TOOLS AND RESOURCES FROM ENERGY STAR

EPA's ENERGY STAR program offers a proven strategy for superior energy management, with tools and resources to help each step of the way. Based on the successful practices of ENERGY STAR partners, EPA's "Guidelines for Energy Management" (available at www.energystar.gov/guidelines*) illustrates how organizations can improve operations and maintenance strategies to reduce energy use and maintain the cost savings that can be realized by financing energy efficiency projects. EPA has sponsored hundreds of presentations (in person and on the internet) on ENERGY STAR tools, resources, and best practices for organizations struggling with the challenge of making their buildings more energy efficient. One of the most common statements from participants, especially those in the public sector, is "We don't have the money needed to do the facility upgrades; in fact, we don't even have enough money to pay for the energy audits needed to determine the size of the savings opportunity." This sentiment is simply not true because the needed funds are currently sitting in their utility operating budgets and being doled out every month to the local utilities. Organizations merely require a way to capture and redirect these "wasted energy" funds to pay for the energy efficiency projects, which will in turn create real savings. For some readers, this may seem to be "circular logic," or what may be called a "Catch 22."

Fortunately, ENERGY STAR has created a number of tools and resources that, when properly used, will allow you to "break the circle" and find a path toward the timely implementation of energy efficiency projects. This section focuses on the tools that tie directly into financing such projects. These include the Guidelines to Energy Management,

---

*Information about ENERGY STAR's tools and resources has been provided by US EPA ENERGY STAR. For more information please refer to their website.

Portfolio Manager, Target Finder, Financial Value Calculator, Building Upgrade Value Calculator, and the Cash Flow Opportunity Calculator. All of these tools are in the public domain and available at www.energystar.gov.

Portfolio Manager

Peter Drucker's famous maxim is, "If you don't measure it, you can't manage it." If your organization wants to use future energy savings to pay for the implementation of energy efficiency projects now, you must start by establishing the baseline of your current energy usage. ENERGY STAR's Portfolio Manager can help you do that. It is an interactive energy management tool that allows you to track and assess energy and water consumption across your entire portfolio of buildings in a secure online environment. Portfolio Manager can help you identify under-performing buildings, set investment priorities, verify effectiveness of efficiency improvements, and receive EPA recognition for superior energy performance.

Any building manager or owner can efficiently track and manage resources through Portfolio Manager. The tool allows you to streamline your portfolio's energy and water data, as well as track key consumption, performance, and cost information portfolio-wide. For example, you can:

- Track multiple energy and water meters for each facility.
- Customize meter names and key information.
- Benchmark your facilities relative to their past performance.
- View percent improvement in weather-normalized source energy.
- Monitor energy and water costs.
- Share your building data with others inside or outside your organization.
- Enter operating characteristics, tailored to each space use category within a building.

For many types of facilities, you can rate energy performance on a scale of 1-100 relative to similar buildings nationwide. Your building is ***not*** compared to the other buildings in Portfolio Manager to determine your ENERGY STAR rating. Instead, statistically representative models are used to compare your building against similar buildings from a national survey conducted by the U.S. Department of Energy's (DOE's) Energy

Information Administration known as the Commercial Building Energy Consumption Survey (CBECS). Conducted every four years, CBECS gathers data on building characteristics and energy use from thousands of buildings across the United States. Your building's peer group for comparison is the group of buildings in the CBECS survey that has similar building and operating characteristics. An EPA rating of 50 indicates that the building, from an energy consumption standpoint, performs better than 50% of all similar buildings nationwide, while a rating of 75 indicates that the building performs better than 75% of all similar buildings nationwide.

EPA's energy performance rating system, based on source energy,* accounts for the impact of weather variations, as well as changes in key physical and operating characteristics of each building. Buildings rating 75 or higher may qualify for the ENERGY STAR label.

Portfolio Manager provides a platform to track energy and water use trends compared against the costs of these resources. This is a valuable tool for understanding the relative costs associated with a given level of performance, helping you evaluate investment opportunities for a particular building, and identifying the best opportunities across your portfolio. It also allows you to track your properties' performance from year to year.

The built-in financial tool within Portfolio Manager helps you compare cost savings across buildings in your portfolio while calculating cost savings for a specific project. Being able to quickly and clearly obtain data showing cumulative investments in facility upgrades or annual energy costs eases the decision making process for best practice management of your buildings nationwide.

From a lender's perspective, a facility with a low rating is more likely to obtain larger energy savings (having more room for improvement) than a facility with a high rating. This becomes important if the energy savings are considered a primary "source of repayment" when financing energy upgrades. Portfolio Manager is also an important tracking mechanism that helps insure that the facilities are being properly maintained and the energy savings are continuing to accrue. As lenders perform due diligence on energy efficiency projects, they will become more aware of the value of Portfolio Manager.

---

*Source energy includes the energy consumed at the building itself—or the *site energy*—plus the energy used to generate, transmit, and distribute the site energy.

Target Finder

Target Finder is another ENERGY STAR interactive energy management tool that helps design teams establish an energy performance target for new projects and major building renovations early in the design process. This target or goal serves as a reference for comparing energy strategies and choosing the best technologies and practices for achieving the performance goal. As the design nears completion, architects enter the project's estimated energy consumption, and Target Finder generates an energy performance rating based on the same underlying database applied to existing buildings in Portfolio Manager.

The energy use intensity generated by Target Finder reflects the distribution of energy performance in commercial buildings derived from CBECS data. The required data inputs (square footage, hours of operation, etc.) were found to be the primary drivers of energy use. The project's ZIP code is used to determine the climate conditions that the building would experience in a normal year (based on a 30-year climate average). The total annual energy use intensity for the target is based on the energy fuel mix typical in the region specified by the ZIP code. Target Finder displays the percent electricity and natural gas defaults used to calculate design targets. Users may enter their own fuel mix, but electricity must be selected as one of the choices. Site and source energy calculations are shown for both energy use intensity and total annual energy. The EPA rating, however, is calculated from source energy use.

EPA's Target Finder helps architects and building owners set aggressive, realistic energy targets and rate a design's intended energy use. Making energy efficiency a design requirement is the most cost-effective way to insure superior energy performance in your properties.

Building Upgrade Value Calculator

The Building Upgrade Value Calculator is built on a Microsoft EXCEL™ platform and is a product of the partnership between EPA's ENERGY STAR, BOMA International, and the BOMA Foundation. The Building Upgrade Value Calculator estimates the financial impact of proposed investments in energy efficiency in office properties. The calculations are based on data input by the user, representing scenarios and conditions present at their properties. Required inputs are limited to general characteristics of the building, plus information on the proposed investments in energy efficiency upgrades.

The calculator's analysis includes the following information:

- Net investment
- Reduction in operating expense
- Energy savings
- Return on investment (ROI)
- Internal rate of return (IRR)
- Net present value (NPV)
- Net operating income (NOI)
- Impact on asset value

In addition to the above outputs, the calculator also estimates the impact the proposed energy efficiency changes will have on a property's ENERGY STAR rating.

The tool provides two ways to use its calculations. Users can save and print a summary of their results, or they can generate a letter highlighting the financial value for use as part of a capital investment proposal. Because energy efficiency projects generally improve net operating income, which is an important consideration when buying and selling commercial properties, the Financial Value Calculator provides strong financial arguments in favor of implementing these projects.

Financial Value Calculator

Investments in energy performance can have a favorable impact on profit margins, earnings per share, and ultimately, shareholder value. The Financial Value Calculator is another tool built on an EXCEL™ platform. It presents energy investment opportunities using the key financial metrics managers need to convey the message of improved energy performance to customers.

Cash Flow Opportunity Calculator

The Cash Flow Opportunity Calculator ("CFO Calculator") has proven to be a very effective tool, especially for public sector projects. This set of spreadsheets helps create a sense of urgency about implementing energy efficiency projects by quantifying the costs of delaying the project implementation. It was developed to help decision makers address three critical questions about energy efficiency investments:

1. How much of the new energy efficiency project can be paid for using the anticipated savings?
2. Should this project be financed now, or is it better to wait and use

cash from a future budget?

3. Is money being lost by waiting for a lower interest rate?

Using graphs and tables, the CFO Calculator is written so that managers who are not financial specialists can use it to make informed decisions, yet it is sophisticated enough to satisfy financial decision makers. This tool works well for projects in both the public and private sectors.

To determine how much of the new project can be paid for using your anticipated savings, the CFO Calculator takes a practical look at your energy efficiency situation and financing opportunities. You can choose to enter either (a) best estimates of how your building currently operates and how much better it could operate, or (b) data generated when you use EPA's energy performance rating system within Portfolio Manager. Either way, the CFO Calculator provides answers to some critical financial questions in just minutes.

The first step in the process is to estimate the amount of the savings that can be captured from the existing utility budget. The working assumption is that these savings will be used to cover the financing costs and that the savings will recur. The savings amount is entered into a "reverse financial calculator," which then asks for an estimated borrowing interest rate, financing term, and the percentage of the savings you wish to use. It then calculates the amount of project improvements that could be purchased by redirecting these energy net savings to pay for the upgrades. Most organizations are surprised to learn how much new equipment and related services are "buried" in their utility bills, all of which could be installed within their existing operating budget and without spending their limited capital budget. The related services often include the initial energy audits that many feel they cannot afford, but are necessary to quantify the savings opportunity. When future energy savings are the main source of the project's repayment, the CFO Calculator becomes an effective sensitivity analysis tool that takes into account the impact of lower interest rates, longer financing terms, and utilization of savings when structuring the project's financing.

A while back, the "See how much money you are leaving on the table" argument was made to the CFO of a large city in the Northeast on behalf of the local electric utility. The CFO responded that the city was fiscally conservative and believed that waiting until funds were available in a future operating budget (thereby avoiding borrowing and paying interest) was in the best interests of the city. The CFO Calculator was used to map the cash flow consequences of these two decision points (financing now or waiting until a future budget) to demonstrate to the city's CFO and city council that *financing now* was a better financial decision than waiting for cash. In most instances, the lost energy savings incurred by waiting for one year are greater than the net present value of all the interest payments of most financing, making "do it now" the better financial decision. This is counter intuitive and surprises most decision makers. Today, this city supports the expeditious implementation of energy efficiency projects.

Another common argument for delay is waiting for a lower interest rate offering rather than financing at a higher rate that is available immediately. This situation may occur when waiting for funds from a future bond issue or for a low-cost specialty fund to replenish itself, versus accepting an immediately available third-party financing offering. The CFO Calculator allows you to compare two different interest rate offerings, and it will compute how long you can wait for the lower interest rate before the lower rate begins to cost more. It does this by including the forfeited

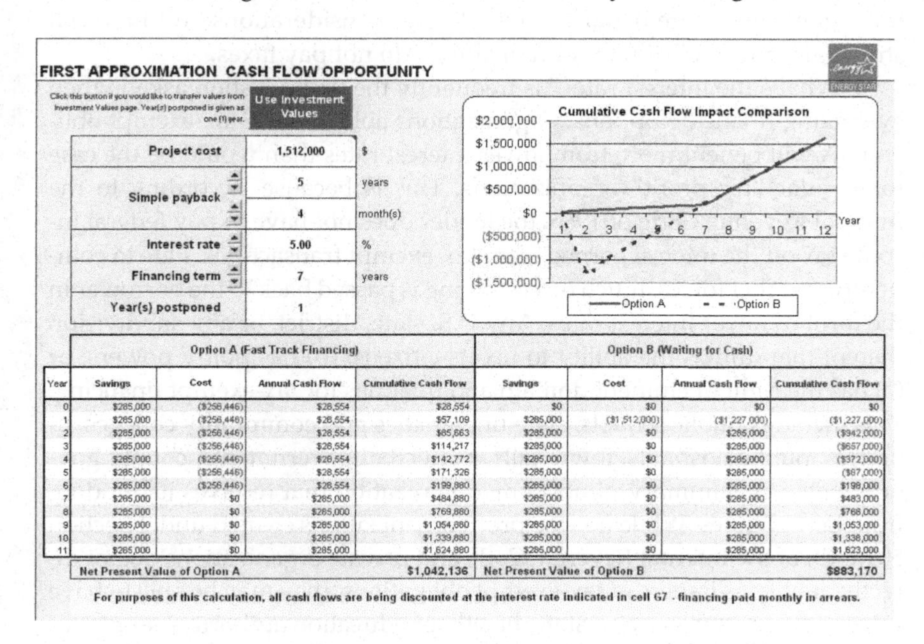

	Option A (Fast Track Financing)				Option B (Waiting for Cash)			
Year	Savings	Cost	Annual Cash Flow	Cumulative Cash Flow	Savings	Cost	Annual Cash Flow	Cumulative Cash Flow
0	$285,000	($256,446)	$28,554	$28,554	$0	$0	$0	$0
1	$285,000	($256,446)	$28,554	$57,109	$285,000	($1,512,000)	($1,227,000)	($1,227,000)
2	$285,000	($256,446)	$28,554	$85,663	$285,000	$0	$285,000	($942,000)
3	$285,000	($256,446)	$28,554	$114,217	$285,000	$0	$285,000	($657,000)
4	$285,000	($256,446)	$28,554	$142,772	$285,000	$0	$285,000	($372,000)
5	$285,000	($256,446)	$28,554	$171,326	$285,000	$0	$285,000	($87,000)
6	$285,000	($256,446)	$28,554	$199,880	$285,000	$0	$285,000	$198,000
7	$285,000	$0	$285,000	$484,880	$285,000	$0	$285,000	$483,000
8	$285,000	$0	$285,000	$769,880	$285,000	$0	$285,000	$768,000
9	$285,000	$0	$285,000	$1,054,880	$285,000	$0	$285,000	$1,053,000
10	$285,000	$0	$285,000	$1,339,880	$285,000	$0	$285,000	$1,338,000
11	$285,000	$0	$285,000	$1,624,880	$285,000	$0	$285,000	$1,623,000
	Net Present Value of Option A			$1,042,136	Net Present Value of Option B			$883,170

For purposes of this calculation, all cash flows are being discounted at the interest rate indicated in cell G7 - financing paid monthly in arrears.

Figure 3-1. This is a sample screen capture of the "cash flow" tab from the CFO Calculator Excel Spread Sheet supporting the "do it now" argument.

energy savings into the decision making process, which is truly another "cost of delay."

CHOOSING THE RIGHT FINANCING

"Financing" should be thought of as a two-step process: financial instruments (or vehicles) and sources of funds. Once you decide which financing vehicle is best for your organization, the next step is to choose the best source of funds. Bear in mind that no one financing alternative is right for everyone. In the world of energy efficiency finance, one size definitely does not fit all! For the purposes of this discussion, our focus is limited to the public and private sectors, not consumer finance options.

Before you can choose the right financial vehicle, however, two other issues must be considered—tax exempt status and interest rates. In general, public sector organizations and some non-profits qualify for tax-exempt financing, while private sector organizations do not. Private sector organizations are usually driven by tax considerations and financial strategies, but public sector organizations do not pay taxes.

What's the interest rate?" is frequently the first question asked when evaluating financing options. Organizations able to issue tax-exempt obligations will benefit more from lower interest rates than would be the case for regular "for profit" organizations. This is because, according to the Internal Revenue Code of 1986, the lender does not have to pay federal income tax on the interest earned from tax-exempt transactions. Due to competitive market forces, much of this saving is passed back to the borrower in the form of lower interest rates. Any U.S. state, district, or any subdivision thereof that (a) has the ability to tax its citizens; (b) has police powers; or (c) has the right of eminent domain and qualifies for tax-exempt financing. This includes public schools, state universities and community colleges, libraries, public hospitals, town halls, municipal governments, county governments—in summary, almost any organization that receives its funding from tax revenues. While not-for-profit organizations created under Section 501(c)(3) of the Internal Revenue Code and private organizations do not directly qualify as issuers of tax-exempt obligations, they may be able to have a "conduit agency"—a city, state, health, or education authority—apply for the financing on their behalf.

For private sector organizations, interest rate alone is rarely the best indicator of the "best deal." To show the importance of proper deal structuring, consider the following question: "Which is the better finance offering, 0% or 6%?" Most people say "0%" until they find out that the 0% obligation must be repaid in 6 months, while the 6% obligation has a term of 5 years, making *cash flow* the deciding factor. The question is further complicated when the 6% obligation requires that the owner(s) personally guarantee the financing; however, the transaction can be done at 12% without a personal guarantee. Once all of the terms and conditions are known, it is easier to understand why many would prefer to pay a 12% financing interest rate rather than 0%. In addition to the term and personal guarantees, the list of structuring points is broad and may include whether the rate is fixed or variable, payment schedules, down payments, balloon payments, balance sheet impact, tax treatment, senior or subordinated debt, and whether additional collateral is required.

Your organization's legal structure, size, credit rating, time in business, sources of income, and profitability or cash flow are also important considerations when choosing the right financing vehicle and source of funds. The type of project, general market conditions, dollar size of the project, and the use of the equipment being financed are also important.

FINANCIAL INSTRUMENTS

There are two basic approaches to funding projects, "pay-as-you-go" and "pay-as-you-use." Pay-as-you-go means paying for the project out of current revenues at the time of expenditure; in other words, paying cash. If you don't have the cash, the project gets postponed until you do. Pay-as-you-use means borrowing to finance the expenditure, with debt service payments being made from revenues generated during the useful life of the project. Because energy efficiency projects generate operating savings over the life of the project, the pay-as-you-use approach makes good sense.

As previously mentioned, public sector organizations can borrow at tax-exempt rates, which are substantially lower than the taxable rates that private sector organizations will pay when financing. While tax-exempt financial instruments can only be used by public sector (and qualifying non profit and private sector) organizations, taxable instruments can be used by all. This section will help identify the financial instruments best suited for public or private sector application. Public sector instruments include bonds and tax-exempt lease purchase agreements. Private sector instruments include a variety of commercial leases. All sectors can benefit from *energy performance contracts* and *power purchase agreements*. Bear in mind that there are exceptions to every rule, and structuring a financing to comply with tax or budget issues is often complicated, which is why working with a financial advisor may prove helpful.

Major capital projects are funded by some form of *debt*, which is categorized as either short term (for periods of less than one year) or long term (for periods greater than one year). Most borrowings by public sector organizations require citizen approval, either directly through referendum or indirectly through actions of an appointed board or elected council. However, revenue bonds and tax-exempt lease-purchase agreements may not require local voter approval. (See details below.)

Most of us are familiar with traditional *loans*, which are debt obligations undertaken by a borrower. The title of the asset being financed is typically in the name of the borrower, and the lender files a lien on the asset being financed, prohibiting the borrower from selling the asset until the lien is lifted. Banks frequently require additional collateral, which may take the form of keeping compensating balances in an account or placing a blanket lien on all other assets of the organization. A *conditional sales agreement* or *installment purchase agreement* is a kind of loan that is secured by the asset being financed; the title of the asset transfers to the borrower after the final payment is made. All loans are considered "on balance sheet" transactions and are common to both the private and public sectors.

Frequently used *short-term debt instruments* include bank loans (term loans or lines of credit), anticipation notes (in anticipation of bond, tax, grant or revenues to be received), commercial paper (taxable or tax-exempt unsecured promissory notes that can be refinanced or rolled over for periods exceeding one year), and floating-rate demand notes (notes that allow the purchaser to demand that the seller redeem the note when the interest rate adjusts).

Long-term debt is frequently in the form of *bonds*. Commercial bonds can be quite complex (asset backed, callable, convertible, debenture, fixed or floating rate, zero coupon, industrial development, etc.) and usually require working with an investment banker. In the public sector, bonds fall into two categories, general obligation (GO) bonds, and revenue bonds. *GO bonds* are backed by the issuer's full faith and credit and can only be issued by units of government with taxing authority. Because the issuer promises to levy taxes to pay for these obligations if necessary, these bonds have the lowest risk of default and, therefore, the lowest cost.

Interest paid on GO bonds is typically exempt from federal income taxes and may be exempt from state income taxes.

Revenue bonds are also issued by local governments or public agencies. However, because they are repaid only from the specific revenues named in the bond, they are considered to be riskier than GO bonds. Revenue bonds may not require voter approval and often contain covenants intended to reduce the perceived risk. Typical covenants include rate formulas, the order of payments, establishing sinking funds, and limiting the ability to issue new debt. Small municipalities that have difficulty issuing debt often add credit enhancements to their bonds in the form of bond insurance or letters of credit.

In the case of most energy efficiency projects, the source of repayment is the actual energy savings (considered part of the operating budget) realized by the project. When the approval process to obtain the necessary debt is a barrier, public sector organizations may be able to limit the repayment of the financing costs to their operating budget by using a tax-exempt lease purchase agreement. This solution may avoid the capital budget process altogether.

Tax-exempt Lease-purchase Agreements

Tax-exempt lease-purchase agreements are the most common public sector financing alternatives that are paid from operating budget dollars rather than capital budget dollars. A tax-exempt lease purchase agreement is an effective alternative to traditional debt financing (bonds, loans, etc.), because it allows a public organization to pay for energy upgrades by using money that is already set aside in its annual utility budget. When properly structured, this type of financing makes it possible for public sector agencies to draw on dollars to be saved in future utility bills to pay for new, energy-efficient equipment and related services today.

A tax-exempt lease-purchase agreement, also known as a municipal lease, is closer in nature to an installment-purchase agreement than a rental agreement. Under most long-term rental agreements or commercial leases (such as those used in car leasing), the renter or lessee *returns* the asset (the car) at the end of the lease term, without building any equity in the asset being leased. In contrast, a lease-purchase agreement presumes that the public sector organization will *own* the assets after the term expires. Further, the interest rates are appreciably lower than those on a taxable commercial lease-purchase agreement because the interest paid is exempt from federal income tax for public sector organizations.

In most states, a tax-exempt lease-purchase agreement usually does **not** constitute a long-term "debt" obligation because of non-appropriation language written into the agreement and, therefore, rarely requires public approval. This language effectively limits the payment obligation to the organization's current operating budget period (typically a 12-month period). The organization will, however, have to assure lenders that the energy efficiency projects being financed are considered of *essential use* (i.e., essential to the operation of the organization), which minimizes the non-appropriation risk to the lender. If, for some reason, future funds are not appropriated, the equipment is returned to the lender, and the repayment obligation is terminated at the end of the current operating period without placing any obligation on future budgets.

Public sector organizations should consider using a lease-purchase agreement to pay for energy efficiency equipment and related services when the projected energy savings will be greater than the cost of the equipment (including financing), especially when a creditworthy energy service company (ESCO) guarantees the savings. If your financial decision makers are concerned about exceeding operating budgets, they can be assured this will not happen, because lease payments can be covered by the dollars to be saved on utility bills once the energy efficiency equipment is installed. Utility bill payments are already part of any organization's normal year-to-year operating budget. Although the financing terms for lease-purchase agreements may extend as long as 15 to 20 years, they are usually less than 12 years and are limited by the useful life of the equipment.

There may be cases, however, when tax-exempt lease-purchase financing is not advisable for public sector organizations; for example, when (1) state statute or charter may prohibit such financing mechanisms; (2) the approval process may be too difficult or politically driven; or (3) other funds are readily available (e.g. bond funding that will soon be accessible or excess money that exists in the current capital or operating budgets).

How is Debt Defined in the Public Sector?

It is important for managers to be aware of the different interpretations of "debt" from three perspectives—legal, credit rating, and accounting. As mentioned above, most tax-exempt lease-purchase agreements are not considered "legal debt," which may prevent the need to obtain voter approval in your locality. However, credit rating agencies, such as

Moody's and Standard & Poor's, do include some or all of the lease-purchase obligations when they evaluate a public entity's credit rating and its ability to meet payment commitments ("debt service"). These two perspectives (legal and credit rating) may differ markedly from the way lease-purchase agreements are treated (i.e., which budget is charged) by your own accounting department and your organization's external auditors.

In general, lease-purchase payments on energy efficiency equipment are small when compared to the overall operating budget of a public organization. This usually means that the accounting treatment of such payments may be open to interpretation. Because savings occur only if the energy efficiency projects are installed, the projects' lease-purchase costs (or the financing costs for upgrades) can be paid out of the savings in the utility line item of the operating budget. Outside auditors may, however, take exception to this treatment if these payments are considered "material" from an accounting perspective. Determining when an expense is "material" is a matter of the auditor's professional judgment.* While there are no strictly defined accounting thresholds, as a practical guide, an item could be considered material when it equals or is greater than 5% of the total expense budget in the public sector (or 5% of the net income for the private sector). For example, the entire energy budget for a typical medium-to-large school district is around 2% of total operating expenses; therefore, so long as the payments stay under 2%, energy efficiency improvements will rarely be considered "material" using this practical guideline.

What are Energy Performance Contracts?

In most parts of the United States, an energy performance contract (EPC) is a common way to implement energy efficiency improvements. It frequently covers financing for the needed equipment, should your organization chose not to use internal funds. In fact, every state except Wyoming† has enacted some legislation or issued an executive directive to deal with energy efficiency improvements. While EPCs are used both in the public and private sectors, 82% of the revenues of ESCOs come

---

*According to Dr. James Donegan, Ph.D. (accounting), Western Connecticut State University, an amount is "considered material when it would affect the judgment of a reasonably informed reader when analyzing financial statements."

†http://www.ornl.gov/info/esco/legislation/

from public sector clients.* Properly structured EPCs can be treated as an operating, rather than a capital expense.

If you search for the phrase "energy performance contract," a variety of definitions appear. The U.S. Department of Housing and Urban Development (HUD) says that a performance contract is "an innovative financing technique that uses cost savings from reduced energy consumption to repay the cost of installing energy conservation measures…" The Energy Services Coalition—a national nonprofit organization composed of energy experts working to increase energy efficiency and building upgrades in the public sector through energy savings performance contracting—states that a performance contract is "an agreement with a private energy service company (ESCO)… [that] will identify and evaluate energy-saving opportunities and then recommend a package of improvements to be paid for through savings. The ESCO will guarantee that savings meet or exceed annual payments to cover all project costs… If savings don't materialize, the ESCO pays the difference…"

Notice that both definitions mention "payment." Let's dig a little deeper.

DOE claims that "Energy performance contracts are generally *financing or operating leases*† provided by an Energy Service Company (ESCO) or equipment manufacturer. What distinguishes these contracts is that they provide a guarantee on energy savings from the installed retrofit measures, and they usually also offer a range of associated design, installation, and maintenance services." The NY State Energy Research and Development Authority (NYSERDA) states that "An EPC is a method of implementation and project financing, whereby the operational savings from energy efficiency improvements is amortized over an agreed-upon repayment period through a *tax-exempt lease purchase arrangement…*" Meanwhile, the Oregon Department of Energy says, "An energy savings performance contract is an agreement between an energy services company (ESCO) and a building owner. The owner uses the energy cost savings to reimburse the ESCO and to pay off the *loan* that financed the energy conservation projects."

So, what is the funding mechanism used in an EPC? Is it a financing or operating lease (two very different structures—see EQUIPMENT

*A Survey of the U.S. ESCO Industry: Market Growth and Development from 2000 to 2006." Lawrence Berkeley National Laboratory, LBNL-62679. May 2007.
†Emphasis in all quotes was done by the author, not the publishers of the websites.

LEASING below), a tax-exempt lease-purchase agreement, or a loan? From a financing perspective, these are all different vehicles with diverse accounting and tax consequences. The answer is ***yes*** to all.

The definition of a performance contract may be found in some state statutes; however, in general it is not clearly defined and usually includes a variety of services such as energy audits, designing, specifying, selling and installing new equipment, providing performance guarantees, maintenance, training, measurement and verification protocols, financing, indoor air quality improvements, and more. One major benefit of using a performance contract is the ability to analyze the customer's needs and craft a custom agreement to address the organization's specific constraints due to budget, time, personnel, or lack of internal expertise. This includes choosing the financing vehicle that best suits the organization's financial and/or tax strategies.

Designed for larger projects, performance contracting allows for the use of energy savings from the operating budget (rather than the capital budget) to pay for necessary equipment and related services. Usually there is little or no up-front cost to the organization benefiting from the installed improvements, which then frees up savings from reduced utility bills that would otherwise be tied up in the operating budget. An energy performance contract is an agreement between the organization and an ESCO to provide a variety of energy saving services and products. Because these improvement projects usually cover multiple buildings and often include upgrades to the entire lighting and HVAC systems, the startup cost when ***not*** using an EPC may be high, and the payback period may be lengthy. Under a well-crafted EPC, the ESCO will be paid based on the verifiable energy savings.

The ESCO will identify energy saving measures through an extensive energy audit and then install and maintain the equipment and other upgrades. This includes low- and no-cost measures which contribute to the projects overall savings. The ESCO works closely with the client throughout the approval process to determine which measures to install, timing of the installations, staffing requirements, etc. The energy savings cover the costs of using the ESCO ***and*** financing for the project.

The most common type of performance contract is called a *"Guaranteed Savings Agreement,"* whereby the ESCO guarantees the savings of the installed energy-efficiency improvements (equipment and services). The ESCO assumes the performance risk of the energy-efficient equipment so that if the promised savings are not met, the ESCO pays the difference between promised savings and actual savings. If the savings allow, a performance contract may include related services, such as the disposal of hazardous waste from the replacement of lighting systems or from the removal of asbestos when upgrading ventilation systems. The ESCO usually maintains the system during the life of the contract and can train staff to assist or continue its care after the expiration of the contract period. The ESCO can also play a major role in educating the customer organization about its energy use and ways to curb it.

A *shared savings agreement* is another type of energy services performance contract under which the ESCO installing the energy-efficient equipment receives a share of the savings during the term of the agreement. In a *fixed shared savings agreement,* the customer agrees to a payment based on stipulated savings, and once the project is completed, the payments usually cannot be changed. After the completion of the project, the savings are verified by an engineering analysis or other mutually agreed upon method. In a *true shared savings agreement,* the savings are verified on a regular basis, with the savings payments changing as the savings are realized.

In summary, performance contracts typically contain three identifiable components: a *project development agreement* indicating which measures will be implemented to save energy (and money); an *energy services agreement* indicating what needs to be done after the installation to maintain ongoing savings; and a *financing agreement*. Organizations may choose to finance the projects independently of the ESCO, especially when they can access lower cost financing on their own (as in the case of public sector organizations when accessing tax-exempt funding). It is important to note that savings are measured in kWh and therms, and then translated into dollars at the current market price for electricity and natural gas.

Regardless of the type of energy services agreement, it is important to remind the reader of two critical components that are needed to ensure that the energy performance and operational goals are met: (1) commissioning, and (2) measurement and verification. Commissioning is the process of making sure a new building functions as intended and communicating the intended performance to the building management team. This usually occurs when the building is turned over for occupancy. Ongoing and carefully monitored measurement and verification protocols are vital to ensure the continuing performance of the improvements, especially when the energy savings are the source of the financing repayment.

Power purchase agreements (PPAs), also know as design-build-own-operate agreements, are ones in which the customer purchases the measurable output of the project (e.g., kilowatt hours, steam, hot water) from the ESCO or a special purpose entity (SPE) established for the project, rather than from the local utility. Such purchases are at lower rates or on better terms than would have been received by staying with the utility. These agreements work well for on-site energy generation and/or central plant opportunities. PPAs are frequently used for renewable energy and cogeneration projects (also known as combined heat and power projects). Due to the complexities of the contracts, projects using PPAs are typically very large. PPAs are frequently considered "off balance sheet" financing and are used in both the public and private sectors.

Commercial Leasing

Energy efficiency equipment that is considered by the Internal Revenue Service (IRS) as personal property (also know as "movable property" or "chattels") may be leased. The traditional equipment lease is a contract between two parties in which one party is given the right to use another party's equipment for a periodic payment over a specified term. Basically, this is a long-term rental agreement with clearly stated purchase options that may be exercised at the end of the lease term. Commercial leasing is an effective financing vehicle and is often referred to as "creative financing." Leases can be written so the payments accommodate a customer's cash flow needs (short-, long-, or "odd-" term; increasing or decreasing payments over time; balloon payments; skip payments, etc.). Leases are frequently used as part of an organization's overall tax and financing strategy and, as such, are mostly used in the private sector.

From a financial reporting perspective, however, commercial leases fall into only two categories, an *operating lease* or a *capital lease*. Each has substantially different financial consequences and accounting treatment. The monthly payments of an *operating lease* are usually lower than loan payments because the asset is owned by the lessor ("lender"), and the lessee's ("borrower's") payments do not build equity in the asset. The equipment is used by the lessee during the term, and the assumption is that the lessee will want to return the equipment at the end of the lease period. This means that the lease calculations must include assumptions that the residual value of the leased asset can be recovered at the end of the lease term. In other words, equipment that has little or no value at the end of the lease term will probably not qualify under an operating lease. For example, lighting systems would not qualify, while a well maintained generator in a cogeneration project might. Operating leases are considered "off balance sheet" financing, and payments are treated as an operating expense.

A common *capital lease* is a "finance lease," which is similar to a conditional sales agreement because the asset must be reflected on the lessee's (borrower's) balance sheet. A finance lease is easily recognized because the customer can buy the equipment at the end of the lease term at a stated price that is less than its fair market value ("a bargain purchase option"). For example, a lease with a one dollar purchase option is clearly a capital lease. Other conditions that define a capital lease deal with the term of the lease, transfer of ownership, and lessor's equity in the asset.*

Not everyone realizes that the tax treatment of a lease may be different from the financial reporting treatment of a lease. A *tax lease* or *guideline lease* is one in which the lessor keeps the tax incentives provided by the tax laws for investment and ownership of equipment (typically depreciation and tax credits). Generally, the lease rate on tax or guideline leases is reduced to reflect the lessor's recognition of this tax incentive. A *true lease,* similar to a long-term rental agreement, gives the lessee the option to buy the equipment at its true fair market value at the end of the lease term and may allow the lessee to deduct the monthly lease payments as an operating expense for income tax purposes. After all, you can't depreciate an asset that you do not own. However, a true lease will be picked up on the balance sheet.

Public sector organizations frequently lease equipment. However, because most public sector organizations are tax-exempt, tax strategies are not usually a consideration when deciding which type of lease to enter into.

FUNDING SOURCES

Once you have determined that internal sources of funds are not available or are insufficient for your energy efficiency project, your options become (a) using third party lenders, (b) postponing the project, or (c) installing part of the upgrades by breaking the project into smaller pieces. Earlier in this chapter, we explained why postponing or delaying

---

*See Financial Accounting Standards Board Statement of Financial Accounting Standards No. 13 for more information. Note that the financial treatment of operating leases is currently under review and may change.

the project can, in fact, be the most expensive alternative. So financing often becomes the best decision. Once you decide to finance the project and identify a preferred financing vehicle, the next step is to evaluate potential funding sources.

Traditional funders include banks, commercial credit companies leasing companies, insurance companies, brokerage houses, and vendors. If you are dealing with a large financial institution, it is important to contact the right department within that organization to obtain the best pricing for your project. For example, when speaking to their bank, most people start with their "regular banker," who may limit their discussions to loans. Keep in mind, however, that larger banks have a public finance department where you will find tax-exempt lease purchasing, and sometimes even a bank-owned leasing company where you may structure a special equipment lease. Large commercial credit companies may divide the market by the size of the transaction—small ticket, middle market, and large ticket. To further complicate matters, companies often define their market divisions differently (i.e., under $25k for micro-ticket, $25k-$500k for small ticket, $500k-$15 million for middle market, and over $15 million for large ticket*). If your project is $500,000, the small ticket and middle-market groups may quote you two different prices, even though they work for the same lender. Add the dimension of public-sector versus private-sector finance, and you may get different pricing again. This is especially true in organizations where the sales staff's compensation plan is commission based, and they do not have a company lead-sharing policy in place.

In almost every state where the electric industry has been restructured (deregulated), legislation has been passed to create a system benefits charge (also know as a public benefits charge) that adds a defined surcharge (fee) to the electricity bills. These fees are used to support energy-related projects that provide public benefits such as renewable energy, energy efficiency, low-income customer programs, energy R&D, or other related activities that may not be available in a competitive market. These fees, usually a per-kilowatt (kWh) hour cost or a fixed charge, are charged to all customers and cannot be bypassed. The amount of the charge varies by state and accumulates in a fund, which is usually administered by the state's energy office or local utility.

---

*There is no industry standard, and individual lenders may set these break points differently than this example.

If you are located in a state with public benefits charges (see map below), you may find that your energy efficiency project qualifies for a low-cost or below-market financing program funded by these charges. A listing of the total funding amounts by category and administrative contact information is available at the American Council for an Energy-Efficient Economy's website (http://www.aceee.org/briefs/aug07_01.htm). In the regulated states, the public utility commissions continue to provide incentives (rebates, loans, grants, etc.) to promote energy efficiency and renewable energy projects.

Examples of some of these programs include Texas' LoanSTAR Revolving Loan Program (available to schools, local governments, state agencies, and hospitals), which will loan up to $5 million at 3% interest (today's rate). In Oregon, the Ashland Electric Utility offers 0% loans to their commercial customers to finance energy efficiency improvements to their facilities. The loans can be used for lighting retrofits and other energy-efficiency measures.

Virtually every state offers some form of incentive, and we recommend starting your funding search by reviewing these special programs.

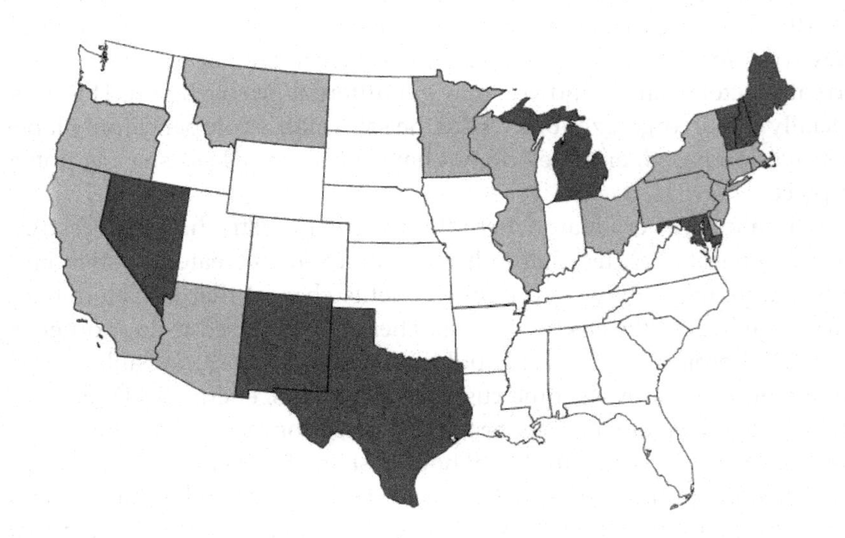

Figure 3-2.

NOTE: Light gray states have PBFs that support energy efficiency and renewable energy. Dark gray states have PBFs that support only energy efficiency (ACEEE 2004; UCS 2004).

A good place to find a list of energy efficiency incentives is the Database of State Incentives for Renewables and Efficiency (www.dsireusa.org).

VISION OF THE FUTURE

In the United States, energy efficiency is big business. A recent Lawrence Berkeley National Laboratory-National Association of Energy Service Companies (LBNL-NAESCO) survey* has ESCOs projecting their 2008 revenues at more than $5 billion and growing at an annual rate of 22% since 2006.

The survey went on to confirm that "MUSH" markets—municipal and state governments, universities and colleges, K-12 schools, and hospitals—have historically represented the largest share of ESCO industry activity, which was 58% of industry revenues in 2006. The federal market represented 22% and public housing represented 2% of industry revenues, while the industrial sector represented 6% and the commercial section 9%. Energy is the single largest operating expense in an office building, representing about 30% of a typical building's costs. This means that the commercial sector represents a growth opportunity for ESCOs.

Historically, however, the commercial real estate sector has faced numerous barriers to increasing their energy efficiency efforts, including:

- Split incentives (when the tenant and not the building owner pays the utility bills).
- Properties are held under complex legal structures that make it difficult and time-consuming to obtain financing (limited liability companies, limited partnerships, etc.).
- Non-recourse financing may already be in place, but new lenders want recourse on traditional energy efficiency project financings, especially when the equity in the property is highly leveraged.
- Difficulty in obtaining a security interest in the new energy assets being installed unless the entire building is refinanced.
- Tenants not wanting to take on debt for long-term leasehold improvements.
- Lenders not wanting to provide secondary financing to building owners for terms longer than the remaining lease term for key tenants.

---

*http://www.ornl.gov/info/esco/legislation/

- Properties are often managed by professional, third-party management companies that generally cannot enter into debt obligations on behalf of owners without special authority.
- Building owners and/or tenants may be unable or unwilling to borrow more money, as they may be concerned about reaching their debt capacity or violating covenants in existing loan agreements.

While many responsible building owners have "stepped up to the plate" and committed to financing new energy efficiency projects, this market sector still represents considerable marketing opportunities and challenges. An innovative financing alternative may help in "getting to yes."

To address this marketing challenge, the adage "the past is the key to the future" comes to mind. Back in 1991, Pacific/Utah Power, the electric generation and distribution divisions of PacifiCorp, offered their retail customers an innovative type of efficiency program in which customers repaid the costs of their efficiency installations through monthly energy service charges on their electric bills. As part of the utility's demand-side management program, the utility had recourse to shutting off power if the customer defaulted. In essence, the electric meter became the "credit." The utility did more than 1,000 transactions, and eventually sold their loan portfolio to a major U.S. bank.

The many benefits of this program included not requiring the tenant or building owner to enter into a debt obligation (thereby overcoming aversions to borrowing), tying the repayment to the use of the equipment, allowing the building owner to acquire new energy saving equipment at no direct cost, and having the tenants reduce their monthly utility bills without incurring debt.

The challenges to making a program such as this work include persuading the appropriate state public utility commissions to authorize a new tariff (energy service charge), finding a lender willing to underwrite the program, and insuring the installed equipment works as promised. Perhaps a "bill-to-the-meter" financing program could be considered to increase participation by this market sector in installing energy efficiency projects.

CONCLUSION

This chapter demonstrated how public and private sector organizations can redirect energy inefficiencies and waste from their current and future operating budgets in order to pay for the needed energy efficiency improvements today. Practical suggestions that support the urgency of implementing these projects were shared, and showing how to quantify the costs of delay was outlined. One section reviewed useful, field-tested ENERGY STAR tools and resources that are in the public domain and yours to use. Finally, potential sources of low- or no-cost funding for your projects were discussed, along with a variety of alternative financing vehicles.

Clearly, a decision not to install more energy-efficient equipment and implement related energy-saving measures is a decision to continue paying higher utility bills to the utility company. Using the captured energy savings to pay for the financing of improvements is recommended, essentially making them "self-liquidating" obligations.

Since these energy efficiency projects pay for themselves over time, the bottom line is that choosing to delay energy improvements may be the wrong decision and cost more in the long term.

Chapter 4

Financing Energy Projects through Performance Contracting

Shirley J. Hansen, Ph.D.

Since soaring energy prices and the growing need for energy efficiency hit the center of our radar screen in the 1970s, there has been growing recognition that using energy more efficiently is good for the economy and the environment. It's just good business. Yet survey after survey reveals that many organizations put off energy-efficiency work for one major reason: money.

Organizations either lacked the money, or those that had the funds were inclined to spend their money elsewhere. The reasons are legitimate: "We must use the money to invest in new production equipment. We must buy new math books. The payback is not short enough," etc., etc. For years many of us thought it was due almost entirely to the discomfort top management felt when the subject of "energy" was introduced. It certainly played a part, but in retrospect we now realize that other concerns were at play.

The horrible truth is that much of top management is not interested in ENERGY! They don't want to hear about gigajoules or British thermal units. CEOs and CFOs do not buy energy; they buy what it can do. How can they worry about the efficiency of something that is virtually non-existent in their lives? The only time it seems to reach their consciousness is when there is a shortage, a sudden power outage, or a spike in prices.

Many years ago, a dear friend who was in top management in a major corporation, gave me some sage advice, that is fundamental to our problem. He said, "Shirley, you folks must learn to fish from the fish's point of view."

To fish from the fish's point of view, we must first realize that top management is interested in delivering promised results, be it education, patient care, or selling widgets. Second, we need to be aware that management is "facility blind." Management can walk the corridors but seldom see the facility until something goes wrong. Third, such details as "energy" are just noise—a small irritating noise for someone else to deal with.

If we are to get the "fish's" attention, we must talk their language and make the case in their terms. For them to bite, the bait on the hook must make energy efficiency (EE) a solid business opportunity.

It is essential that we provide our "fish": (1) an effective cost/benefit analysis procedure, which compares the net benefits of energy efficiency to increased production; (2) a new perspective of energy savings as a percentage of the bottom line; and (3) energy efficiency and conservation as a very cost-effective delivery system for meeting environmental mandates—a way to make money while reducing emissions.

Probably our biggest challenge is to remind top management, as forcefully as possible, that EE can be a self-funding endeavor. CEOs and CFOs have a tendency to compare energy investments to other business investments and fail to appreciate that no new money is required to do energy efficiency work. The money needed for energy investments is already in the budget—and it's being spent on wasted energy. The financing source for the EE investment is right there in avoided utility costs—money that will go up the smokestack creating more pollution every day that the energy efficiency measures are not taken.

All this, however, becomes much more palatable if the initial expenditures are from someone else's pocketbook—even better if there is someone out there that will guarantee that the funds to do the energy work will come out of avoided utility costs (the future energy savings from the project).

With such a backdrop, it is not surprising that performance contracting emerged as an attractive financing mechanism for energy work. Imagine the jubilation when someone figured out that future energy costs could finance the rebuilding or replacement of decrepit heating and cooling systems. And it just got better, because the money being set aside to fix those behemoths could now be used for something else.

The concept of performance contract is the same idea that stands behind commercial paper: the fact that a future obligation to pay already has value today. The future obligation in performance contracting is not derived from serving a debt but rather from the known and unavoidable cost of heating, cooling and illuminating buildings or fueling industrial processes.

Administrators around the world are burdened with obsolete, money-hungry schools, hospitals, etc. built in the low-energy-cost era of the 1950s and 1960s. Unfortunately, they cannot see affording to re-equip these structures with the immensely improved energy-use technologies that have emerged in recent years. They feel compelled to continue to throw good money after bad to keep buildings in service.

By a kind of financial judo, performance contracting turns this grim prospect into an asset. Future utility bills can be discounted, while the commercial debt and the resulting cash can be used today for retrofits.

PERFORMANCE CONTRACTING IN RETROSPECT

In the late 1970s, Scallop Thermal, a division of Royal Dutch Shell, introduced the concept of using third-party financing to improve energy efficiency and cut operating costs in North America. Scallop offered to meet a Philadelphia hospital's existing energy services for 90 percent of its current utility bill. By upgrading the mechanical system and implementing energy-efficient practices, Scallop was able to bring consumption well below the 90 percent level. Scallop both paid for its services and made a profit from this difference.

These early energy financing agreements generally were based upon each party receiving a percentage of the energy cost savings. The energy service company (ESCO) received a share to cover its costs and make a profit. The owner also received a share (as well as capital improvements) as an incentive to participate. Since each party received a share of the energy cost savings, this procedure became known as "shared savings."

During the life of the contract, the ESCO expected its percentage of the cost savings to cover all costs it had incurred, plus deliver a profit. This concept worked quite well as long as energy prices stayed the same or increased.

In the mid-1980s, energy prices began to drop. With lower prices, it took longer than predicted for the firm to recover its costs. Some firms could not meet their payments to their suppliers or financial backers. Companies closed their doors and, in the process, defaulted on their commitments to their shared savings customers. Some suppliers tried to recover costs from the building owners. Lawsuits were filed and "shared savings" nearly died a painful death.

From this tenuous thread, the "shared savings" industry in North America survived, but its character changed dramatically. Those supplying the financial backing and/or equipment recognized the risks of basing contract on future energy prices. Higher risks meant higher interest rates, if the money could be found. Insurance, which had been available to ESCOs for "shared savings," became a scarce commodity. By the late 1980s, shared cost savings agreements had shrunk to approximately 5 percent of the market.

From the end user's point of view, through the "shared savings" boom, it soon became apparent that the energy payments to the ESCO were unpredictable. All too often, the customers found themselves paying far more than expected for the opportunity. A customer, who had accepted a shared savings deal with $3.5 million in equipment, who was asked to pay 70 percent of a predicted $1 million annual savings for five years, assumed the total payment would be about $3.5 million for the acquired equipment and services. If, however, the savings were greater than expected or the price of energy went up, the costs could easily become $5-7 million for the same equipment. Payment procedures became confrontational. In the final analysis, too often the only real benefit of the shared savings transactions were their "off balance" sheet feature. This was attractive to customers who did not wish to incur more debt.

Out of the "shared savings" confusion of the 1970s-1980s evolved a new financing mechanism with the guarantee centered on the reduction in the amount of energy consumed and the value of the energy in dollar savings calculated at current billing rates. Typically, the projected dollar savings were guaranteed to cover any of the associated debt service obligations of the owner. To avoid the risks associated with falling energy prices, ESCOs began setting an energy floor price below which money guarantees would not apply.

A new term, "performance contracting," emerged that embraced all guaranteed energy efficiency financing schemes, including shared savings and guaranteed savings.

The initial attraction to performance contracting was the financing help. Through the years, the ESCO expertise has become equally attractive. For example, a 1988 survey found that the majority of US school administrators still thought work on the building envelope, (e.g., added insulation, double glazing of windows, etc.) was the most cost-effective energy savings measure to take. However, a U.S. Department of Energy analysis at the time showed that building envelope measures averaged the longest payback (over eight years) of those measures studied. In contrast, control measures paid back in roughly two years. In attempts to cut energy costs, school administrators were not investing money the most effective place. If fact, they were waiting eight years to get the return on investment that they could have achieved in two years.

ESCOs have learned, sometimes at great expense, what works—and what doesn't!! The earlier temptation to invest in elaborate control systems, for example, no longer exists. Today, experienced ESCOs know exactly how sophisticated a system should be in a given facility. That expertise can keep owners from investing valuable money in the wrong measures.

Performance contracting has matured into a viable and reliable way of doing business in North America and many countries around the world. In the United States and Canada, federal law allows—even encourages—federal agencies to use performance contracting to cut operating costs. Over 4,000 of the US school systems have had some type of performance contract, and these school systems have as many as 800 buildings each. Major manufacturers of energy-related equipment have ESCO services and divisions, and many regional ESCOs have grown from engineering firms, distributorships and mechanical contractors. It is conservative to say that in the United States there are at least 10 major national ESCOs and another 40 operating on broad, regional bases. Local firms offering some ESCO services undoubtedly exceed 100. In fact, the concept has become so popular there is great fear in the industry that "wannabe ESCOs," often dubbed WISHCOs, are offering services before they fully understand the complexity of the process and are creating credibility difficulties for the industry.

In the mid-1990s, the International Energy Agency went on record at the Conference on Energy Efficiency in Latin America to support and encourage the ESCO industry. More recently the European Union has been actively promoting the concept through papers and conferences. The ESCO Europe Conference, sponsored in part by the EU, has become an annual event.* More recently, the ESCO concept has become accepted in Asia. In late 2007, the Japanese and Chinese ESCO associates put together the second Asia ESCO Conference in Beijing. Some version of performance contracting is now being practiced in all industrialized nations and most developing countries. The US Agency for International Development, the World Bank, Asian Development Bank, and the European Bank for Reconstruction and Development are all very active in fostering ESCO industries in developing countries.

TYPICAL ESCO SERVICES

Through the years, ESCO services have become more varied. It has become a customer-driven industry, and the customer typically has a selection of ESCO services from which to choose. Services offered by an ESCO usually include:

- An investment grade energy audit to identify energy and operational savings opportunities, assess risks, determine risk management/mitigating strategies, and calculate cost-effectiveness of proposed measures over time;
- Financing from its own resources or through arrangements with banks or other financing sources;
- The purchase, installation and maintenance of the installed energy-efficient equipment (and possibly maintenance on all energy-consuming equipment);
- New equipment training of operations and maintenance (O&M) personnel;
- Training of O&M personnel in energy-efficient practices;
- Monitoring of the operations and energy savings so reduced energy consumption and operation costs persist;

*See "Latest Development of Energy Service Companies across Europe," authored by Paolo Bertoldi, Beneigna Boza-Kiss, and Silvia Rezessy, published in 2007 by the IES of the Joint Research Center, European Commission. ISBN 978-92-79-06965-9.

- Measurement and savings verification; and
- A guarantee of the energy savings to be achieved.

The popularity of performance contracting rests on the many benefits it delivers. The highlights of the benefits it offers the customer or the business which offers ESCO services are addressed below.

CUSTOMER ADVANTAGES

The recipient of ESCO services can achieve many benefits, including:

- an immediate upgrade of facilities and reduced operating costs—without any initial capital investment;
- access to the ESCO's energy efficiency expertise;
- positive cash flow (most projects generate savings that exceed the guarantee);
- the opportunity to use the money that would have been used for required upgrades or replacements to meet other needs;
- improved and more energy-efficient operations and maintenance;
- several normal business risks are assumed by the ESCO, including the guaranteed performance of the new equipment for the life of the contract (not just through a warranty period);
- a more comfortable, productive environment;
- services paid for with the money that the customer would otherwise have paid to the utilities for wasted energy.

Performance contracting is probably the best and quickest way to be sure an organization is operating as efficiently as possible. It offers the customer a risk shedding opportunity; however, risks do still exist.

In 1995-96 we had the privilege of working with the City and County of San Francisco on its "Kilowatts to Megawatts" research and development project funded by the Urban Consortium. A key component of this effort, ably led by Mr. John Deakin and Ms. Christine Vance, was to develop a decision model to aid other cities and counties in determining when to use their limited financial resources to do energy efficiency work, when to outsource the work, and which of the many outsourcing options would best meet the municipalities' needs. A particularly challenging aspect of this work was the identification of specific risks associated with a broad range of energy efficiency financing opportunities and the extent of those risks. Table 4-1 offers the general magnitude of those risks by option. Clearly, the specific magnitude of these risks will vary with the customer's local conditions.

Risk shedding always carries a price tag. It can be presumed, therefore, that the customer will typically pay more for the services provided on the right side of Table 4-1. It should be noted, however, that, as the customer moves to the more comprehensive option, he, or she, also receives more services, has less administrative burden, and will probably achieve greater savings persistence. The costs associated with the "right-hand side of the menu" can also be mitigated by controlling the contributing factors listed. The greatest factor, however, could be the relative speed with which projects can be implemented, thus avoiding valuable dollars going up the smokestack while the customer tries to get more of the energy efficiency work out of his or her organization's limited resources. (Options are shown on the left in the table.)

In performance contracting, the risks shed by the owner are largely assumed by the ESCO. Effective performance contracting then becomes a matter of risk assessment and management by the ESCO. The more accurately the ESCO assesses the risks and the more effectively it manages them, the greater the benefits to all parties. Since performance contracting risks are primarily managed through the project's financial structure, effective risk management presents a major point of differentiation among ESCOs.

ESCO RISK ASSESSMENT AND MANAGEMENT

Effective, experienced ESCOs have found a series of critical risk assessment components and management techniques, which include:

Table 4-1. Risk Assessment Matrix

Risk Factors	Relative Risk Level Associated with Financing Options — Option #1	#2	#3	#4	#5	#6	Major Variables	Mitigating Strategies
Equipment performance warranty	3	3	3	3	3	1	Quality of specs Selection process Contract conditions	In-house/consultants' expertise Legal ability available
Technical expertise	4	4	4	3	2	1	Consultant qualifications	Selection process Negotiation
Audit quality; accuracy	3	3	3	3	3	2	Auditor ability Review capability Vender bias	Selection process Review a sample audit 3rd party validation Establish procedural criteria & scope of audit
Project 3 management inadequate	3	3	3	3	2		Ability of PM or CM provided in-house and through project delivery option	Careful planning with DPW and clear definition of roles Service provider selection process
Construction/ installation	5	4	3	2	2	1	Vendor/subcontractor qualifications Contract provisions	Selection process Performance/payment bonds Legal ability
Maintenance & operations	5	5	5	5	3	1	Manpower In-house staff qualifications Training quality	Outsourcing Training/experience Selection process
Savings persistence	5	4	4	4	3	1	Varies by measure Administrative commitment O&M attitudes, Training/experience	Contractual obligations Vendor selection ESCO selection Guarantees offered

Legend: N/A not applicable; 1 low risk; 2 low-medium risk; 3 medium risk; 4 medium-high risk; 5 high risk

Table 4-1. Risk Assessment Matrix (*Continued*)

Risk Factors	Relative Risk Level Associated with Financing Options — Option #1	#2	#3	#4	#5	#6	Major Variables	Mitigating Strategies
Savings verification instruments approach	3	3	3	3	3	1	Needs vary by measure/guarantee Accuracy	Costs paid for accuracy 3rd party validation
Establishing base year; procebaseline adjustments	3	3	3	3	3	4	Availability of historical data Clarity of formulas in the contract	Organizations energy management dures; Combined legal& tech. expertise
Provisions & cost for M&V	3	3	3	3	3	5	Procedures accuracy	Instruments & approach used Cost
Cost of delay/ schedule adherence	5	4	3	2	2	1	Audit delays Financing/ESA Design delays Construction delays	Recognize costs incurred by delay; Consider penalty clauses when completion delayed; City expedite reviews, etc. when possible; Check past practice of firms during selection process
Hidden project costs	1	1	2	4	3	5	Margins, mark-ups Profit	Option 4-6; Reserve right to bid equipment; Open book pricing
Facility control problems	1	1	1	1	1	5	Contract provisions	Specify acceptable parameters in contract
Quality of maintenance training	5	5	5	5	2	1	Specification for maintenance trainer's abilities; cost	Selection of service provider to provide training
Intrusion/ interruption	4	4	4	4	4	5	Contract language—stipulate hours of installation	Close coordination with maintenance staff; careful attention to operation needs

Legend: N/A not applicable; 1 low risk; 2 low-medium risk; 3 medium risk; 4 medium-high risk; 5 high risk

Pre-qualify the customer. Prequalification criteria should be established; the customer's organizational, technical and financial data should be carefully collected, validated, and evaluated.

Conduct an investment-grade audit. The typical energy audit assumes that conditions observed in the audit will stay the same for the life of the equipment and/or the project. An investment-grade audit assesses administrative risks, operation and maintenance risks and the impact these risks will have on the project's savings over time. An investment-grade audit also considers the time value of money. Projected payback calculations, for example, discount the value of the dollars saved each year for the life of the project. A typical four-year payback, for example, can easily become 5+ years in the real value of dollars saved, thus changing the dynamics of the cost effectiveness calculations. Further, the investment-grade audit incorporates the cost of risk mitigation in payback calculations. It also recognizes that facilities and processes are critical portions of an organization's investment portfolio. The investment grade audit, therefore, offers the owner a guide in ways to enhance the work environment and the value of the portfolio. A list of measures that save energy but do not address the workplace conditions is no longer sufficient.

Establish sound baseyear data. Baseyear data are more than the average energy consumption over the last two or three years. They also consider existing conditions and what was happening in the facility or industrial process during that period. Hours of occupancy, level of occupancy, run times, etc., all become critical issues that must be verified prior to project implementation and signed off on by both parties. How these variables are adjusted to establish an annually adjusted baseline for reconciliation purposes is critical.

Secure a solid contract fair to all parties. Sometimes ESCOs get a little too zealous in managing their risks. For example, the following paragraph, nearly a deal stopper, was put into a draft contract in caps:

IN NO EVENT SHALL [THE ESCO] BE LIABLE FOR ANY SPECIAL, INCIDENTAL, INDIRECT, SPECULATIVE, REMOTE OR CONSEQUENTIAL DAMAGES ARISING FROM, RELATING TO, OR CONNECTED WITH THE WORK, THE SUPPORT SERVICES, EQUIPMENT, MATERIALS, OR ANY GOODS OR SERVICES PROVIDED HEREUNDER.

After months of discussion, the paragraph was finally dropped and the project moved ahead.

Implement Quality Measures

Customers too often insist on quantity instead of quality. One federal procurement officer reportedly told an ESCO the military base would rather have "15 Tempos than 10 Volvos." The agency may have reasons for needing 15 sets of wheels, but when the focus is energy, the ESCO must look at parts, repairs, maintenance and savings persistence.

Operations and Maintenance (O&M)

Too often overlooked, energy efficient O&M practices are absolutely vital to the success of a project. An evaluation of the schools and hospitals federal energy grants program (Institutional Conservation Program) after eight grant cycles revealed that in effective energy management programs found that up to 80 percent of the savings could be attributed to energy efficient O&M practices. If ESCOs are to guarantee the savings from a measure, they must look beyond the equipment to its "care and feeding" over the life of the contract. If the ESCO does not perform the maintenance, it must; (a) train the owner's staff; (b) police the O&M practices; and/or (c) discount the predicted level of savings. In all cases the associated risk level must be evaluated.

Measurement and Savings Verification (M&V)

Once the recommended measures are known and approved, the ESCO and owner should cooperatively establish the desired level of accuracy and the associated costs for M&V, that they are willing to have the project carry. The question is: How much accuracy are the parties willing to pay for? When M&V costs are included in the project, they cut into the savings and thereby diminish the amount of equipment and services that can be funded. But anytime money changes hands based upon savings, those savings need to be quantified in an acceptable manner. The International Performance Measurement and Verification Protocol (IPMVP) is the most broadly accepted means of measuring and verifying savings. The IPMVP guidelines can be downloaded from the organization's website: evo-world.org. (Accessibility to the guidelines requires membership in EVO.)

Project Management

The "business end" of the project, securing the savings, doesn't begin until the equipment is installed. Weak ESCOs tend to think of the project as "complete" once it is in the ground. Customers often think the ESCO is in business to sell equipment. ESCOs typically lose money during equipment installation but make up for it through the service and savings over the life of the contract. A good project manager is an incredibly valuable part of the ESCOs project delivery during the project implementation and in all the contract years.

When ESCOs have completed a thorough risk assessment and determined the procedures needed to manage those risks, they then discount the savings to adjust for risk exposure. In other words, they never guarantee 100 percent of the expected savings.

PERFORMANCE CONTRACTING: A FINANCIAL TRANSACTION

Risk is managed through the financial structure. The ESCOs investment grade audit will provide the information necessary to establish how much can be guaranteed and how much must be held back as a risk cushion. The financial structure has essentially three components:

- the guarantee, which covers design, acquisition, installation and the cost of money;
- the ESCO fee for services performed;
- the positive cash flow (the savings in excess of the guarantee and ESCO fee).

The size of the ESCO fee will vary with the services provided and the risks perceived and assumed by the ESCOs. Some ESCOs put their fee in the guarantee package, others take it off the top of any savings that exceed the guaranteed amount, and still others do some of each. Owners should realize that an ESCO that offers the customer "all of the positive cash flow" has all its fee imbedded in the guarantee package. In such cases, the customer will pay finance charges and interest on the ESCO fees for the life of the project.

Performance contracting has had its problems. Many of them in recent years, however, can be traced to the assumption by fledgling ESCOs and customers alike, that performance contracting is a technical procedure. It is not. Technical issues are important; they are the foundation upon which the project is based. On the other hand, lawyers would have you believe it's all managed through the contract. It is not. A good contract is simply the basis of a good project.

Make no mistake: Performance contracting is primarily a financial transaction. The ultimate performance contract decisions cannot be made in the boiler room; they must be made in the business office. Engineers and lawyers must also be in that business office when the decisions are made, but a strong financial voice is crucial.

Performance contracting is a simple idea but a complex process. If we are to get the maximum benefit out of performance contracting, it helps if financial officers have some understanding of engineering and legal issues; however, it is absolutely critical that lawyers and engineers understand and appreciate the fundamental role that finance plays in performance contracting.

Chapter 5

The Power Purchase Agreement (PC for Solar)

Ryan Park

INTRODUCTION

The installation of a solar electricity project is a great way to green your business, and the financial return can be very attractive if your business is positioned in a state or utility district offering rebates and/or tax credits for solar installations. Due to the fact that solar electricity panels will last many decades and have a significant amount of embedded energy, the upfront cost of a solar project can be significant. At this time there are several ways to proceed with a solar electricity project: cash purchase, project finance, capital or operational lease, or power purchase agreement (PPA). Due to the many benefits of a PPA, over 60 percent of all commercial solar projects are financed using this structure. In this chapter we will explain the power purchase agreement and its advantages over other forms of financing. Then we will proceed into design criteria to maximize your return on investment if pursuing a PPA.

The primary benefit for using a PPA is to get a solar system on your roof with zero upfront costs, while also having a known price for energy for 20 years. (The energy produced from the solar system reduces risk from energy supply price spikes.)

WHAT IS A POWER PURCHASE AGREEMENT (PPA)?

A power purchase agreement is a long-term agreement to buy power from a company that produces electricity. A third-party financier will provide the capital to build, operate, and maintain a solar electricity installation for 15 to 20 years. The host customer is only responsible for purchasing the electricity produced by the solar system. It is the responsibility of the

PPA Advantages

	FINANCE LEASE	TRUE LEASE	CAPITAL PURCHASE	POWER PURCHASE AGREEMENT
Upfront Capital?	NONE	NONE	YES	NONE
Performance Risk?	YES	YES	YES	NONE
System Expertise Required?	YES	YES	YES	NONE
Maintenance Required?	YES	YES	YES	NONE
Purchase Required?	YES	YES – but option to re-lease	YES	NONE

Figure 5-1

PPA provider to assume all risks and responsibilities of ownership. A PPA provider will own, operate, maintain, and clean the system for optimal performance. The PPA provider will also have sophisticated real-time monitoring services to verify the system is working properly.

The host customer will run their business as usual, without any concerns about how to best operate the solar power system. At the end of the PPA term, the system can be purchased by the host at fair market value, or the PPA can be renewed. Overall, a PPA enables a host customer (and our world) to benefit from the use of "green" energy, while still receiving some of the benefits of ownership through lower electricity costs and an improved public image. It also allows the host customer to spend their capital budget on a core business instead of on a solar electricity system.

HOW IS IT DIFFERENT FROM OTHER FORMS OF FINANCING?

A PPA is unique because there are zero upfront costs and the host only pays for the power produced, with no responsibility for maintaining the system. If a system were financed through project financing or a lease, the host would be required to make payments on the loan regardless of the functionality of the system. While solar electricity systems are not prone to maintenance problems, few businesses want to add solar

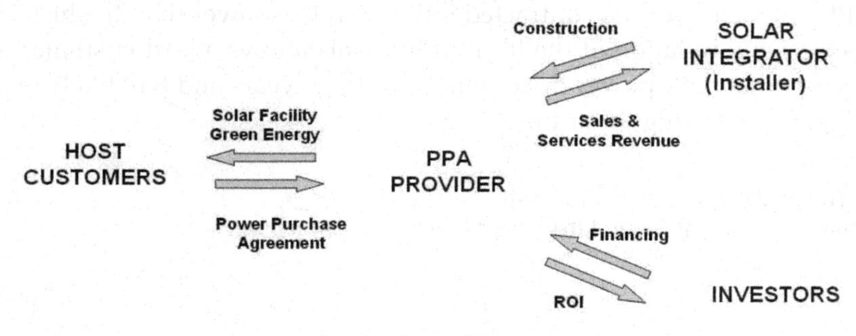

Figure 5-2

electricity system oversight to an already overloaded list of issues to deal with for their core business.

TYPICAL PPA TERMS

Zero Upfront Cost

15- 20 year PPA Agreement

This is the term during which you will be required to purchase all electricity generated by the solar system. PPA providers realize that solar electricity systems will last decades. They are willing to purchase the system to own, operate, and maintain over 15-20+ years while traditional financing is for only a third to half that time frame. The result is that many PPA agreements can be cash flow positive from the start, which means the price you will pay for solar power is actually less than you would pay for traditional electricity from the utility. Keep in mind that when entering a PPA you will still receive power from the utility company to provide electricity when solar is not able to provide all of your electricity needs, as may be the case with night or heavy power usage facilities.

Starting kWh Price

Depending on the location of the solar installation, size of the project, and other parameters affecting power production, a PPA will have a specific kWh starting price.

PPA kWh Price Inflation

In order to reduce the starting kWh price as much as possible, the PPA agreement will have a set kWh price inflation per year. For virtually all PPA agreements, the contracted inflation rate is lower than the historical electricity inflation of the local utility. This allows a host customer to forecast what their power prices will be in 10-20 years and will likely lead to significant savings over time.

Comparative Cost of Electricity:
Utility vs. Solar Over Life of System

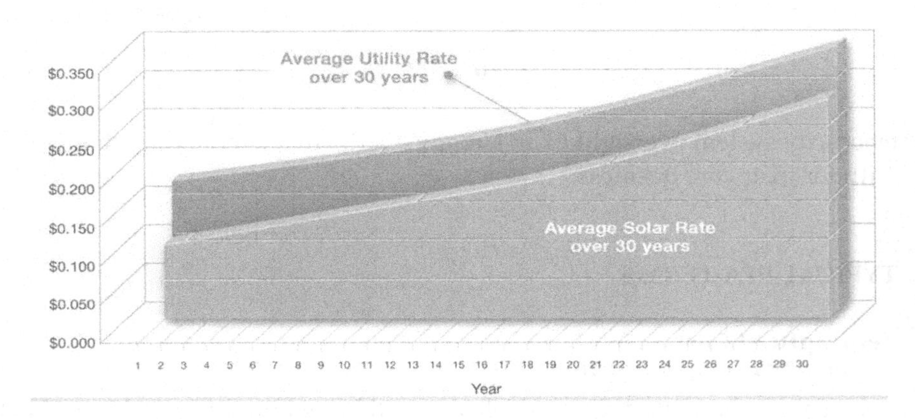

Figure 5-3

An Example of PPA Financing

The following example is from a 1 MW Southern California project. While the numbers below are not representative for all PPAs, they still can be used to show what is possible with a PPA.

Criteria for a Low Cost PPA

Stable Host

Not all solar systems or hosts are viewed equally by a PPA provider. The key evaluation criterion for a PPA provider is stability. Ideally, they want the host customer to have been in business for at least five years with strong credit. They also want the business to own its building; however, some PPA providers will still enter an agreement with a long term lease in place.

PPA Example – 1 MW

Estimated Savings Analysis: SCE GS2 TOUa Electricity Rate
Forecasted Savings and Cash Flow Summary - Current Solar Rebate

Year	Electricity Cost Without Solar	Equivalent Solar Power Cost	Savings With Solar	Annual % Savings
1	$396,000	$285,600	$110,400	28%
2	$413,664	$296,214	$117,450	28%
3	$432,116	$307,222	$124,894	29%
4	$451,391	$318,639	$132,752	29%
5	$471,526	$330,480	$141,046	30%
6	$492,559	$342,762	$149,797	30%
7	$514,530	$355,500	$159,031	31%
8	$537,482	$368,711	$168,771	31%
9	$561,457	$382,413	$179,044	32%
10	$586,501	$396,624	$189,877	32%
11	$612,663	$411,364	$201,299	33%
12	$639,991	$426,651	$213,340	33%
13	$668,539	$442,507	$226,032	34%
14	$698,360	$458,951	$239,408	34%
15	$729,511	$476,007	$253,504	35%
16	$762,052	$493,697	$268,355	35%
17	$796,044	$512,044	$284,000	36%
18	$831,553	$531,073	$300,480	36%
19	$868,645	$550,809	$317,836	37%
20	$907,392	$571,278	$336,114	37%
Total	**$8,206,289**	**$5,599,644**	**$2,606,645**	

Figure 5-4

Projects at Least $1,000,000 in Value

There are substantial costs associated with each PPA deal, due to all the legal, tax structure, and engineering fees. With a larger project, those fixed costs are spread throughout a longer period of solar power and have less of an impact on the PPA price.

Excellent Solar Location

The more electricity a given solar installation can produce, the lower the resulting PPA price will be. The reason is that a PPA provider will purchase the solar system and require a certain return on investment for the PPA term. The more a system produces over that term, the lower the resulting kWh price will need to be to cover the investment return requirements.

How to Receive a Quote for a PPA

There are a number of PPA firms available. Below you will find a short list of several reputable PPA providers, but there are many others available. Not all providers will offer the same price for a PPA for a given project so it is good to ask for a PPA quote from more than one. In order to receive a PPA quote, the PPA provider will need to know:

- Location of the solar project
- Product specifications (panels, inverters, and racking)
- Installation parameters (system size, azimuth [S, W, etc.], tilt)

Keep in mind that many solar system integrators can evaluate your facility, develop the project details, and source a PPA provider for you. Many solar system integrators can handle everything in-house, and some PPA providers will only work with a few solar installers for their projects because they are confident the installation will be quality.

PPA Providers

- MMA Renewable Ventures: http://www.mmarenewableventures.com/
- Recurrent energy: http://www.recurrentenergy.com/
- Solar Power Partners: http://www.solarpowerpartners.com/
- Tioga energy: http://www.tiogaenergy.com/

Reputable Solar Integrators

- REC Solar: http://www.recsolar.com/cm/Home.html
- Sunpower: http://www.sunpowercorp.com/
- Sun Edison: http://www.sunedison.com/

Chapter 6

Selling Projects to Financiers

Financing is an art, not a science. The best way to appreciate the process is to put yourself in the lender's shoes. If you were considering lending someone several million of your hard-earned dollars, what sort of questions would you ask? How concerned would you be about certain details of your borrower's financial history that were unclear? What weight, if any, would you give to the borrower's reputation?

Traditionally, lenders evaluate a possible loan from several perspectives. These are known as the "C's of Lending":

- *Capital*. Is there capital at risk on the part of the borrower? How motivated is the borrower to pay back the loan if its own capital is not involved in the transaction?

- *Credit/Cash flow*. What are the revenues and expenses of the borrower and/or the project? Are there noticeable trends in past financial performance? What are expected trends in the energy industry or the industry of the borrower?

- *Collateral*. What is the security for the loan? If the loan is not paid back as scheduled, what items of value remain for the lender to own and/or sell? Is it the project equipment or another corporate assets, such as stock?

- *Conditions*. What are the desired terms of the loan?

- *Character*. How experienced is the management team? What is the reputation of the company?

Regardless of lending policies and rules, developing the trust of your lender will go a long way when you need to secure a loan approval. Many borrowers approach the lending process as a "torture chamber" of difficult questions and paperwork. This type of attitude is certainly not going to make a lot of friends in the lending community.

A lender relationship is well worth cultivating. The best way to create a foundation for a long-term lending relationship is to offer accurate information about the project and the financial status of the parties involved, in addition to presenting your references and project history.

Lenders have what is called a "due diligence period." This is a time when, even though a loan approval may have been issued, they are looking for ways to verify everything disclosed in the initial application package. They want to expose all the weaknesses of the project in order to properly evaluate the risks of the loan.

Never try to hide or manipulate information, as this will land you in court or bankruptcy proceedings. Lenders should have all the relevant information, both good and bad. You should make the effort to meet face-to-face with the bankers and, if possible, make a presentation to one or more members of the loan committee. This "relationship" aspect of the loan decision-making process is an opportunity for you to turn what could be a marginal loan application into an approval.

If you are preparing a package of information about your energy project to send to a lender, this is what you should include:

- Your company's financial statements (two years, audited)
- ESCO financial statements (two years, audited)
- Project pro formas (revenue and expense projections)
- Energy services agreement between your company and the ESCO
- Contract between your company (or the ESCO) and the utility (if a rebate program is available)
- Description of equipment to be installed
- Energy audit results summary
- Measurment and verification plan
- Construction schedule
- Management team summary
- References from suppliers and customers

Once you have submitted this information to a lender, it's your turn to start asking for information. Here are some important questions to ask a lender:

- Are you interested in a lease on the equipment, a loan to my company, or a loan to the ESCO?
- Is construction financing being offered as well as a long-term loan? What factors will affect the rate and terms of the loan?
- How soon can you send me a proposal to finance the project?
- What is the timeline for loan approval, due diligence, and funding?
- How often does the loan committee meet? Can I make a presentation to them?
- Do you require the ESCO to have a construction bond?
- What will be your course of action if the project fails to generate the expected savings?
- Is there a prepayment penalty on the loan? If so, under what conditions?
- What happens if utility rates go down and the dollar amount of savings is insufficient to pay for the loan?
- Can payments be made quarterly or annually instead of monthly?

Lastly, and most importantly, remember these words of wisdom: *Everything is negotiable*. Don't be afraid to ask for a lower rate or better terms than initially offered by a lender. As long as you have clearly demonstrated to the loan committee that your project is well-planned and has a high likelihood of success, you will be surprised at how flexible lenders can be.

THE FINANCIAL ANALYSIS OF A PROJECT

When lenders are making a loan based on the full faith and credit of a company, they tend to focus on the financial statements and general business performance of the proposed borrower. With project financing, however, it is the project itself which receives much attention. Even though the ESCO is the borrower, the project must perform in order for the loan to be paid back on schedule.

The best way for the ESCO and lender to analyze a project is to view it as a stand-alone company. This "company" has revenues, expenses, assets and debt which will vary over time. The revenues are the estimated share of savings payments your organization agrees to make to the ESCO or to the single-purpose entity. The expenses may include equipment maintenance, measurement and verification ("M&V"), or even the cost of raw materials to produce sold "goods," such as chilled water.

Lenders will be most interested in the debt service coverage analysis. How do the net revenues (net income after expenses) compare to the project loan payments? If the net revenues are $25,000 per month and the lease payments are $20,000 per month, the Debt Service Coverage Ratio is 1.25. Most lenders prefer debt service ratios above 1.10 in order to provide some protection against fluctuations in project revenues and expenses.

If the debt service coverage ratio is too low for your project, you may want to run some alternative scenarios. First, try to extend the term of the loan. This will have the effect of lowering the payment per month and increasing the coverage ratio. Second, if the expenses of the project are high, relative to other projects, you may want to consider restructuring the project. Eliminating a costly piece of equipment with high maintenance expenses may well lower the simple payback of the project and increase the chances of the project's overall success. A third option is for the ESCO to use its own cash to pay for a small portion of the project, thus decreasing the loan amount.

In terms of a presentation to your organization's finance team, any project needs to be compared to other projects which are vying for management's attention. The tools of comparison are Net Present Value (NPV) and Internal Rate of Return (IRR). The NPV is the value in current dollars of the positive and negative cash flows over time related to the project. In order to perform this calculation, an interest rate must be selected; usually it is equivalent to the opportunity cost of your company. This cost is the rate of return that the company receives on average in most of its relatively safe projects. The IRR is the rate at which the NPV is zero. The project has a high likelihood of being accepted by your management if its NPV or IRR is higher than other projects. For financing success stories, see Chapter 2.

Chapter 7

Key Risk and Structuring Provisions for Bankable Transactions

Jim Thoma

Editor's Note: Chapter 7 provides greater detail (and some different perspectives) than what was covered in Chapter 6. The key message is to start "on the right foot" with your lender *early* in the process.

INTRODUCTION

Energy services projects, in all their varying forms, are technically complex undertakings, requiring significant design, engineering and development efforts to create a transaction that is economically compelling to the end user customer. However, all these efforts could be for naught if third party financing is required and the transaction is not structured to satisfy the requirements of the capital markets. Few things are more frustrating for the customer and the energy services company (ESCO) than having high expectations for a project, only to learn that funding is unobtainable due to an improperly structured transaction. This chapter is intended to help ESCOs and their customers avoid this problem by identifying and explaining specific risk and structural provisions that are key to creating a "bankable" energy services transaction.

Developing a bankable energy services transaction requires a thorough understanding and balancing of the needs and objectives of the three main parties to the deal—the customer (end user or obligor), the ESCO, and the lender. Although each party's perspective must be considered when developing the deal, this chapter is intended to primarily represent the lender's perspective and, by doing so, explain the critical components of a bankable transaction. Once a structure is determined to be bankable, the capital markets should welcome it with virtually *unlimited* capacity.

Energy services transactions can take many different forms, as long as they're properly structured and documented. The initial challenge is to accurately identify a structure that meets the needs of all three parties to the transaction. In most cases, the parties begin with a common structure but work through a series of negotiated modifications to achieve the final bankable result. Many ESCOs have learned the hard way that leaving the lender out of these structural negotiations usually results in unmet expectations and frustration for the customer.

The matrix below provides a sample of commonly used project structures and corresponding financing solutions.

RISK ASSESSMENT

Risk assessment and allocation is the fundamental principal that

Project Structures	**Financing Structures**
Performance Contracts	Capital Lease Loan & Security Agreement Operating or Synthetic Lease Tax Exempt Lease Municipal Lease
Construction/Installation Contracts with or without Operations & Maintenance Agreements	Capital Lease Loan & Security Agreement Synthetic Lease Tax Exempt Lease Municipal Lease
Shared Savings Agreements	Embedded Capital Lease Embedded Operating Lease Purchase & Assignment Agreement
Energy Services Agreements Customized Debt & Equity	Purchase & Assignment Agreement
Power Purchase Agreements Customized Debt & Equity	Purchase & Assignment Agreement

guides a successful transaction development process and serves as the structural foundation for all bankable transactions. In order to attract capital markets investors, a transaction must present a risk profile commensurate with the nature of the financing (i.e. debt or equity) and anticipated investor returns. This concept is frequently and most commonly described in terms of a basic risk/reward analysis—higher risk requires a greater return to the investor. This is the foundational structuring concept it its simplest form, but risk analysis and mitigation in energy management transactions is significantly more complex than simply pricing the transaction to provide attractive investor returns.

In underwriting an energy management transaction, the capital markets analyze four broad categories of risk:

- Obligor Credit Risk
- Construction/Installation Risk
- ESCO Credit Risk
- Structural Risk

In order to structure a bankable transaction, each of these risks must be considered and mitigated to the extent required by the (debt) capital markets. To do so, one must understand the nature of these risks and the capital markets' tolerance for each of them.

Obligor Credit Risk

Obligor credit risk is the core risk for all financing transactions and simply translates to the end user's ability to satisfy all of its obligations (payment and otherwise) under the terms of the transaction. Despite the wide variety of financing structures for energy services transactions, their common foundation is the end user's obligation to make payments in exchange for the products or services received. In whatever form it takes (i.e. lease payment, loan payment, services payment, usage payment, etc.), that payment obligation provides the lender's primary source of debt service and return on the equity investment, if any. As a result, the bulk of the lender's underwriting activities are focused on a detailed analysis of the end-user's overall credit profile. Operating performance, cash flow, debt service capacity, liquidity, balance sheet strength, market position, management capabilities, and future projections are among the many factors thoroughly examined during the underwriting process. In order to achieve the desired end result, i.e.

transaction approval, the credit analysis must conclude that the obligor is capable of fully satisfying its obligations throughout the full term of the transaction.

Construction/Installation Risk

Construction/installation risk is defined as the risk of loss to the investor during the period of time the necessary facilities and equipment are being constructed or installed at the customer's site. Construction/installation risk is a function of the project specifications, the length of the construction period, the nature of the customer, and ESCO obligations during the construction period, as well as the ESCO's technical capabilities and financial condition.

The assessment of construction/installation risk begins with identifying the party responsible for guaranteeing repayment to investors should the project not be completed. In most cases, construction/installation risk is borne by the ESCO and/or its subcontractors (in the form of repayment guarantees to the lender in the event that the project is not completed to the end user's satisfaction and there is no commencement of debt service payments). Depending on the credit strength of the party or parties guaranteeing the repayment of funds advanced during the construction period, the lender may also require performance bonds from a creditworthy surety.

ESCO Credit Risk

ESCO credit risk is a function of the ESCO's role in the construction and ongoing operations and maintenance of the project, typically taking three main forms: construction/installation risk, performance risk, and general risk of insolvency. Because these ESCO credit risks stem from different circumstances, the lender must analyze them individually and mitigate each to the maximum extent possible.

As previously discussed, the lender must have a reliable source of repayment for all funds disbursed during the construction/installation process. Direct recourse to the ESCO is the most common vehicle for sufficiently mitigating construction/installation risk. In order to rely on its recourse to the ESCO, the lender will require a thorough review and approval of the ESCO's credit profile. The ESCO should be prepared to provide at least three years audited financial statements, historical performance data, and any other information necessary to underwrite the transaction. Depending on the credit strength of the ESCO, the lender may impose additional credit conditions such as performance bonds and parent guarantees.

Performance Risk

Performance risk is measured in terms of the end user's payment obligations in the event of a performance deficiency and / or performance default. If the end user has an absolute and unconditional payment obligation to the lender regardless of ESCO performance problems, then the lender's exposure to performance risk is minimized. On the other hand, if the end user's payments are conditional upon satisfactory performance, then the lender must have recourse to the ESCO for non-payment resulting from performance deficiencies. In that case, the lender will closely scrutinize performance risk and the ESCO's ability to satisfy its obligation in the event of actual or alleged non-performance. As mentioned above, additional structural and credit enhancements may be required.

Structural Risk

Structural risk stems from the contractual nature of the transaction and the specific terms and conditions therein. In order to develop a bankable energy services transaction, it is critically important to consider and properly address these risks early in the transaction development process. Failure to clearly specify these risks and allocate them to the appropriate parties is probably the most common mistake that leads to a non-bankable transaction. The most important of these risks and the most effective ways to mitigate them are identified and discussed in detail below.

PERFORMANCE RISK VS. CREDIT RISK

An important general rule of thumb in financing energy services projects is that the capital markets are in the business of taking credit risk, not performance risk. As a result, transactions must be structured with bright line separation of these risks, with credit risk clearly allocated to the lender and performance risk clearly assumed by the ESCO or the end user obligor. This bright line separation enables the lender to evaluate and underwrite the transaction according to standard credit procedures, without focusing on technical specifications, savings estimates, energy costs, equipment reliability and a host of other factors that affect the ESCO's ability to perform as required under the terms of the contract. The bottom line is that the lender must be protected from any risk of non-payment due to an actual or alleged performance deficiency or default.

IMPORTANT PROVISIONS

Payment Obligation(s)

Payment obligations are closely related to and sometimes a function of the separation of performance and credit risk. The lender's preferred structure is to have a "hell or high water" payment obligation directly from the end user obligor regardless of ESCO performance. This is typically the case with basic debt financing and enables the lender to perform a standard analysis of the obligor's creditworthiness and to underwrite the transaction if that analysis indicates a high likelihood of repayment. In a structure where the end user obligor does have the right to suspend payment due to alleged or actual performance deficiencies, the lender will look to the ESCO to guarantee repayment. In all cases, the clear separation of performance and credit risk is critical to the lender's ability to preserve the hell or high water nature of the repayment obligation. If done properly, the lender's risk will be limited solely to the creditworthiness of the end user obligor or, in a performance guarantee scenario, the technical capabilities and creditworthiness of the ESCO.

Funding During Construction/Installation

There are a number of ways to provide for funding of the project during the construction/installation period (e.g. escrow funding, direct disbursements from the lender, bridge loans, etc.). Regardless of the funding mechanism, the related documents must address specific issues such as mechanics and contractor liens; procedures for invoice submittal, approval and payment; escrow or payment agent costs; title and security interest in the equipment; and waivers and consents, if necessary.

Payment and Cash Flow Logistics

Not only do the debt capital markets require a strict repayment obligation, but they also require certainty as to the timing and logistics of those payments. In most cases, the lender will assume all responsibilities for billing, collection and distribution of payments. When debt service payments are bundled with payments to the ESCO for its operations and maintenance services, the lender will require a lock-box agreement. The lock-box would be controlled by the lender under the terms of an agreement that clearly states that all payments received will be applied first to debt service, next to ESCO receivables, and last to equity, if any.

Taxes, Insurance and Maintenance

Whether the financing transaction is documented as a standard capital lease or incorporated into an executory services agreement, the capital markets will require a "triple net" arrangement. In other words, the lender will require the end user obligor or the ESCO to:

- Pay all applicable property, sales and use taxes.
- Acquire property and liability insurance in amounts determined by the lender and name the lender as a loss payee and/or additional insured under those policies.
- Properly maintain the equipment. The ESCO's operations and maintenance obligations normally cover this requirement.

Title/Security Interest in Assets

Depending on the financing structure used, the lender may or may not hold title to the financed equipment. However, in all cases, the lender will require a first priority, perfected security interest in the project assets. Three key elements must be included in the documents to satisfy this requirement. They are:

- Explicit identification of the assets as personal property and agreement by the parties that the installed equipment will never be deemed fixtures or any other type of real property.

- Unequivocal rights of the lender to perfect its first priority security interest in the equipment by making any and all UCC and fixture filings necessary to document its lien(s) on the equipment. Furthermore, the end user obligor and ESCO will be required to keep the equipment free and clear of liens other than those asserted by the lender.

- Reasonably limited rights of the lender to access, inspect and/or remove the equipment without liability to the end user or the ESCO.

Termination Provisions

Many energy services transactions allow for early termination at the option of one or more of the parties, or upon an event of default. Optional early termination provisions do not pose a problem for the capital markets as long as the mechanics of the termination process provide for written notice 60 to 180 days in advance and protect the lender from any loss on its investment. To achieve this acceptable end result, the transaction must contain unambiguous termination language that provides for payment of an early termination value equal to the lender's outstanding principal balance plus customary early termination fees. The fee is usually calculated according to a pre-negotiated sliding scale based on when the early termination takes place—the earlier in the term, the higher the fee. Depending on the preferences of the parties, the early termination value can either be calculated at the time of early termination or be predetermined and documented in an early termination table within the documents.

In the event of early termination resulting from a performance default, the lender will still require payment of either the defined early termination value or a casualty loss value, as defined in the documents. The structure of the financing will dictate which party is obligated to pay the early termination value. In standard financing transactions directly between the end user obligor and the lender, the obligor will be required to pay the early termination payment. Depending on the terms and conditions of its contract with the ESCO, the obligor may or may not have recourse to the ESCO if the early termination was caused by the ESCO's performance default. In bundled service agreement structures, the defaulting party is usually obligated to make the early termination or casualty loss payment, which is assigned to the lender.

Equity/Residual Risk

In some structures, such as operating leases and synthetic leases, the lender or third party investor(s) will make an equity investment in the project. As a result, the obligor's payments will not fully amortize the project cost. The investor must rely on the residual value of the equipment at the end of the term to recoup his original investment plus, ideally, a reasonable return on the invested capital. Due to the limited secondary market for the assets used in energy services projects, most investors take very limited residual risk or simply avoid it altogether.

Assignment Rights

The lender's unlimited right of assignment is a critical component of a bankable transaction. Without this feature, the deal will be completely isolated from the capital markets and extremely difficult, if not impossible, to fund. This reality stems from the fact that all lenders must maintain liquidity in their investment portfolios by having the right to sell (i.e. assign) their paper to other investors in the secondary market. Conversely, liquidity is further enhanced by prohibiting or precisely limiting the assignment rights of the other party or parties to the transaction. This is a necessary recognition of the fact that the lender's investment decision is based on the creditworthiness of the original obligor. By controlling the obligor's right of assignment, the lender can prevent credit and portfolio deterioration that would result from assumption of the borrower's obligations by an entity with weaker credit profile than the original obligor.

Contracts & Documentation

Once the parties have agreed upon the basic business and structural terms of the transaction, the deal must be documented in compliance with the standards and expectations of the capital markets, most of which have been identified above. An experienced legal team is key to achieving this requirement. Among the most common mistakes that must be avoided are:

- Lack of Clarity. The parties' rights and obligations must be stated clearly and unequivocally so as not to be open to interpretation, either by the parties themselves or in a court of law. Something as intuitively simple as the customer's payment obligation can create significant funding problems if not written with the clarity necessary to ensure consistent interpretation by all parties.

- Co-mingling of Lender & ESCO Risks. The documents must achieve a bright line separation between the lender's credit risk and the ESCO's performance risk.

- Inseparability of Financing and Services. In a bundled services agreement, the terms and conditions of the financing must be separable from those of the technical project in order to successfully monetize the transaction.

- Failure to Include Key Capital Markets Provisions. Documents without the previously discussed critical provisions, such as lender assignment rights, security interests in the equipment, hell or high water payment obligations, etc. will be poorly received by the capital markets.

CONCLUSION

The capital markets have an unlimited capacity to fund creditworthy, properly structured energy services transactions. By being aware of investor requirements and working collaboratively with a knowledgeable and experienced financing partner from the earliest stages of a project, energy services companies can avoid costly project delays, maintain competitive advantages, meet or exceed customer expectations, avoid leveraging their balance sheets, and gain broad access to the capital markets to fund their bankable projects.

This material is the work of the individual author and is presented for informational purposes only. This work does not represent the formal opinions or positions of Banc of America Leasing & Capital, Bank of America, N.A. or Bank of America Corporation including any of its subsidiaries and affiliates. Readers are strongly encouraged to consult their own independent tax, legal and accounting advisors when entering into any financing or investment transaction.

Chapter 8

When Firms Publicize Energy Management Projects Their Stock Prices Go Up:

How high? As much as 21.33% within 150 days of announcement.

John R. Wingender,
Eric A. Woodroof, Ph.D.

Editor's Note: **Energy managers need all the help they can get to implement energy projects. Authors Wingender and Woodroof describe a new way to catch top management's attention, via an objective close to their hearts: How to improve the price of the company's stock.**

The potential for increased profits via *cost-reducing* energy management projects (EMPs) exists in nearly all firms. However, when allocating capital, priority is often given to *revenue-enhancing* projects, such as starting new product lines or joint ventures.

Frequently, these projects are perceived to be superior to EMPs, even though they may yield the same increased profit and present value. A justification is that *revenue-enhancing* projects are more likely to attract publicity and investor attention. Investor speculation and reaction to announcements can increase the firm's stock price. Most EMPs do not generate as much publicity as joint ventures or new product lines.

If "publicity-gaining" potential is a decision factor during project selection, then a new product line or joint-venture will usually be selected over an EMP. But is this a fair comparison? There has not been any research to determine if an EMP announcement increases a firm's stock price. In theory, it should—because most EMPs increase profits (via *cost reduction* instead of *increased revenues*). From a cash flow perspective, an EMP is equivalent to any other profit-enhancing project.

This article seeks to determine whether an EMP announcement correlates with an abnormal increase in a firm's stock price. If such announcements positively impact stock price, then the firm has one more incentive to implement EMPs.

LITERATURE REVIEW

The purpose of this literature review is three-fold:

1. To demonstrate that EMPs are credible investments, with relatively low risk;

2. To present some background on stock price reaction to announcements of typical capital investments; and

3. To show that abnormal increases in stock prices from EMP announcements have not been measured.

Public announcements (such as mergers, joint ventures or new product lines) correlate with abnormal stock price returns.[1,2] When a firm announces a joint venture (or other revenue-enhancing project) it is trying to attract publicity, which can raise the stock price based on expected future profits. However, since such projects can also be unprofitable, the anticipated cash flows are at risk.

When firms implement EMPs, they also expect improved profits by becoming more cost-competitive. EMPs and equipment replacement projects usually have more predictable cash flows (less risk) than many other types of capital investments, especially new product lines or joint ventures.[3] Today, the risk from most EMPs is so low there are many third party lenders who are eager to locate and finance EMPs.[4] In 1995, leasing (including third-party leasing and performance contacting) accounted for nearly one third of all equipment utilization.[5]

Thus, EMPs and other facilities improvement projects are recognized as credible investments; however, they are frequently put on the "back burner" relative to *revenue-enhancing* projects.

Maximizing stock price should be a goal of the corporation. Increasing productivity, offering new product lines, and increasing profits are examples of tangible factors that can increase the firm's stock price. However, stock price may also increase due to intangible factors, such as investor speculation and reaction immediately following an announcement. Executives may incorporate this investor reaction when deciding which projects to implement.

Although investor reaction has not been assessed for EMP announcements, there has been some research in this area. It has been shown that firms increasing expenditures on general facility and equipment improvements had a 1.98% abnormal stock increase immediately after the announcement.[6]

Announcements of joint ventures correlated with a positive abnormal return of 1.95% over a 21-day interval (–10 days to +10 days) centered on the announcement day (day 0).[7] However, not all joint ventures correlate with positive abnormal stock returns.[8]

The correlation between EMP announcements and stock price has not previously been investigated. This article examines whether EMP announcements correlate with positive abnormal stock price returns. If they do, then perhaps the capital budgeting process should incorporate this benefit.

METHODS

Using the Nexis/Lexis Data Base, the world wide web, and other resources, a search for EMP announcements resulted in over 5,500 citations. Of the 5,500 citations, only 23 announcements fit the following criteria:

1. The firm announcing the EMP was publicly traded and its returns were available on the data files of the Center for Research in Security Prices (CRSP).

2. The announcement was the first public information released about the EMP.

3. The EMP was large enough to represent a significant investment for the company. *For example, if a large fast-food chain was announcing an EMP at only one restaurant, it was deleted from the sample. However, if the EMP was company-wide, it was included in the sample.*

4. The announcement was made in a major US newspaper, newswire service or a monthly trade magazine between 1986 and 1995.

The 23 announcements which represented the "Complete Sample," consisted of several subsamples:

- 16 Announcements made within daily newspapers or via electronic wire ("daily")
- 3 Announcements within monthly trade magazines ("monthly")
- 4 Announcements of post-implementation results ("post-implementation")

An additional sub-sample was created ("daily + monthly") to maximize the sample size of preimplementation announcements.

For each announcement, the firm's name, CRSP identification number and announcement date were entered into the *Eventus* computer program, which tracks daily stock price performance for every U.S. stock listed within the CRSP.[9] *Eventus* uses a common "event-study" methodology to calculate abnormal stock returns, and it indicates statistical significance levels. Statistically, a null hypothesis (H_0) was proposed and tested against an alternative hypothesis (H_a).

H_0 = EMP announcements have no impact on stock price.

H_a = EMP announcements have a positive impact on stock price.

Analysis Intervals

Usually, a firm's stock price performance is analyzed over several time intervals around the announcement date. Often, a stock price improvement can be noticed between the day of the announcement (day 0) and the next trading day (day 1). In this interval, the range is represented by the following notation: (day 0, day I) or (0,1). Another typical interval for event-study analysis is a two-day interval, one day before the announcement to the announcement date (–1,0).[10]

EMP announcements frequently appear in monthly trade magazines, which are distributed to readers on different days in different geographic locations. In this case, an exact announcement date cannot be determined. Thus, analysis of the stock performance over a wider interval is appropriate. The (–10,10) interval represents the period at which EMP announcements would most likely be noticed through monthly trade magazines.

In addition, since EMP announcements may not capture as much publicity as other announcements, it is reasonable to expect a longer period for the market to "learn" about an EMP.

The (–10,10) interval is useful for identifying if an abnormal stock price increase correlates with an EMP announcement. However, to observe the long-term stock impact, the sub-samples were analyzed over additional intervals, such as (1,100), (1,150), etc.

Applying the aforementioned hypothesis tests to the sub-samples yielded the results which are presented in the following section of this article. A more detailed explanation of the "event-study" methodology (statistical analysis) is included in the Appendix.

RESULTS

Tables 8-1 and 8-2 present the short-term and long-term abnormal returns. The returns are categorized by interval around the announcement date. The level of significance at which H_0 was rejected is also indicated.

The daily + monthly sub-sample is the most appropriate sample because it is the largest sample possible that excludes post-implementation announcements. The post-implementation subsample is substantially different in nature because it represents firms announcing cost savings (increased profits) from projects already implemented.

Using the daily + monthly sub-sample, EMPs correlate with a 3.90% increase in stock price, measured from ten days prior to the announcement to ten days after the announcement, (–10,10). The level of significance was 0.01. See Table 8-1.

Table 8-2 shows the long-term performance, where the daily + monthly sub-sample correlated with a 21.33% abnormal return over the (1,150) interval, at the 0.001 significance level. Figure 8-1 is a graphical illustration of the abnormal returns over the long-term interval.

Samples	# of Firms in Sample	Day Range From Announcement Date			
		(-5,5)	(-10,10)	(-15,15)	(-20,20)
Complete Sample	23	2.21%, *	2.80%, **	1.06%, *	1.92%, *
Announcements from Daily Newspapers	16	2.89%, **	3.75%, ***	2.31%, **	3.18%, **
Announcements from Monthly Magazines	3	2.31%	6.91%	8.23%	8.50%
Post Implementation Announcements	4	-0.57%	-2.70%	-5.86%	-7.12%
Daily + Monthly Subsamples Combined	19	2.43%, *	3.90%, ***	2.83%, *	3.64%, **

Significant at: * =0.10
** = 0.05
*** = 0.01

Table 8-1. Abnormal Return of Firms Announcing Energy Management Projects

Samples	# of Firms in Sample	Day Range From Announcement Date					
		(1,100)	(1,120)	(1,150)	(1,200)	(1,220)	(1,240)
Complete Sample	23	8.23%, *	11.00%, **	13.93%, ***	12.09%, **	13.75%, **	12.71%, **
Announcements from Daily Newspapers	16	10.67%, **	16.29%, ***	17.66%, ***	13.38%, **	17.10%, **	17.11%, **
Announcements from Monthly Magazines	3	27.05%	21.38%	31.54%	44.89%	42.44%	53.02%
Daily + Monthly Subsamples Combined	19	14.07%, **	18.03%, ***	21.33%, ****	21.73%, ***	23.42%, ***	24.95%, ***

Significant at: * = 0.10
** = 0.05
*** = 0.01
****= 0.001

Table 8-2. Long-Term Abnormal Return of Firms Announcing Energy Management Projects

CONCLUSION

The results from this study indicate that EMP announcements do correlate with significant abnormal increases in a firm's stock price. On average, an EMP announcement correlated with a 21.33% abnormal increase in the firm's stock price. This increase was experienced from the day after the announcement to 150 days after the announcement. This increase is *in addition to* the risk-adjusted return the firms would normally experience.

For example, during a "bull market" a firm 's expected return was 10%. After the announcement, the return increased by 21.33%, for a net return of 31.33%. Because these EMPs were announced by a diverse group of firms at various periods over a ten-year time span, the significance of these results is impressive. In other words, the EMP is probably the only event that all firms within the sample have in common.

From these results, it appears that shareholders recognize EMPs as investments that should increase profits and add value to the firm. With the new information presented here, firms may have an additional strategic incentive to implement EMPs.

DISCUSSION AND RECOMMENDATIONS FOR FURTHER RESEARCH

Despite the small scale of this study, the significance of the results is impressive. This study could serve as a first step to understanding investor reaction to EMP announcements. Additional studies with increased sample size and greater stratification would yield more information.

It is interesting to note that detailed cost savings estimates were not always included in the EMP announcements. *For example, many firms simply announced that they were going to retrofit a portion of their facilities, without an estimate of dollar savings.* Perhaps more detailed information was released after the announcement date, triggering greater stock price increases in the long-term intervals. However, it is more likely that shareholders associate EMPs as effective profit enhancing projects that are almost always good for the bottom line.

It would be interesting to determine if there is a relationship between an EMP's potential profits and the value of the abnormal return. The value of the abnormal return should predictably be related to the

Figure 8-1. Abnormal Return of Firms Announcing Energy Management Projects

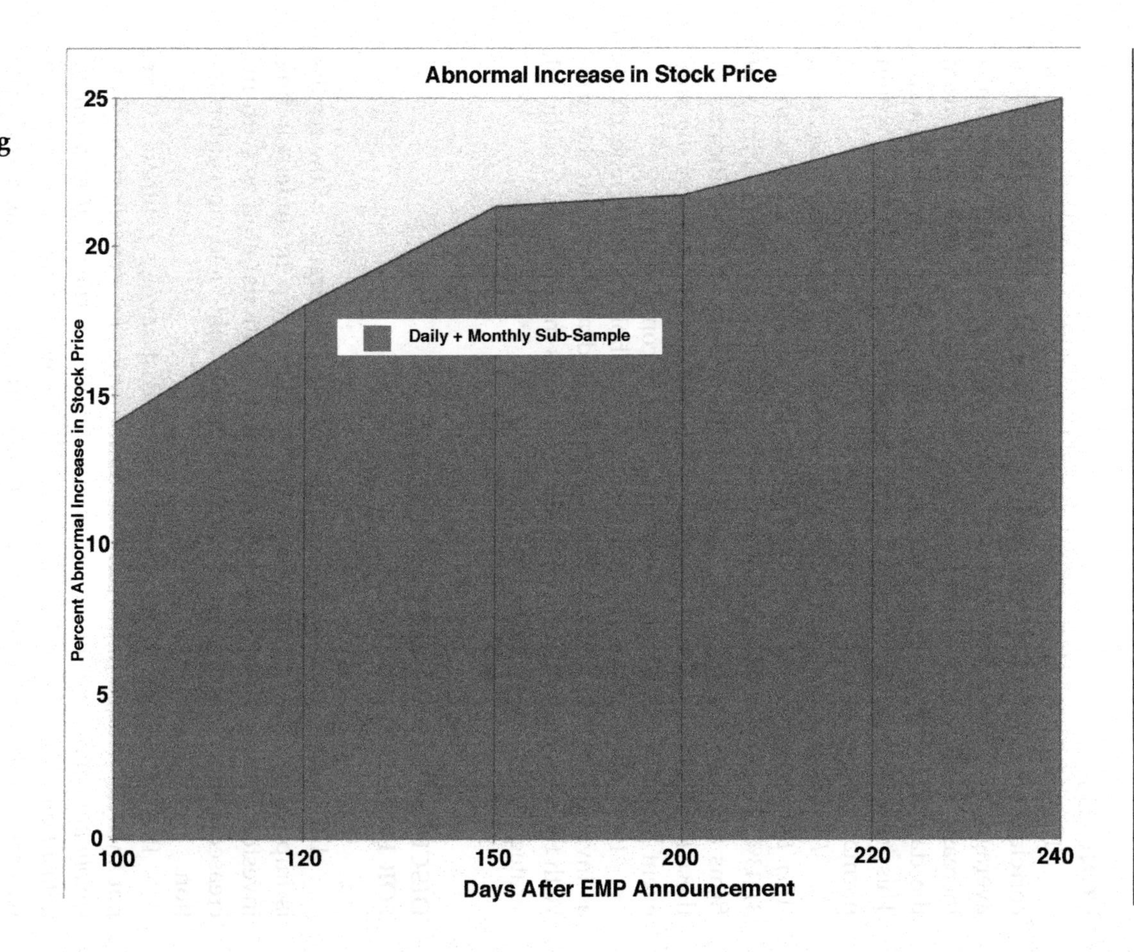

amount of increased profit from the EMP. Identifying these values could indicate whether the investor reaction is proportional to the potential added value of the EMP. Calculating these values would require additional information about each firm as well as each project. This could be a focus of additional research.

It was recognized that the type of EMP could influence the magnitude of the abnormal stock increase. Thus, the Complete Sample was further stratified into two sub-samples: EMPs that were lighting retrofits and EMPs that were installations of other types of energy efficient equipment (such as HVAC upgrades, chiller upgrades, etc.).

Both sub-samples were analyzed by the *Eventus* software. The energy efficient equipment sub-sample correlated with a 1.42% increase over the (–10,10) interval, at the 0.01 significance level; the lighting retrofit sub-sample yielded no significant returns over the (–10,10) interval.

However, because this sub-sample only contained seven firms, this comparison needs to be re-evaluated with a greater sample size. In addition, this analysis was tainted because the post-implementation announcements were included.

It was also recognized that the finance method for each EMP could influence the magnitude of the abnormal stock increase. Since off-balance sheet financing (leasing) is common for EMPs, a comparison was made between EMPs that utilized leasing versus EMPs where the equipment was purchased by the firm (and the debt was carried on their balance sheet). Assuming that stock analysts frequently look at balance sheets to assess a company's performance, it is reasonable to hypothesize that off-balance sheet financed EMPs would correlate with higher abnormal stock returns than EMPs where equipment was purchased by the firm.

The complete sample was further stratified and analyzed by the *Eventus* software. The EMPs that were purchased directly by the firm did show a 3.74% increase over the (–10,10) interval, at the 0.01 significance level. The sub-sample of leased EMPs yielded no significant returns over the (–10,10) interval, although it should be noted this sub-sample included only three firms.

Again, the samples were tainted with the post-implementation announcements. Therefore, this comparison needs to be re-evaluated with a larger sample size, and with the post-implementation announcements removed.

Although the post-implementation sub-sample was small, it did not yield any significant positive abnormal returns. In fact, the returns were negative, although not significant. This is intriguing because a post-implementation announcement is basically a statement of increased profits already realized.

A more extensive study on the effects of post-implementation announcements could reveal if these types of announcements yield different abnormal returns than announcements prior to implementation.

All of the sub-samples should be analyzed over a longer time interval. In addition, increasing the sample size would also improve the validity of the results.

References

1. McConnell, J. and Nantell, T. (1985), "Corporate Combinations and Common Stock Returns: The Case of Joint Ventures," *The Journal of Finance*, **40** (2), 519-536.
2. Pettway, R. and Yamada, T. (1986), "Mergers in Japan and their Impacts on Stockholders' Wealth," *Financial Management*, **15** (4), 43-52 (Winter 1986).
3. Pohlman, R., Santiago, E., and Markel, E. (1988), "Cash Flow Estimation Practices of Large Firms," *Financial Management*, **17** (2), 71-79.
4. Zobler, N. (1995), "Lenders Stand Ready to Fund Energy Projects," *Energy User News*, **20** (3), 19.
5. Sharpe, S. and Nguyen, H. (1995) "Capital Market Imperfections and Incentive to Lease," *Journal of Financial Economics*, **39** (2), 271-294.
6. McConnell, J. and Muscarella, C. (1985), "Corporate Capital Expenditure Decisions and the Market Value of the Firm," *Journal of Financial Economics*, **40** (3), 399-422. *Abnormal return was measured over a two-day announcement period that encompassed the day on which the announcement appeared in print, plus the previous day (–1,0). The stock performance over a 21-day interval (–10,10) was not reported. This abnormal return value can be different than the value over the two-day interval. The result was different from zero at the 0.01 statistical significance.*
7. McConnell, J. and Nantell, T. (1985), "Corporate Combinations and Common Stock Returns: The Case of Joint Ventures," *The Journal of Finance* **40** (2), 519-536.
8. Lee, I. And Wyatt, S. (1990) "The Effects of International Joint Ventures on Shareholder Wealth," *The Financial Review*, **25** (4), 641-649.
9. Cowan, A. (1996), *Eventus Version 6.2 Users Guide*, Cowan Research.
10. Brown, S. and Warner, J. (1985) "Using Daily Stock Returns," *Journal of Financial Economics*, **14** (1), 3-31.

APPENDIX

Event-Study Methodology

An "event study" is a popular analysis tool for analyzing stock price reactions to particular events. In this study, the "event" is the announcement of an EMP by one of the sample firms. The event date is the first trading day that the market could react to the announcement. The impact of EMP announcements on stock price is tested by calculating risk-adjusted abnormal returns on and around the announcement date.

For this study, we use the market model event-study method and test the results for significance with the standardized residual method. The market model event-study method uses a linear regression to predict stock returns; then it compares the predicted value to its actual returns.

The abnormal return (ABR_{jt}) is the difference between the actual return (R_{jt}) on a specific date and the expected return ($E(R_{jt})$) calculated for the firm on that specific date. The expected return is calculated using the parameters of a single index regression model during a pre-event estimation period. The regression model parameters are determined by the following equation:

$$R_{jt} = a_j + b_j R_{mt} + e_{jt}$$

where

R_{jt} = the return on security j for period t,
a_j = the intercept term,
b_j = the covariance of the returns on the jth security with those of the market portfolio's returns,
R_{mt} = the return on the CRSP equally-weighted market portfolio for period t, and
e_{jt} = the residual error term on security j for period t.

To calculate the market model parameters (a_j and b_j) a 220-day estimation period was used that begins 260 days before the announcement date. For each sample firm, the event period begins 30 days before the announcement date and ends 30 days after the announcement date. The expected return ($E(R_{jt})$) is then calculated using the return on the market (R_{mt}) for the specific event period date:

$$(E(R_{jt}) = a_j + b_j R_{mt}$$

The abnormal return *(ABRjt)* for an event date is then calculated by subtracting the expected return (which uses the parameters of the firm from the estimation period and the actual market return for a particular date in the event period) from the actual return (R_{jt}) on that date. The equation is as follows:

$$\text{ABRjt} = \text{Rjt} - E(R_{jt})$$

The average abnormal return (AAR_t) for a specific event date is the mean of all the individual firms' abnormal returns for that date:

$$AAR_t = \sum_{j=1}^{N} \left(ABR_{jt}\right)/N$$

where N is the number of firms used in the calculation.

The cumulative average abnormal return ($CAAR$) for each interval is calculated as follows

$$CAAR_{T_1T_2} = \sum_{T_1}^{T_2} (AAR_t)$$

The standardized residual method is used to determine whether the abnormal return is significantly different from zero. The standardized abnormal return (SAR_{jt}) is calculated as follows:

$$SAR_{jt} = ABK_{jt}/s_{jt}$$

where

s_{jt} = the standard deviation of security j's estimation period variance of its ABR_{jt}'s.

The estimation period variance $s^2{}_{jt}$, is calculated as follows

$$s_{jt}^2 = s_j^2\left[1 + 1/D_j + \left[(R_{mt} - \bar{R}_m)^2 / \sum_{k=1}^{D_j} (R_{mk} - \bar{R}_m)^2\right]\right]$$

$$s_j^2 = \left[\sum_{k=1}^{D_j} \left(ABR_{jk}^2\right)\right] / \left(D_j - 2\right)$$

Where

$\bar{R}_m$ = the mean market return over the estimation period, and

$D_j =$ the number of trading day returns (220) used to estimate the parameters of firm *j*.

Finally, the test statistic for the null hypothesis (H_0) that $CAAR_{T_1, T_2}$ equals zero is defined as follows:

$$Z_{T_1,T_2} = (1/\sqrt{N}) \sum_{j=1}^{N} \left(Z^j_{T_1,T_2}\right)$$

where

$$Z_{T_1,T_2} = \left(1/\sqrt{Q^j_{T_1,T_2}}\right) \sum_{t=T_1}^{T_2} \left(SAR_{jt}\right)$$

and

$$Q^j_{T_1,T_2} = (T_2 - T_1 + 1)\,\frac{D_j - 2}{D_j - 4}$$

Chapter 9

Overcoming the Three Main Barriers to Energy Efficiency or "Green" Projects*

Eric A. Woodroof, Ph.D., CEM

ABSTRACT

Although the popularity of energy management and "green" projects is improving, there are *many* good projects that are postponed or cancelled due to common barriers. This article discusses these barriers and problems, as well as effective, proven strategies to overcome them. These timeless, cutting-edge strategies involve marketing, educational resources, and financing approaches to make your projects *irresistible*. The goal of this article is to help organizations get more good energy management/green projects approved and implemented, thereby helping to slow global warming.

INTRODUCTION

The polar ice is melting and, as far as the planet is concerned, engineers are *wasting their time* if the projects they so carefully develop are not implemented and deliver no value. This article refers to "good" projects as those with a three-year payback or less. Why don't good proj-

---

*Published in *Strategic Planning for Energy and the Environment*, 2008

ects get implemented? There are a variety of reasons and a few common barriers:

- *Marketing* (under-marketing a project's value)
- *Education and collaboration* (not expanding the value of a project)
- *Money* (not having a positive cash flow solution)

If a project can't satisfy these criteria, it probably won't be implemented anyway… so focus on the ones that will!

PROBLEM #1: MARKETING

People often ask me why marketing is first on the list. Answer: because NOTHING HAPPENS WITHOUT A SALE. For example, your first job (or your first date) began with you "selling yourself." This is the goal of an interview. In fact, the development of every product/service begins with someone selling a solution to some type of problem. I am saying that selling/convincing is neither "bad" nor unethical. Convincing others when it improves their lives is good—it can be done with passion! When something (like an energy management project) is great, we should sell the benefits *with all the passion in the world*. You would do the same when talking to your kids about "getting a good education," or "learning good manners." Passion can also emerge from fear, such as from the chaos and violence that occurs during an electrical blackout. *Most of the time, humans are more passionate and action-oriented when they are at risk of losing something versus the possibility of gaining something.*

So, we must communicate in a way that the audience (the buyer or project endorser) can understand the problem/pain that they are in now. After they agree that they are "in pain," then they will want to hear about potential solutions.

Attention

It starts with getting the buyer's attention on the problem, the pain it is causing, and a sense of urgency to solve the problem. Only then will a solution seem to be logical. In addition, after the problem/pain is understood, they will be able to become passionate about the solution.

If you fail to get the attention of the endorsers, you are actually doing them a disservice; they won't know they are in trouble and are wasting money. It's like allowing someone to bleed to death when they don't even know they are cut. So, don't be shy—you have a duty to perform.

Warning! Some endorser's personalities' won't like to hear about problems/pain. They may "put their head in the sand" like an ostrich when problems are discussed. Don't blame them; it's part of their personality (which has strengths in other areas). Discover ways to communicate in a way that they will respond. FYI, it can take *seven* impressions (explanations/presentations) before some people will agree on the problem and take action on a solution. Don't give up and don't be surprised or depressed when they don't take action after the first impression.

Below are a few examples of effective headlines* that can help get the attention of an endorser. Feel free to use these in executive summaries:

- "How will the shareholders feel about us throwing money away every month?"
- "A way to make money while reducing emissions..."
- "What will we do with the yearly savings?"
- "We are paying for energy-efficiency projects, whether or not we do them!"
- "Guaranteed, high-yield investments..."
- "If you enjoy throwing money away every month, don't read this..."
- "4.6 billion years of reliability... solar energy."
- "This project could improve our stock price by over 20%!"**
- "Good planets are hard to find."

There are many other great proven examples that are available.† However, you can experiment by looking around for "marketing copy" in magazine advertisements, commercials, etc. There is a reason they call it "copy"—some of the principles are thousands of years old, and they still work! Just change the words to relate to your problem/solution. Try a few versions and test, test, test to see which ones are most effective. Go for it!

---

*Ultimate Marketing for Engineers Course, www.ProfitableGreenSolutions.com

**Wingender, J. and Woodroof, E., (1997) "When Firms Publicize Energy Management Projects: Their Stock Prices Go Up"—How much—21.33% on Average! *Strategic Planning for Energy and the Environment,* Summer Issue 1997.

†The "Vault Files," www.ProfitableGreenSolutions.com

Benefits*

After you have their attention, be sure that you include compelling benefits that "take away the pain" the audience is feeling. As engineers, we are good at mentioning the typical benefits:

- Saves energy, money, waste and emissions;
- Offsets the cost of a planned capital project;
- Improves cost-competitiveness, productivity, etc.;
- Is a relatively low-risk, high-profit investment that directly impacts the bottom line**.

In today's green-minded economy, we could also demonstrate that "green" projects are a very effective marketing tool—which could get the client's marketing department behind your project—because these benefits have also been proven†:

- Improves the client's "green" image;
- Differentiates the client from the competition††;
- Introduces them to new markets, suppliers and clients‡¶;
- Helps them grow sales/revenue.

However, we should also mention the passionate, global and moral reasons behind a good "green" project:

- Slows global warming, reduces acid rain;
- Reduces mercury pollution, which allows us to eat healthy;
- Improves our national energy independence;
- Reduces security/disaster risks, etc.

---

*Download the FREE emissions calculator from www.ProfitableGreenSolutions.com

**For Example: an energy-efficient project that saves $100,000 in operating costs is equivalent to generating $1,000,000 in new sales (assuming the company has a 10% profit margin). It can be more difficult to add $1,000,000 in sales, and would require more infrastructure, etc.

†Several examples include: Patagonia, Google, GE, Home Depot, etc. Other examples can be downloaded from the "Resource Vault" at www.ProfitableGreenSolutions.com

††For Example: A construction firm switched to hybrid vehicles, which offsets the carbon emissions. The firm's name is prominently displayed on each vehicle. They get tons of new business because they are seen and known as the "greenest construction firm" in the city. Plus, they charge a premium for their services!

‡For Example: A law firm renovated their office in a "green" manner and attracted a new client (who chose the firm due to its "green" emphasis). The new client was worth an extra $100,000 in revenue in the first month.

¶Additional Examples: "Green" networking groups such as "greendrinks.org" can supplement the traditional business of networking clubs like Rotary Club, Kiwanis, Chamber of Commerce, etc. Also, when joining groups such as the Climate Action Registry, companies are exposed to other members, who could be superior suppliers, clients and partners.

Dollar values for these benefits can often be calculated and should be included in your proposal. To calculate the "green benefit equivalencies," such as "number of trees planted" (from reduced power plant emissions), see the "Money" section of this article.

The list above can be expanded, refined and optimized for any project. To build a list like the one above, one technique is "WSGAT"—"What is So Good About That?" Ask that question for every project feature and you will develop a long list of passionate benefits. By the way, this approach has been used in TV sales and has helped sell billions of dollars of material*. If they can sell this much junk on TV, we should have no problem selling green projects that are factually saving the planet! Add the emotional benefits of going "green," and you will have a project that touches the hearts of leaders in your organization.

Call to Action

The "call to action" becomes easy and logical when all of the benefits have been quantified and they are aligned with the client's strategic objectives. Tell endorsers what you want them to do and why. Be sure to include the "cost of delay" in your executive summary. Remind them that they are "in pain" and the project/solution will solve it. Visual aids can be helpful. For example, during one presentation, buckets of dollars were shoveled out a window to demonstrate the losses that were occurring every minute. The executives were literally in pain watching those dollars fly away. They couldn't stand it, and they took action. It is OK to get creative and have some fun in your presentation!

But wait... there is more!

"Configuring" your presentation can make the difference between immediate approval and further delay. There are many ways to configure or "package" your product/project so that it is IRRESISTIBLE. One way to do this is to find a way to make a project's performance guaranteed or "risk-free." Another way is to separate (or add) one part of the project and introduce it as a "free bonus." Everyone likes a "FREE" bonus—it helps them understand that they are getting a good deal. For example, on a recent "green," facility-related project, carbon offsets for a company's fleet were included as a free bonus. The bonus delighted the client and distinguished the project (adding extra value), yet the additional costs were less

*Marketing to Millions Manual, Bob Circosta Communications, LLC.

than $1,000.

Engineers can be two, three or ten times more productive by developing sales and marketing skills. However, there is another reason for developing these sales/marketing skills: Your Career! The skills you learn will be valuable to your organization (as well as other organizations). These skills are transferable to other industries too. So keep this in mind when you are investing in yourself; there will almost always be a fantastic pay-off.

Finally, there are two prerequisites that a buyer must see in you before any sale is made: "Trust" and "Value." As far as trust goes, it must be earned, and once earned… it must be cherished. To accelerate the buyer's trust in you, be an advocate for the client and put their needs ahead of your own. Assume the role of their "most trusted advisor," and then deliver. Value comes from applying knowledge, tools, resources, partners, etc., in the best way for the client, which is why education and collaboration is such an important component of success. This is discussed further in the next section. Be sure to read the sub-sections on reciprocal business agreements, joint ventures, and incentives/rebates… great ideas!

PROBLEM #2: Education & Collaboration

Knowing how to deliver the value is an area that requires continuous updating. Today, with the proliferation of energy/green technologies, it is impossible for one person to know all the ways to add value to a project. Green specialties are expanding every day, for example: energy efficiency, water efficiency, green janitorial, LEED*, recycling, transportation, etc.

Learn all you can, then collaborate with other professionals who are also actively learning, and the value available to your clients' increases exponentially. It is important to be open to new ideas and fresh perspectives in this process. "Mind-sharing" or brain-storming techniques can facilitate the process and maximize the number of useful ideas.*

Fortunately, education is a low-cost investment. Collaboration and even joint-ventures/partnering can be done inexpensively as well—and the returns can be huge!

---

*LEED = Leadership in Energy & Environmental Design
*Results from the Profitable Green Strategies Course, www.ProfitableGreenSolutions.com

Free sources of green/energy efficiency education:
- https://www.aeecenter.org/seminars/
- http://www.eere.energy.gov/
- http://www.ashrae.org/education/
- www.usgbc.org
- www.ase.org
- www.energystar.org
- http://greeninginterior.doi.gov

In addition, there are many innovative ways to bring more value to a project. Some include:
- Reciprocal business agreements
- Joint ventures
- Free tax and utility incentives/rebates

Reciprocal Business Agreements

For example: After presenting a $1,000,000 service contract for a global car rental company, the deal was sweetened with an agreement on our part to choose that car company while traveling, which generated over $1,000,000 in extra car rentals for them. To the client, they were getting an extra $1,000,000 in revenue by working with us versus the competitors. With suppliers, partners, colleagues, professionals, etc. could you develop reciprocal business agreements? How could you help two clients (or suppliers) benefit from each other? How could you help them become more "green"?

Another example: We helped client #1 supply green solutions to client #2. Both clients were extremely happy to generate more sales/and save money. When it was time to approve our next round of projects, there was little resistance because we had helped them earn/save far more than the costs of the proposed projects. This illustrates the value of being the "trusted advisor."

Joint Ventures

For example: A "green" travel agency gives 50% of its commissions back to its clients in exchange for their travel business.* The client can use this extra, free money to fund "green" initiatives or scholarships, or other

* www.GreenTravelPartners.com

social programs. The travel agency guarantees the lowest prices and easily doubles its business because it delivers more value to its clients via joint ventures.

Free Tax and Utility Incentives/Rebates

For example: In California, 50% of a solar project was funded by federal and state rebates. Utility incentives lowered the installation costs even further. There are numerous free tax and utility incentives available, and some are discussed in the next section.

In addition to the options above, many utilities and third parties are offering "green power purchase agreements," which are essentially "wind and solar performance contracts." For example, if you want to put solar panels on your roof, a third party (often a utility or solar contractor) finances the project installation and then sells you the renewable energy produced from your roof (at a known price) for 15-25 years. So you get "green" power at no up-front cost, as well as a known future energy cost (lowers your risk to energy price volatility). The financier wins because the project will pay back the investment within 10 years and the rest is profit.

There are an unlimited number of creative "win-win" contracts available. However, before finalizing or even developing your solution, be sure that you understand the client's strategic and financial goals, then align the value to support the client's larger objectives.

PROBLEM #3: MONEY

If you do a good job tapping into the passion behind the project and are satisfying the emotional, financial and other approval criteria, you should have enough benefits to get the project approved, especially if the project is above the client's MARR.* However, if your organization is capitally constrained, you can finance a project and have positive cash flow. CFOs like positive cash flow projects! Cash flow constraints (not having the up-front capital to install a project) represent over 35% of the reasons why projects are not implemented†.

---

*MARR= Minimum Attractive Rate of Return. For more info on this topic see: Woodroof, E., Thumann, A. (2005) *Handbook for Financing Energy Projects*, Fairmont Press, Atlanta.

†U.S. Department of Energy, (1996) "Analysis of Energy-Efficiency Investment Decisions by Small and Medium-Sized Manufacturers," U.S. DOE, Office of Policy and Office of Energy Efficiency and Renewable Energy, pp. 37-38.

Financing does not have to be complicated. In fact, financing energy efficiency/green projects can be very similar to your mortgage or car payment—fixed payments for a length of time. However, with a good project, you can finance the project such that the annual savings are greater than the finance payments, which means the project becomes "cash flow positive" and does not impact the capital budget! This can allow the endorser to move forward without sacrificing any other budget line item.

Table 9-1 shows the cash flow for a non-financed project*. Assume the project costs $100,000 and saves $28,000 per year for 15 years. This project could get approved IF the client has $100,000 in cash to fund it. The project has a net present value of $ 102,700 and an internal rate of return of 27%.

Now, let's look at financing the project with a simple loan. Let's say the client finances the $100,000 for 15 years at 10% per year. That means that instead of investing $100,000 up front (the bank provides these funds),

Table 9-1. Project Cash Flow (paid with cash)

EOY	Savings	Cost	Cash Flow
0	-	(100,000)	$ (100,000)
1	28,000		$ 28,000
2	28,000		$ 28,000
3	28,000		$ 28,000
4	28,000		$ 28,000
5	28,000		$ 28,000
6	28,000		$ 28,000
7	28,000		$ 28,000
8	28,000		$ 28,000
9	28,000		$ 28,000
10	28,000		$ 28,000
11	28,000		$ 28,000
12	28,000		$ 28,000
13	28,000		$ 28,000
14	28,000		$ 28,000
15	28,000		$ 28,000
NPV i=10%			**$102,700**
IRR			**27%**

*Advanced Project Financing Course, www.ProfitableGreenSolutions.com

the client pays $13,147 each year to the bank for 15 years. At the end of 15 years, the bank loan is paid off (just like a mortgage or car payment—just a different time period). To keep this simple, ignore interest tax deductions as well as depreciation, which would likely improve the financials even further.

Table 9-2. Financed Project Cash Flow

EOY	Savings	Finance Cost	Cash Flow
0	-	-	$ -
1	28,000	13,147	$ 14,853
2	28,000	13,147	$ 14,853
3	28,000	13,147	$ 14,853
4	28,000	13,147	$ 14,853
5	28,000	13,147	$ 14,853
6	28,000	13,147	$ 14,853
7	28,000	13,147	$ 14,853
8	28,000	13,147	$ 14,853
9	28,000	13,147	$ 14,853
10	28,000	13,147	$ 14,853
11	28,000	13,147	$ 14,853
12	28,000	13,147	$ 14,853
13	28,000	13,147	$ 14,853
14	28,000	13,147	$ 14,853
15	28,000	13,147	$ 14,853
NPV i=10%			**$112,970**
IRR			**n/a**

In this case, the project generates $14,853 each year for the client. Because there is no up-front investment required, the IRR value becomes infinity.

Adding in the "Green Benefits" could further illustrate the project's benefits. Table 9-3 shows what some of these benefits could include. Note that it can be easier for the audience to visualize equivalencies ("car miles not driven," or "trees planted") instead of lbs of CO_2.

However, there are even more benefits when you consider the following impacts the project could have on:

- Shareholders in the annual report,
- Community morale and "green image,"

Table 9-3. Green Benefits*

kWh Saved per Year	260,000	
# of Years	15	
kWh Saved during Project	3,900,000	
Barrels of Oil Not Consumed	7,917	
Car Miles Not Driven	6,924,450	
Acid Rain Emission Reduction	29,250	lbs of Sox
Smog Emission Reductions	14,040	lbs of Nox
GreenHouse Gas Reduction	6,045,000	lbs of CO2
Mature Trees Planted	13,260	

- Productivity improvements,
- Legal risk reduction,
- LEED points, white certificates, RECs†,
- FREE public press¶.

FREE Money

In addition, there are utility rebates, tax refunds, credits, and other sources of free money that will improve a project's financial return. Here are some useful web sites that allow you to see utility and tax benefits in your state:

- www.dsireusa.org/
- www.energytaxincentives.org
- http://www.efficientbuildings.org
- http://www.lightingtaxdeduction.org/

But don't just rely on web sites. Use professionals; they should know what techniques, technologies, and rebates are best for your geographic area.

SUMMARY

This article has described the three common barriers (marketing, education, and money), as well as provided a start on how to overcome them. To get a project approved:

*Download the FREE emissions calculator from www.ProfitableGreenSolutions.com
†REC = Renewable energy credit
¶Press release samples from the "Vault" at www.ProfitableGreenSolutions.com

- Articulate the problem/pain.
- Collaborate to add value in the solution.
- Quantify all the benefits.
- Minimize financial risk.
- Develop/configure an executive summary that "sings" to the heart of the endorser.

Hopefully, these techniques will help you get your next project approved. Why is this important? Because the ice is melting! We are counting on you.

Chapter 10

Basics of Energy Project Financing

The main job of the chief financial officer of your organization is to reduce risks. The risk analysis of an energy project is not complete unless all technical and financial options are explored and understood.

For the purposes of this chapter, it is assumed that an energy services company (ESCO) will be responsible for providing the energy conservation measures (ECMs) and managing your project. These ECMs may include audits, design, construction, installation, monitoring, and maintenance.

It is also assumed that a separate party, such as a bank or investment company, is providing the capital to purchase the equipment. Sometimes ESCOs market themselves as being financiers, but most often they have a financing source such as a corporate parent or bank in the background. For simplicity's sake, the ultimate source of capital for the project will be considered a separate lender with its own guidelines and its own decision-making process.

Some of the most common financing alternatives are:

COMMERCIAL BANK LOAN

The obvious first alternative is approaching your local business bank and applying for a loan. The bank will review your company's credit history and financial statements in order to make a decision. Interest rates are usually based on the prime rate (the rate offered by the largest U.S. banks to their best customers) plus a margin. For example, as of this writing, a company with established, above-average credit can expect to pay 1% over prime. Most banks prefer to lend to businesses in the form of credit lines which can be used by the company as needed, paid back, and reused again without time-consuming reapplication procedures.

There is a well-known "catch-22 effect" in banking: If you are sufficiently well-qualified to be approved for a loan, you probably don't need one. If your company's financial statements show a $1,000,000 net worth and positive net income for the past few years, chances are that you will be approved for a loan for a $200,000 energy project. On the other hand, if you have a similar net worth and income and you ask your banker to help you with a $5,000,000 project, your banker will help you find the door. Commercial banks understand financial statements, but they don't understand kilowatts.

GENERAL OBLIGATION BOND

A bond is a security instrument representing an obligation to pay by the issuer (the "borrowing entity") to the buyers (the public or investment companies). Bonds can be secured by certain assets or by the good faith and credit of the issuer. General obligation bonds (known as "GOBs" to bankers) are specialized bonds issued by local and state government entities in order to raise money for general business operations. The interest rate paid by these bonds is based on the current overall interest rate market, as well as the credit quality of the state or local government issuer. The process of preparing and issuing a general obligation bond is long and complicated, but the interest rate that the issuer will have to pay is relatively low. As of this writing, bonds are being issued with interest rates in the range of 5%-7%. Cheap money is good as long as it's not too difficult or too time-consuming to get!

LEASE

Leasing has gained tremendous popularity in the 1980s and 1990s because of its flexibility and practicality. For example, many car buyers are choosing to use leases instead of bank loans, because they can get a better quality car for the same payment. They also prefer to return their car to the dealership after a few years and get a newer model, instead of having to go through the hassle and uncertainty of selling the car themselves. As you may know, the dealer is more than willing to take the car back, because he can estimate the car's value fairly accurately and resell it at a profit.

For energy equipment, leases have the added benefit of offering 100% financing; that is, your company does not have to make a substantial down payment and can thus conserve its cash for daily operations. There are numerous types of leases. Operating, capital, and municipal leases deserve special attention.

A *capital lease* is one which appears on the balance sheet of the lessee (the borrower) as a debt in much the same way as a commercial bank loan. A typical commercial or industrial company, such as a steel manufacturer, is a good candidate for lease financing. The rates for such leases are usually 1%-5% above prime.

Most leased energy equipment uses the capital lease structure, which may include any of the following: transfer of ownership of equipment at the end of the lease, a dollar buyout clause at the end of the lease, a lease term equaling 75% or more of the economic life of the equipment, the net present value of the lease payments equaling or being greater than 90% of the value of equipment.

An *operating lease* is designed to fulfill certain requirements of the Financial Accounting Standards Board (FASB) and the Internal Revenue Service. A well-structured operating lease does not have to be reported on your company's balance sheet as debt. That is, it does not appear as a liability like most bank loans or credit lines but is merely a regular expense obligation on the income statement, such as office rent. Oftentimes this can be very important for a company or institution that does not want to affect its borrowing capacity or its cost of capital by adding more debt to its financial structure.

Operating leases are difficult to structure for energy equipment because it is not easy to estimate the useful life and the fair market value of the equipment several years after installation. What is the value of an energy management system or compact fluorescent lights five years after they have been installed? There are stories of energy equipment leases that were structured as operating leases but which, upon closer scrutiny by auditors, had to be converted to capital leases. This could cause a substantial effect to a company's financial structure, or worse, trigger default clauses under bank credit lines that are vital to the company for day-to-day business.

Another kind of lease is the *municipal lease*. Any entity that raises revenue through taxation and qualifies according to the IRS as tax-exempt (under Section 103(c) of the Internal Revenue Code of 1986) can have access to a municipal lease. This form of lease is often used by state and local governments because of its low rates (below prime and comparable to GOB rates).

It is important to note that in all of the categories listed above, your company will be obligated to make the loan or lease payments regardless of the performance of the energy-saving equipment or the ESCO. You may have recourse for repairs under a separate energy services contract with the ESCO, or through an equipment manufacturer's warranty, but the loan or lease payments have to be made on time every month. The interest rate and terms of the loan or lease are almost exclusively based on the credit quality of your organization. Figure 10-1 shows the structure of a traditional commercial bank loan or equipment lease.

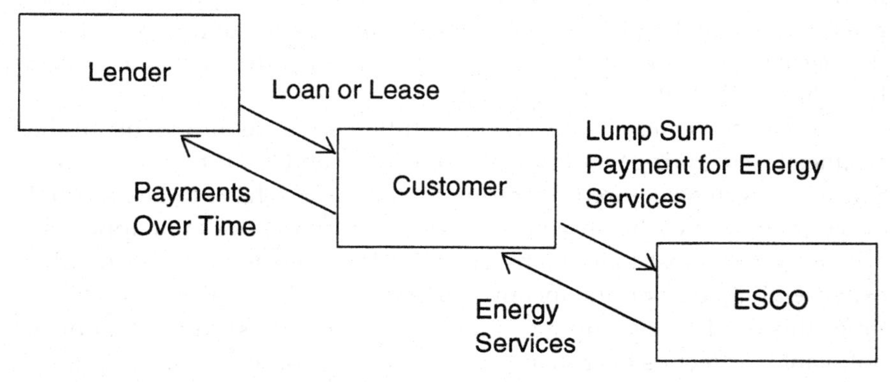

Figure 10-1. Commercial Bank Loan or Lease

PROJECT FINANCING/SHARED SAVINGS/ PERFORMANCE CONTRACTING

The United States government spends over $8 billion a year on its utility bills. The Energy Independence and Security Act of 2007 requires that total energy use in federal buildings (relative to the 2005 level) be reduced 30% by 2015. Yet the government continues to suffer from decaying and outdated energy systems in many of its facilities that cannot be repaired without the issuance of additional government debt. This problem is not isolated solely to the government, as many institutions and corporations find themselves facing budget cutbacks due to downsizing or increased competition.

The solution to this problem is known as energy savings performance contracting (ESPC). In essence, the government (or any other organization) can contract with an ESCO for energy efficiency projects throughout its facilities, but it will be the ESCO that will have to incur the cost of implementing the energy-savings measures. In exchange for this, both the ESCO and the government will derive the benefits from the savings generated by the more efficient equipment. For example, for every $10,000 saved, the ESCO will get paid $8,000 and the government keeps the other $2,000. This is a win-win situation, since the ESCO will be very motivated to keep the equipment in peak operating condition in order to generate the greatest amount of revenue over time. Another benefit of this methodology is that the government or your organization will consider the ESCO payments an operating expense rather than a debt obligation. Your chief financial officer will appreciate that!

With ESPC, the ESCO, not the customer, has to find the financing. For small to mid-size energy projects, the ESCO may choose to use commercial banks or leasing firms that will lend money based on the ESCO's creditworthiness. The federal projects mentioned above are so large (in the tens and hundreds of millions of dollars) that few, if any, ESCOs will qualify for traditional bank loans.

Bankers and specialized investment companies are increasingly becoming involved in ESPC. They are offering an innovative loan program known as full-recourse project financing, which is shown in Figure 10-2. They will review the ESCO's financial statements, but they will also review the customer's financial statements and the project specifics prior to making a lending decision. The term "full-recourse" is derived from the right which the lender will have to take back any asset of the borrower should the loan not be paid on schedule.

As some lenders become more comfortable with energy projects, they may be willing to structure what is called "non-recourse project financing." In this case, the loan is made to a single-purpose entity (SPE), which then owns the equipment of the project and contracts with the ESCO to perform the energy services. Figure 10-3 will show how non-recourse project financing is different than a full recourse arrangement. If the project fails to perform as expected, the lender only has recourse to the project equipment in the SPE. The lender cannot reach the assets of either the customer or the ESCO. The single-purpose entity concept is also used in the asset sale financing programs.

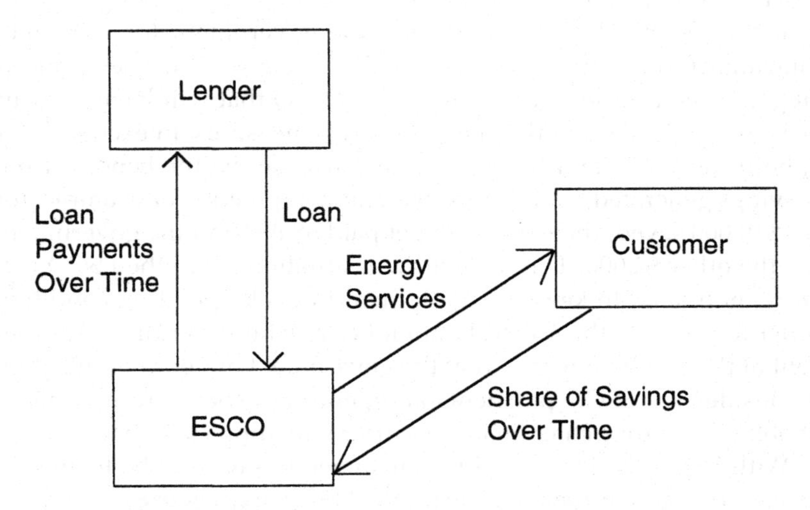

Figure 10-2. Full-recourse Project Financing

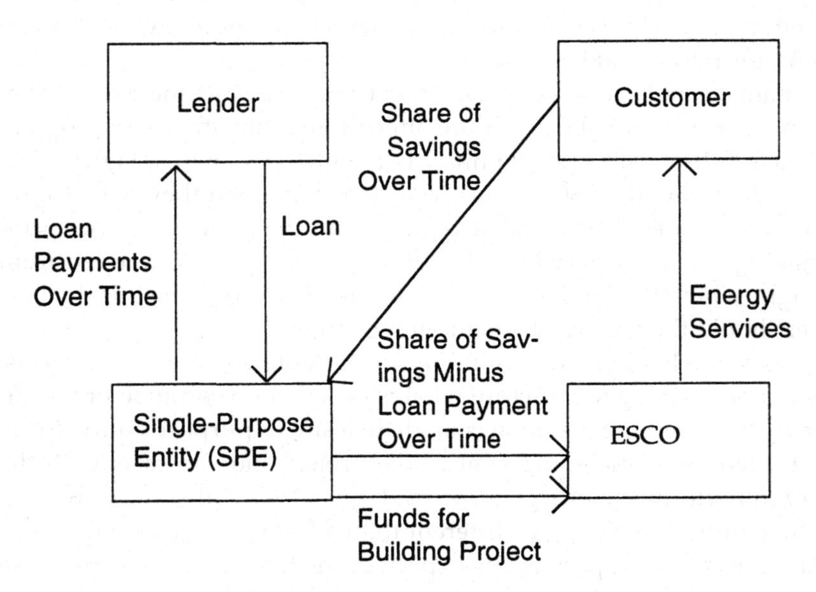

Figure 10-3. Non-recourse Project Financing

Chapter 11

Codes, Standards And Legislation

This chapter presents a historical perspective on key codes, standards, and regulations, all of which have impacted energy policy and are still playing a major role in shaping energy usage. The context of past standards and legislation must be understood in order to properly implement the proper systems and to be able to impact future codes. The Energy Policy Act, for example, has created an environment for retail competition. Electric utilities will drastically change the way they operate in order to provide power and lowest cost. This in turn will drastically reduce utility-sponsored incentive and rebate programs, which have influenced energy conservation adoption. The chapter attempts to cover a majority of the material that currently impacts the energy related industries, with respect to their initial writing.

The main difference between standards, codes and regulations is an increasing level of enforceability of the various design parameters. A group of interested parties (vendors, trade organizations, engineers, designers, citizens, etc.) may develop a standard in order to assure minimum levels of performance. The standard acts as a suggestion to those parties involved, but is not enforceable until it is codified by a governing body (local or state agency) that makes the standard a code. Not meeting this code may prevent continuance of a building permit or the ultimate stoppage of work. Once the federal government makes the code part of the federal code, it becomes a regulation. Often this progression involves equipment development and commercialization prior to codification in order to assure that the standards are attainable.

THE ENERGY INDEPENDENCE AND SECURITY ACT OF 2007 (H.R.6)

The Energy Independence and Security Act of 2007 (H.R.6) was enacted into law December 19, 2007. Key provisions of the law are summarized below.

Title I Energy Security through Improved Vehicle Fuel Economy

- The Corporate Average Fuel Economy (CAFE) sets a target of 35 miles per gallon for the combined fleet of cars and light trucks by 2020.
- The law establishes a loan guarantee program for advanced battery development, a grant program for plug-in hybrid vehicles, incentives for purchasing heavy-duty hybrid vehicles for fleets, and credits for various electric vehicles.

Title II Energy Security through Increased Production of Biofuels

- The law increases the Renewable Fuels Standard (RFS), which sets annual requirements for the quantity of renewable fuels produced and used in motor vehicles. RFS requires 9 billion gallons of renewable fuels in 2008, increasing to 36 billion gallons in 2022.

Title III Energy Savings Through Improved Standards for Appliances and Lighting

- The law establishes new efficiency standards for motors, external power supplies, residential clothes washers, dishwashers, dehumidifiers, refrigerators, refrigerator freezers, and residential boilers.
- It contains a set of national standards for light bulbs. The first part of the standard would increase energy efficiency of light bulbs 30% and phase out most common types of incandescent light bulb by 2012-2014.
- It requires the federal government to substitute energy efficient lighting for incandescent bulbs.

Title IV Energy Savings in Buildings and Industry

- Increases funding for the Department of Energy's Weatherization Program, providing 3.75 billion dollars over five years.
- Encourages the development of more energy efficient "green" commercial buildings, creating an Office of Commercial High Performance Green Buildings at the Department of Energy.
- Sets a national goal to achieve zero-net energy use for new commercial buildings built after 2025. A further goal is to retrofit all pre-construction 2025 buildings to zero-net energy by 2050.
- Requires that total energy use in federal buildings (relative to the 2005 level) be reduced 30% by 2015.
- Requires federal facilities to conduct a comprehensive energy and water evaluation for each facility at least once every four years.
- Requires new federal buildings and major renovations to reduce fossil fuel energy use 55% (relative to 2003 level) by 2010 and be 100% eliminated by 2030.
- Requires that each federal agency ensure that major replacements of installed equipment (such as heating and cooling systems) or renovation/expansion of existing space employ the most energy efficient designs, systems, equipment, and controls that are life cycle cost effective. For the purpose of calculating life-cycle cost calculations, the time period will increase from 25 years in the prior law to 40 years.
- Directs the Department of Energy to conduct research to develop and demonstrate new process technologies and operating practices to significantly improve the energy efficiency of equipment and processes used by energy-intensive industries.
- Directs the Environmental Protection Agency to establish a recoverable waste energy inventory program. The program must include an ongoing survey of all major industry and large commercial combustion services in the United States.
- Includes new incentives to promote new industrial energy efficiency through the conversion of waste heat into electricity.
- Creates a grant program for Healthy High Performance Schools that aims to encourage states, local governments and school systems to build green schools.
- Creates a program of grants and loans to support energy efficiency and energy sustainability projects at public institutions.

Title V Energy Savings in Government and Public Institutions

- Promotes energy savings performance contracting in the federal government and provides flexible financing and training of federal contract officers.
- Promotes the purchase of energy efficient products and procurement of alternative fuels with lower carbon emissions for the federal government.
- Reauthorizes state energy grants for renewable energy and energy efficiency technologies through 2012.
- Establishes an energy and environmental block grant program to be used for seed money for innovative local best practices.

Title VI Alternative Research and Development

- Authorizes research and development to expand the use of geothermal energy.
- Improves the cost and effectiveness of thermal energy storage technologies that could improve the operation of concentrating solar power electric generation plants.
- Promotes research and development of technologies that produce electricity from waves, tides, currents, and ocean thermal differences.
- Authorizes a development program on energy storage systems for electric drive vehicles, stationary applications, and electricity transmission and distribution.

Title VII Carbon Capture and Sequestration

- Provides grants to demonstrate technologies to capture carbon dioxide from industrial sources.
- Authorizes a nationwide assessment of geological formations capable of sequestering carbon dioxide underground.

Title VIII Improved Management of Energy Policy

- Creates a 50% matching grants program for constructing small renewable energy projects that will have an electrical generation capacity less than 15 megawatts.

- Prohibits crude oil and petroleum product wholesalers from using any technique to manipulate the market or provide false information.

Title IX International Energy Programs

- Promotes US exports in clean, efficient technologies to India, China and other developing countries.
- Authorizes US Agency for International Development (USAID) to increase funding to promote clean energy technologies in developing countries.

Title X Green Jobs

- Creates an energy efficiency and renewable energy worker training program for "green collar" jobs.
- Provides training opportunities in the energy field for individuals who need to update their skills.

Title XI Energy Transportation and Infrastructure

- Establishes an office of climate change and environment to coordinate and implement strategies to reduce transportation related energy use.

Title XII Small Business Energy Programs

- Loans, grants and debentures are established to help small businesses develop, invest in, and purchase energy efficient equipment and technologies.

Title XIII Smart Grid

- Promotes a "smart electric grid" to modernize and strengthen the reliability and energy efficiency of the electricity supply. The term "Smart Grid" refers to a distribution system that allows for flow of information from a customer's meter in two directions, both inside the house to thermostats, appliances, and other devices; and from the house back to the utility.

THE ENERGY POLICY ACT OF 2005

The first major piece of national energy legislation since the Energy Policy Act of 1992, EPAct 2005 was signed by President George W. Bush on August 8, 2005 and became effective January 1, 2006. The major thrust of EPAct 2005 is energy production. However, there are many important sections of the act that do help promote energy efficiency and energy conservation. There are also some significant impacts on federal energy management. Highlights are described below:

Federal Energy Management

- The United States is the single largest energy user, with about a $10 billion energy budget. Forty-four percent of this budget was used for non-mobile buildings and facilities. The United States is also the single largest product purchaser, with $6 billion spent for energy using products, vehicles, and equipment.

Energy Management Goals

- An annual energy reduction goal of 2% is in place from fiscal year 2006 to fiscal year 2015, for a total energy reduction of 20%.
- Electric metering is required in all federal building by the year 2012.
- Energy efficient specifications are required in procurement bids and evaluations.
- Energy efficient products to be listed in federal catalogs include Energy Star and FEMP recommended products by GSA and Defense Logistics Agency.
- Energy Service Performance Contracts (ESPCs) are reauthorized through September 30, 2016.
- New federal buildings are required to be designed 30% below ASHRAE standard or the International Energy Code (if life-cycle cost effective.) Agencies must identify those that meet or exceed the standard.
- Renewable electricity consumption by the federal government cannot be less than: 3% from fiscal year 2007-2009, 5% from fiscal year 2010-2012, and 7.5% from fiscal year 2013-present. Double credits are earned for renewables produced on the site or on federal lands and used at a federal facility, as well as renewables produced on Native American lands.

- The goal for photovoltaic energy is to have 20,000 solar energy systems installed in federal buildings by the year 2012.

Tax Provisions

- Tax credits will be issued for residential solar photovoltaic and hot water heating systems. Tax deductions will be offered for highly efficient commercial buildings and highly efficient new homes. There will also be tax credits for improvements made to existing homes, including high efficiency HVAC systems and residential fuel cell systems. Tax credits are also available for fuel cells and microturbines used in businesses.

THE ENERGY POLICY ACT OF 1992

The Energy Policy Act of 1992 is far-reaching, and its implementation is impacting electric power deregulation, building codes, and new energy efficient products. Sometimes policy makers do not see the extensive impact of their legislation. This comprehensive legislation significantly impacts energy conservation, power generation, and alternative fuel vehicles, as well as energy production. The federal and private sectors are impacted by this comprehensive energy act. Highlights are described below:

Energy Efficiency Provisions

Buildings

- Requires states to establish minimum commercial building energy codes and to consider minimum residential codes based on current voluntary codes.

Utilities

- Requires states to consider new regulatory standards that would require utilities to undertake integrated resource planning, allow efficiency programs to be at least as profitable as new supply options, and encourage improvements in supply system efficiency.

Equipment Standards

- Establishes efficiency standards for commercial heating and air-conditioning equipment, electric motors, and lamps.

- Gives the private sector an opportunity to establish voluntary efficiency information/labeling programs for windows, office equipment and luminaries (or the Department of Energy will establish such programs).

Renewable Energy

- Establishes a program for providing federal support on a competitive basis for renewable energy technologies. Expands program to promote export of these renewable energy technologies to emerging markets in developing countries.

Alternative Fuels

- Gives Department of Energy authority to require a private and municipal alternative fuel fleet program, starting in 1998. Provides a federal alternative fuel fleet program with phased-in acquisition schedule; also provides state fleet program for large fleets in large cities.

Electric Vehicles

Establishes comprehensive program for the research and development, infrastructure promotion, and vehicle demonstration for electric motor vehicle.

Electricity

- Removes obstacles to wholesale power competition in the Public Utilities Holding Company Act by allowing both utilities and non-utilities to form exempt wholesale generators without triggering the PUHCA restrictions.

Global Climate Change

- Directs the Energy Information Administration to establish a baseline inventory of greenhouse gas emissions and establishes a program for the voluntary reporting of those emissions. Directs the Department of Energy to prepare a report analyzing the strategies for mitigating global climate change and to develop a least-cost energy strategy for reducing the generation of greenhouse gases.

Research and Development

- Directs Dept. of Energy to undertake research and development on a wide range of energy technologies, including energy efficiency technologies, natural gas end-use products, renewable energy resources, heating and cooling products, and electric vehicles.

CODES AND STANDARDS

Energy codes specify how buildings *must* be constructed or perform and are written in a mandatory, enforceable language. State and local governments adopt and enforce energy codes for their jurisdictions. Energy standards describe how buildings *should* be constructed to save energy cost effectively. They are published by national organizations such as the American Society of Heating, Refrigerating, and Air Conditioning Engineers (ASHRAE). Such standards are not mandatory but serve as national recommendations, with some variation for regional climate. State and local governments frequently use energy standards as the technical basis for developing their energy codes. Some energy standards are written in a mandatory, enforceable language, making it easy for jurisdictions to incorporate the provisions of the energy standards directly into their laws or regulations. The requirement for the Federal sector to use ASHRAE 90.1 and 90.2 as mandatory standards for all new Federal buildings is specified in the Code of Federal Regulations (10 CFR 435).

Most states use the ASHRAE 90 standard as their basis for the energy component of their building codes. ASHRAE 90.1 is used for commercial buildings, while ASHRAE 90.2 is used for residential buildings. Some states have quite comprehensive building codes (for example, California Title 24).

ASHRAE Standard 90.1

- Sets minimum requirements for the energy efficient design of new buildings so they may be constructed, operated, and maintained in a manner that minimizes the use of energy without constraining the building function and productivity of the occupants.
- Addresses building components and systems that affect energy usage.
- Sections 5-10 are the technical sections specifically addressing components of the building envelope, HVAC systems and equipment, service water heating, power, lighting, and motors. Each technical

section contains general requirements and mandatory provisions. Some sections also include prescriptive and performance requirements.

ASHRAE Standard 90.2

- Sets minimum requirements for the energy efficient design for new low-rise residential buildings.

When the Department of Energy determines that a revision would improve energy efficiency, each state has two years to review the energy provisions of its residential or commercial building code. For residential buildings, a state has the option of revising its residential code to meet or to exceed the residential portion of ASHRAE 90.2. For commercial buildings, a state is required to update its commercial code to meet or exceed the provision of ASHRAE 90.1.

ASHRAE standards 90.1 and 90.2 are developed and revised through voluntary consensus and public hearing processes that are critical to widespread support for their adoption. Both standards are continually maintained by separate Standing Standards Projects Committees. Committee membership varies from ten to sixty voting members. Committee membership includes representatives from many groups to ensure balance among all interest categories. After the committee proposes revisions, the standard undergoes public review and comment. When a majority of the parties substantially agree, the revised standard is submitted to the ASHRAE Board of Directors. This entire process can take anywhere from two to ten years to complete. ASHRAE Standards 90.1 and 90.2 are automatically revised and published every three years. Approved interim revisions are posted on the ASHRAE website (www.ashrae.org) and are included in the next published version.

The energy cost budget method permits tradeoffs between building systems (lighting and fenestration, for example) if the annual energy cost estimated for the proposed design does not exceed the annual energy cost of a base design that fulfills the prescriptive requirements. Using the energy cost budget method approach requires simulation software that can analyze energy consumption in buildings and model the energy features in the proposed design. ASHRAE 90.1 sets minimum requirements for the simulation software; suitable programs include BLAST, eQUEST, and TRACE.

CLIMATE CHANGE

Kyoto Protocol

The goal of the Kyoto Protocol is to stabilize green house gases in the atmosphere that would prevent human impact on global climate change. The nations that signed this treaty come together to make decisions at meetings called "Conferences of the Parties." The 38 parties are grouped into two groups, developed industrialized nations and developing countries. The Kyoto Protocol, an international agreement reached in Kyoto in 1997 by the third Conference of the Parties (COP-3), aims to lower emissions from two groups of three green house gases: (1) carbon dioxide, methane, and nitrous oxide; and (2) hyrdofluorocarbon (HFC): sulfur hexafluoride and perfluorocarbons.

INDOOR AIR QUALITY STANDARDS

Indoor air quality (IAQ) is an emerging issue of concern to building managers, operators, and designers. Recent research has shown that indoor air is often less clean than outdoor air and federal legislation has been proposed to establish programs to deal with this issue on a national level. This, like the asbestos issue, will have an impact on building design and operations. Americans today spend long hours inside buildings, so building operators, managers, and designers must be aware of potential IAQ problems and how they can be avoided.

IAQ problems, sometimes termed "sick building syndrome," have become an acknowledged health and comfort problem. Buildings are characterized as sick when occupants complain of acute symptoms such as headache; eye, nose, and throat irritations; dizziness; nausea; sensitivity to odors; and difficulty in concentrating. The complaints may become more clinically defined so that an occupant may be said to have developed an illness believed to be related to IAQ problems.

The most effective means to deal with an IAQ problem is to remove or minimize the pollutant source, when feasible. If not, dilution and filtration may be effective.

The purpose of ASHRAE Standard 62 is to specify minimum ventilation rates and indoor air quality that will be acceptable to human occupants and thereby minimize the potential for adverse health effects. ASHRAE defines acceptable indoor air quality as the air in which there are no known contaminants at harmful concentrations (as determined by cognizant authorities) and with which a substantial majority of those exposed (80% or more) do not express dissatisfaction.

ASHRAE Standard 55, which sets environmental conditions for human occupancy, covers several environmental parameters including temperature, radiation, humidity, and air movement. The standard specifies conditions in which 80% of the occupants will find the environment thermally acceptable. This applies to healthy people in normal indoor environments for winter and summer conditions. Adjustment factors are described for various activity levels and clothing levels.

The International Performance Measurement and Verification Protocol (IPMVP) is used for commercial and industrial facility operators. It offers standards for measurement and verification of energy and water efficiency projects. The IPMVP volumes are used to: (1) Develop a measurement and verification strategy and plan for quantifying energy and water savings in retrofits and new construction, (2) Monitor indoor environmental quality, and (3) Quantify emissions reduction. (www.evo-world.org)

REGULATORY AND LEGISLATIVE ISSUES IMPACTING COGENERATION AND INDEPENDENT POWER PRODUCTION

Public Utilities Regulatory Policies Act (PURPA)

This legislation was part of the 1978 National Energy Act and has had perhaps the most significant effect on the development of cogeneration and other forms of alternative energy production in the past decade. Certain provisions of PURPA also apply to the exchange of electric power between utilities and cogenerators. It provides a number of benefits to those cogenerators who can become qualifying facilities (QFs) under this act. Specifically, it:

- Requires utilities to purchase the power made available by cogenerations at reasonable buy-back rates. These rates are typically based on the utilities' cost.

- Guarantees the cogeneration or small power producer interconnection with the electric grid and the availability of backup service from the utility.

- Dictates that supplemental power requirements of cogeneration must be provided at a reasonable cost.

- Exempts cogenerations and small power producers from federal and state utility regulations and associated reporting requirements of these bodies.

In order to assure a facility the benefits of PURPA, a cogenerator must become a qualifying facility. To achieve this status, a cogenerator must generate electricity and useful thermal energy from a single fuel source. In addition, a cogeneration facility must be less than 50% owned by an electric utility or an electric utility holding company. Finally, the plant must meet the minimum annual operating efficiency standard established by the Federal Energy Regulatory Commission (FERC) when using oil or natural gas as the principal fuel source. The standard is that the useful electric power output, plus one half of the useful thermal output of the facility, must be no less than 42.5% of the total oil or natural gas energy input. The minimum efficiency standard increases to 45% if the useful thermal energy is less than 15% of the total energy output of the plant.

Natural Gas Policy Act

The Natural Gas Policy Act created a deregulated natural gas market for natural gas. The major objective of this regulation was to create a deregulated national market for natural gas. It provides for incremental pricing of higher cost natural gas to fluctuate with the cost of fuel oil. Cogenerators classified as qualifying facilities under PURPA are exempt from the incremental pricing schedule established for industrial customers.

Public Utility Holding Company Act of 1935

The Public Utility Company Holding Act of 1935 authorized the Securities and Exchange Commission (SEC) to regulate certain utility "holding companies" and their subsidiaries in a wide range of corporate transactions.

The utility industry and would-be owners of utilities lobbied

Congress heavily to repeal PUHCA, claiming that it was outdated. On August 8, 2005, the Energy Policy Act of 2005 passed both houses of Congress and was signed into law, repealing PUHCA—despite consumer, environmental, union and credit rating agency objections. The repeal became effective on February 8, 2006.

SUMMARY

The dynamic process of revisions to existing codes, plus the introduction of new legislation, will impact the energy industry and bring a dramatic change. Energy conservation and creating new power generation supply options will be required to meet the energy demands of the twenty-first century.

Chapter 12

The Energy Audit

Barney L. Capehart and Mark B. Spiller
Scott Frazier, Ph.D.

Editor's Note: The energy audit is often one of the first steps to identifying potential savings projects. Beyond finding energy savings, it is important to estimate the maintenance savings and avoided capital costs associated with potential projects. Also consider "secondary costs" such as "down-time," damages, and emergency repair costs that may be likely if a project is not implemented. Ultimately, these "cash flows" help you and the CFO better understand and justify the economics of the project.

INTRODUCTION

Saving money on energy bills is attractive to businesses, industries, and individuals alike. Customers with large energy bills have a strong motivation to initiate and continue an on-going energy cost-control program. "No-cost" or very low-cost operational changes can often save a customer or an industry 10-20% on utility bills. Capital cost programs with payback times of two years or less can often save an additional 20-30%. In many cases these energy cost control programs will also result in both reduced energy consumption and reduced emissions of environmental pollutants.

The energy audit is one of the first tasks to be performed in the accomplishment of an effective energy cost control program. An energy audit consists of a detailed examination of how a facility uses energy and what the facility pays for that energy. It also includes a recommended program for changes in operating practices or energy-consuming equipment that will cost-effectively save dollars on energy bills. The energy audit is sometimes called an energy survey or an energy analysis so that it is not hampered with the negative connotation of an audit in the sense of an IRS "audit." The energy audit is a positive experience with significant benefits to the business or individual, and the term "audit" should be avoided if it clearly produces a negative image in the mind of a particular business or individual.

ENERGY AUDITING SERVICES

Energy audits are performed by several different groups. Electric and gas utilities throughout the country offer free residential energy audits. A utility's residential energy auditors analyze the monthly bills, inspect the construction of the dwelling unit, and inspect all of the energy-consuming appliances in a house or an apartment. Ceiling and wall insulation is measured, ducts are inspected, appliances such as heaters, air conditioners, water heaters, refrigerators, and freezers are examined, and the lighting system is checked.

Some utilities also perform audits for their industrial and commercial customers. They have professional engineers on their staff to perform the detailed audits needed by companies with complex process equipment and operations. When utilities offer free or low-cost energy audits for commercial customers, they usually provide only walk-through audits rather than detailed audits. Even so, they generally consider lighting, HVAC systems, water heating, insulation and some motors.

Large commercial or industrial customers may hire an engineering consulting firm to perform a complete energy audit. Other companies may elect to hire an energy manager or set up an energy management team whose job is to conduct periodic audits and to keep up with the available energy efficiency technology.

The U.S. Department of Energy (U.S. DOE) funds a program where universities around the country operate industrial assessment centers that perform free energy audits for small and medium sized manufacturing companies. There are currently 26 IACs funded by the Industrial Division of the U.S. DOE.

The State Energy Program (SEP) is another energy audit service funded by the U.S. Department of Energy. It is administered through state energy offices. This program pays for audits of schools, hospitals, and other institutions, and has some funding assistance for energy conservation improvements.

BASIC COMPONENTS OF AN ENERGY AUDIT

An initial summary of the basic steps involved in conducting a successful energy audit is provided here, and these steps are explained more fully in the sections that follow. This audit description primarily addresses the steps in an industrial or large-scale commercial audit; however, not all of the procedures described in this section are required for every type of audit.

The audit process starts by collecting information about a facility's operation and about its past record of utility bills. These data are then analyzed to get a picture of how the facility uses—and possibly wastes—energy, as well as to help the auditor learn what areas to examine to reduce energy costs. Specific changes—called energy conservation opportunities (ECOs)—are identified and evaluated to determine their benefits and their cost-effectiveness. These ECOs are assessed in terms of their costs and benefits, and an economic comparison is made to rank the various ECOs. Finally, an action plan is created where certain ECOs are selected for implementation, and the actual process of saving energy and saving money begins.

The Auditor's Toolbox

To obtain the best information for a successful energy cost control program, the auditor must make some measurements during the audit visit. The amount of equipment needed depends on the type of energy-consuming equipment used at the facility, as well as the range of potential ECOs that might be considered. For example, if waste heat recovery is being considered, then the auditor must take substantial temperature measurement data from potential heat sources. Tools commonly needed for energy audits are listed below:

Tape Measures

The most basic measuring device needed is the tape measure. A 25-foot tape measure (1″ wide) and a 100-foot tape measure are used to check the dimensions of walls, ceilings, windows and distances between pieces of equipment for purposes such as determining the length of a pipe for transferring waste heat from one piece of equipment to the other.

Lightmeter

One simple and useful instrument is the lightmeter, which is used to measure illumination levels in facilities. A lightmeter that reads in footcandles allows direct analysis of lighting systems and comparison with recommended light levels specified by the Illuminating Engineering Society. A digital lightmeter that is portable and can be hand carried is the most useful. Many areas in buildings and plants are still significantly over-lighted, and measuring this excess illumination then allows the auditor to recommend a reduction in lighting levels either through lamp removal programs or by replacing inefficient lamps with high efficiency lamps that may not supply the same amount of illumination as the old inefficient lamps.

Thermometers

Several thermometers are generally needed to measure temperatures in offices and other worker areas, as well as to measure the temperature of operating equipment. Knowing process temperatures allows the auditor to determine process equipment efficiencies and identify waste heat sources for potential heat recovery programs. Inexpensive electronic thermometers with interchangeable probes are now available to measure temperatures in both these areas. Some common types include an immersion probe, a surface temperature probe, and a radiation shielded probe for measuring true air temperature. Other types of infrared thermometers and thermographic equipment are also available. An infra-red "gun" is valuable for measuring temperatures of steam lines that are not readily reached without a ladder.

Infrared Cameras

Infrared cameras have come down in price substantially, but they are still rather expensive pieces of equipment. An investment of at least $10,000-15,000 is needed to have a good-quality infrared camera. However, these are very versatile pieces of equipment and can be used to find overheated electrical wires, connections, neutrals, circuit breakers, transformers, motors, and other pieces of electrical equipment. They can also be used to find wet insulation, missing insulation, roof leaks, and cold spots. Thus, infrared cameras are excellent tools for both safety related diagnostics and energy savings diagnostics. A good rule of thumb is that if one safety hazard is found during an infrared scan of a facility, then that has paid for the cost of the scan for the entire facility. Many insurers require infrared scans of buildings for facilities once a year.

Voltmeter

An inexpensive digital voltmeter is useful for determining operating voltages on electrical equipment, and especially useful when the nameplate has worn off of a piece of equipment or is otherwise unreadable or missing. The most versatile instrument is a digital combined volt-ohm-ammeter with a clamp-on feature for measuring currents in conductors that are easily accessible. This type of multi-meter is convenient and relatively inexpensive. Any newly purchased voltmeter (or multimeter) should be a true RMS meter for greatest accuracy where harmonics might be involved.

Clamp-on Ammeter

These are very useful instruments for measuring current in a wire without having to make any live electrical connections. The clamp is opened up and put around one insulated conductor, and the meter reads the current in that conductor. New clamp on ammeters can be purchased rather inexpensively that read true RMS values. This is important because of the level of harmonics in many of our facilities. An idea of the level of harmonics in a load can be estimated from using an old non-RMS ammeter, and then a true RMS ammeter can be used to measure the current. If there is more than a five to ten percent difference between the two readings, there is a significant harmonic content to that load.

Wattmeter/Power Factor Meter

A portable hand-held wattmeter and power factor meter is very handy for determining the power consumption and power factor of individual motors and other inductive devices. This meter typically has a clamp-on feature which allows an easy connection to the current-carrying conductor, and has probes for voltage connections. Any newly purchased wattmeter or power factor meter should be a true RMS meter for greatest accuracy where harmonics might be involved.

Combustion Analyzer

Combustion analyzers are portable devices capable of estimating the combustion efficiency of furnaces, boilers, or other fossil fuel burning machines. Electronic digital combustion analyzers perform the measurements and read out in percent combustion efficiency. Today these instruments are hand-held devices that are very accurate, and they are also quite inexpensive. $800-$1000 will buy an analyzer for most residential and commercial heaters and boilers.

Airflow Measurement Devices

Measuring air flow from heating, air conditioning or ventilating ducts, or from other sources of air flow is one of the energy auditor's tasks. Airflow measurement devices can be used to identify problems with air flows, such as whether the combustion air flow into a gas heater is correct. Typical airflow measuring devices include a velometer, an anemometer, or an airflow hood. See page 121 for more detail on airflow measurement devices.

Blower Door Attachment

Building or structure tightness can be measured with a blower door attachment. This device is frequently used in residences and in office buildings to determine the air leakage rate or the number of air changes per hour in the facility. This is often helps determine whether the facility has substantial structural or duct leaks that need to be found and sealed. See page 140 for additional information on blower doors.

Smoke Generator

A simple smoke generator can also be used in residences, offices and other buildings to find air infiltration and leakage around doors, windows, ducts and other structural features. Care must be taken in using this device, since the chemical "smoke" produced may be hazardous, and breathing protection masks may be needed. See page 139 for additional information on the smoke generation process, and use of smoke generators.

Safety Equipment

The use of safety equipment is a vital precaution for any energy auditor. A good pair of safety glasses is an absolute necessity for almost any audit visit. Hearing protectors may also be required on audit visits to noisy plants or areas with high horsepower motors driving fans and pumps. Electrical insulated gloves should be used if electrical measurements will be taken, and asbestos gloves should be used for working around boilers and heaters. Breathing masks may also be needed when hazardous fumes are present from processes or materials used. Steel-toe and steel-shank safety shoes may be needed on audits of plants where heavy materials, hot or sharp materials, or hazardous materials are being used. (See pages 110 and 124 for an additional discussion of safety procedures.)

Miniature Data Loggers

Miniature—or mini—data loggers have appeared in low-cost models in the last five years. These are often devices that can be held in the palm of the hand and electronically record measurements of temperature, relative humidity, light intensity, light on/off, and motor on/off. If they have an external sensor input jack, these little boxes are actually general purpose data loggers. With external sensors they can record measurements of current, voltage, apparent power (kVA), pressure, and CO_2.

These data loggers have a microcomputer control chip and a memory chip, so they can be initialized and then record data for periods of time from days to weeks. They can record data on a 24-hour-a-day basis, without any attention or intervention on the part of the energy auditor. Most of these data loggers interface with a digital computer PC and can transfer data into a spreadsheet of the user's choice, or the software provided by the suppliers of the loggers can be used.

Collecting audit data with these small data loggers gives a more complete and accurate picture of an energy system's overall performance, because some conditions may change over long periods of time, or when no one is present.

Vibration Analysis Gear

Relatively new in the energy manager's tool box is vibration analysis equipment. The correlation between machine condition (bearings, pulley alignment, etc.) and energy consumption is related, and this equipment monitors such machine health. This equipment comes in various levels of sophistication and price. At the lower end of the spectrum are vibration pens (or probes) that simply give real-time amplitude readings of vibrating equipment in in/sec or mm/sec. This type of equipment can cost under $1,000. The engineer compares the measured vibration amplitude to a list of vibration levels (ISO2372) and is able to determine if the vibration is excessive for that particular piece of equipment.

The more typical type of vibration equipment will measure and log the vibration into a database (on-board and downloadable). In addition to simply measuring vibration amplitude, the machine vibration can be displayed in time or frequency domains. The graphs of vibration in the frequency domain will normally exhibit spikes at certain frequencies. These spikes can be interpreted by a trained individual to determine the relative health of the machine monitored.

The more sophisticated machines are capable of trend analysis so that facility equipment can be monitored on a schedule and changes in vibration (amplitudes and frequencies) can be noted. Such trending can be used to schedule maintenance based on observations of change. This type of equipment starts at about $3,000 and goes up depending on features desired.

Preparing for the Audit Visit

Some preliminary work must be done before the auditor makes the actual energy audit visit to a facility. Data should be collected on the facility's use of energy through examination of utility bills, and some preliminary information should be compiled on the physical description and operation of the facility. These data should then be analyzed so that the auditor can do the most complete job of identifying energy conservation opportunities during the actual site visit to the facility.

Energy Use Data

The energy auditor should start by collecting data on energy use, power demand and cost for at least the previous twelve months. Twenty-four months of data might be necessary to adequately understand some types of billing methods. Bills for gas, oil, coal, electricity, etc. should be compiled and examined to determine both the amount of energy used and the cost of that energy. These data should then be put into tabular and graphic form to see what kind of patterns or problems appear from the tables or graphs. Any anomaly in the pattern of energy use raises the possibility for some significant energy or cost savings by identifying and controlling that anomalous behavior. Sometimes an anomaly on the graph or in the table reflects an error in billing, but generally the deviation shows that some activity is going on that has not been noticed or completely understood by the customer.

Rate Structures

To fully understand the cost of energy, the auditor must determine the rate structure under which that energy use is billed. Energy rate structures may go from the extremely simple ones (such as $1.00 per gallon of Number 2 fuel oil) to very complex ones (such as electricity consumption that may have a customer charge, energy charge, demand charge, power factor charge, and other miscellaneous charges that vary from month to month). Few customers or businesses really understand the various rate structures that control the cost of the energy they consume. The auditor can help here because the customer must know the basis for the costs in order to control them successfully.

- Electrical demand charges: The demand charge is based on a reading of the maximum power in kW that a customer demands in one month. Power is the rate at which energy is used, and it varies quite rapidly for many facilities. Electric utilities average the power reading over intervals from fifteen minutes to one hour, so that very short fluctuations do not adversely affect customers. Thus, a customer might be billed for a monthly demand based on a maximum value of a 15-minute integrated average of their power use.

- Ratchet clauses: Some utilities have a rachet clause in their rate structure which stipulates that the minimum power demand charge will be the highest demand recorded in the last billing period or some percentage (i.e., typically 70%) of the highest power demand recorded in the last year. The rachet clause can increase utility charges for facilities during periods of low activity or where power demand is tied to extreme weather.

- Discounts/penalties: Utilities generally provide discounts on their energy and power rates for customers who accept power at high voltage and provide transformers on site. They also commonly assess penalties when a customer has a power factor less than 0.9. Inductive loads (e.g., lightly loaded electric motors, old fluorescent lighting ballasts, etc.) reduce the power factor. Improvement can be made by adding capacitance to correct for lagging power factor, and variable capacitor banks are most useful for improving the power factor at the service drop. Capacitance added near the loads can effectively increase the electrical system capacity. Turning off idling or lightly loaded motors can also help.

- Water and wastewater charges: The energy auditor also looks at water and wastewater use and costs as part of the audit visit. These costs are often related to the energy costs at a facility. Wastewater charges are usually based on some proportion of the metered water use, since the solids are difficult to meter. This can needlessly result in substantial increases in the utility bill for processes which do not contribute to the wastewater stream (e.g., makeup water for cooling towers and other evaporative devices, irrigation, etc.). A water meter can be installed at the service main to supply the loads not returning water to the sewer system. This can reduce the charges by up to two-thirds.

Energy bills should be broken down into the components that can be controlled by the facility. These cost components can be listed individually in tables and then plotted. For example, electricity bills should be broken down into power demand costs per kW per month, and energy costs per kWh. The following example illustrates the parts of a rate structure for an industry in Florida.

Example: A company that fabricates metal products gets electricity from its electric utility at the following general service demand rate structure.

Rate structure:

Customer cost	=	\$21.00 per month
Energy cost	=	\$0.051 per kWh
Demand cost	=	\$6.50 per kW per month
Taxes	=	Total of 8%
Fuel adjustment	=	A variable amount per kWh each month

The energy use and costs for that company for a year are summarized below:

The auditor must be sure to account for all the taxes, the fuel adjustment costs, the fixed charges, and any other costs so that the true cost of the controllable energy cost components can be determined. In the electric rate structure described above, the quoted costs for a kW of demand and a kWh of energy are not complete until all these additional costs are added. Although the rate structure says there is a basic charge of \$6.50 per kW per month, the actual cost, including all taxes, is \$7.02 per kW per month. The average cost per kWh is most easily obtained

Summary of Energy Usage and Costs

Month	kWh Used (kWh)	kWh Cost ($)	Demand (kW)	Demand Cost ($)	Total Cost ($)
Mar	44960	1581.35	213	1495.26	3076.61
Apr	47920	1859.68	213	1495.26	3354.94
May	56000	2318.11	231	1621.62	3939.73
Jun	56320	2423.28	222	1558.44	3981.72
Jul	45120	1908.16	222	1558.44	3466.60
Aug	54240	2410.49	231	1621.62	4032.11
Sept	50720	2260.88	222	1558.44	3819.32
Oct	52080	2312.19	231	1621.62	3933.81
Nov	44480	1954.01	213	1495.26	3449.27
Dec	38640	1715.60	213	1495.26	3210.86
Jan	36000	1591.01	204	1432.08	3023.09
Feb	42880	1908.37	204	1432.08	3340.45
Totals	569,360	24,243.13	2,619	18,385.38	42,628.51
Monthly Averages	47,447	2,020.26	218	1,532.12	3,552.38

by taking the data for the 12-month period and calculating the cost over this period of time. Using the numbers from the table, one can see that this company has an average energy cost of $0.075 per kWh.

These data are used initially to analyze potential ECOs and will ultimately influence which ECOs are recommended. For example, an ECO that reduces peak demand during a month would save $7.02 per kW per month. Therefore, the auditor should consider ECOs that would involve using certain equipment during the night shift, when the peak load is significantly less than the first shift peak load. ECOs that save both energy and demand on the first shift would save costs at a rate of $0.075 per kWh. Finally, ECOs that save electrical energy during the off-peak shift should be examined too, but they may not be as advantageous. They would only save at the rate of $0.043 per kWh because they are already used off-peak, so there would not be any additional demand cost savings.

Physical and Operational Data for the Facility

The auditor must gather information on factors likely to affect energy use in the facility. Geographic location, weather data, facility layout and construction, operating hours, and equipment can all influence energy use.

- **Geographic Location/Weather Data:** The geographic location of the facility should be noted, together with the weather data for that location. Contact the local weather station, the local utility, or the state energy office to obtain the average degree days for heating and cooling for that location for the past twelve months. These degree-day data will be very useful in analyzing the need for energy for heating or cooling the facility. Bin weather data would also be useful if a thermal envelope simulation of the facility were going to be performed as part of the audit.

- **Facility Layout:** Next the facility layout or plan should be obtained and reviewed to determine the facility size, floor plan, and construction features such as wall and roof material and insulation levels, as well as door and window sizes and construction. A set of building plans could supply this information in sufficient detail. It is important to make sure the plans reflect the "as-built" features of the facility, since many original building plans do not get used without alterations.

- **Operating Hours:** Operating hours for the facility should also be obtained. Is there only a single shift? Are there two shifts? Three? Knowing the operating hours in advance allows some determination as to whether some loads could be shifted to off-peak times. Adding a second shift can often be cost effective from an energy cost view, since the demand charge can then be spread over a greater amount of kWh.

- **Equipment List:** Finally, the auditor should get an equipment list for the facility and review it before conducting the audit. All large pieces of energy-consuming equipment such as heaters, air conditioners, water heaters, and specific process-related equipment should be identified. This list, together with data on operational uses of the equipment, allows a good understanding of the major energy-consuming tasks or equipment at the facility. As a general rule, the largest energy and cost activities should be examined first to see what savings could be achieved. The greatest effort should be devoted to the ECOs that show the greatest savings, and the least effort to those with the smallest savings potential.

The equipment found at an audit location will depend greatly on the type of facility involved. Residential audits for single-family dwellings generally involve smaller-sized lighting, heating, air conditioning, and refrigeration systems. Commercial operations such as grocery stores, office buildings and shopping centers usually have equipment similar to residences but much larger in size and in energy use. However, large residential structures such as apartment buildings have heating, air conditioning and lighting that is very similar to many commercial facilities. Business operations is the area where commercial audits begin to involve equipment substantially different from that found in residences.

Industrial auditors encounter the most complex equipment. Commercial-scale lighting, heating, air conditioning, and refrigeration, as well as office business equipment, is generally used at most industrial facilities. The major difference is in the highly specialized equipment used for the industrial production processes. This can include equipment for chemical mixing and blending, metal plating and treatment, welding, plastic injection molding, paper making and printing, metal refining, electronic assembly, and making glass, for example.

Safety Considerations

Safety is a critical part of any energy audit. The audit person or team should be thoroughly briefed on safety equipment and procedures, and they should never place themselves in a position where they could injure themselves or other people at the facility. Adequate safety equipment should be worn at all appropriate times. Auditors should be extremely careful making any measurements on electrical systems or on high temperature devices such as boilers, heaters, cookers, etc. Electrical gloves or asbestos gloves should be worn as appropriate.

The auditor should be careful when examining any operating piece of equipment, especially those with open drive shafts, belts or gears, or any form of rotating machinery. The equipment operator or supervisor should be notified that the auditor is going to look at that piece of equipment and might need to get information from some part of the device. If necessary, the auditor may need to come back when the machine or device is idle in order to safely get the data. The auditor should never approach a piece of equipment and inspect it without the operator or supervisor being notified first.

Safety Checklist

1. Electrical:
   a. Avoid working on live circuits, if possible.
   b. Securely lock off circuits and switches before working on a piece of equipment.
   c. Always keep one hand in your pocket while making measurements on live circuits to help prevent cardiac arrest.

2. Respiratory:
   a. When necessary, wear a full face respirator mask with adequate filtration particle size.
   b. Use activated carbon cartridges in the mask when working around low concentrations of noxious gases. Change the cartridges on a regular basis.
   c. Use a self-contained breathing apparatus for work in toxic environments.

3. Hearing:
   a. Use foam insert plugs while working around loud machinery to reduce sound levels up to 30 decibels.

Conducting the Audit Visit

Once the information on energy bills, facility equipment, and facility operation has been obtained, the audit equipment can be gathered up, and the actual visit to the facility can be made.

Introductory Meeting

The audit person (or team) should meet with the facility manager and the maintenance supervisor to briefly discuss the purpose of the audit and indicate the kind of information that is to be obtained during the visit to the facility. If possible, a facility employee who is in a position to authorize expenditures or make operating policy decisions should also be at this initial meeting.

Audit Interviews

Getting the correct information on facility equipment and operation is important if the audit is going to be most successful in identifying ways to save money on energy bills. The company philosophy towards investments, the impetus behind requesting the audit, and the expectations from the audit can be determined by interviewing the general manager, chief operating officer, or other executives. The facility manager or plant manager is one person that should have access to much of the operational data on the facility and equipment. The finance officer can provide any necessary financial records (e.g.; utility bills for electric, gas, oil, other fuels, water and wastewater, plus expenditures for maintenance and repair, etc.).

The auditor must also interview the floor supervisors and equipment operators to understand the building and process problems. Line or area supervisors usually have the best information on the times their equipment is used. The maintenance supervisor is often the primary person to talk to about types of lighting and lamps, sizes of motors, sizes of air conditioners and space heaters, and electrical loads of specialized process equipment. Finally, the maintenance staff must be interviewed to find the equipment and performance problems.

The auditor should write down these people's names, job functions and telephone numbers, since it is frequently necessary to get additional information after the initial audit visit.

Walk-through Tour

A walk-through tour of the facility or plant should be conducted by the facility/plant manager, and should be arranged so the auditor or audit team can see the major operational and equipment to obtain general information. More specific information should be obtained from the maintenance and operational people after the tour.

Getting Detailed Data

Following the facility or plant tour, the auditor or audit team should acquire the detailed data on facility equipment and operation that will lead to identifying the significant energy conservation opportunities (ECOs) that may be appropriate for this facility. This includes data on lighting, HVAC equipment, motors, and water heating, as well as specialized equipment such as refrigerators, ovens, mixers, boilers, heaters, etc. These data are most easily recorded on individualized data sheets that have been prepared in advance.

What to Look for

- **Lighting:** Making a detailed inventory of all lighting is important. Data should be recorded on the numbers of each type of light fixtures and lamps, wattages of lamps, and hours of operation of groups of lights. A lighting inventory data sheet should be used to record these data. Using a lightmeter, the auditor should also record light intensity readings for each area. Taking notes on types of tasks performed in each area will help the auditor select alternative lighting technologies that might be more energy efficient. Other items to note are the areas that may be infrequently used and may be candidates for occupancy sensor controls of lighting. Areas where daylighting may be feasible should also be considered.

- **HVAC Equipment:** All heating, air conditioning and ventilating equipment should be inventoried. Prepared data sheets can be used to record type, size, model number, age, electrical specifications, or fuel use specifications, as well as estimated hours of operation. The equipment should be inspected to determine the condition of the evaporator and condenser coils, the air filters, and the insulation on the refrigerant lines. Air velocity measurement may also be made and recorded to assess operating efficiencies or to discover conditioned air leaks. These data will allow later analysis to examine alternative equipment and operations that would reduce energy costs for heating, ventilating, and air conditioning.

- **Electric Motors:** An inventory of all electric motors over one horsepower should also be taken. Prepared data sheets can be used to record motor size, use, age, model number, estimated hours of operation, other electrical characteristics, and possibly the operating power factor. Measurement of voltages, currents, and power factors may be appropriate for some motors. Notes should be taken on the use of motors, particularly recording those that are infrequently used and might be candidates for peak load control or shifting use to off-peak times. All motors over one hp and with times of use of 2000 hours per year or greater are likely candidates for replacement by high efficiency motors—at least when they fail and must be replaced.

- **Water Heaters:** All water heaters should be examined, with data recorded on their type, size, age, model number, electrical characteristics or fuel use. What the hot water is used for, how much is used, and what time it is used should all be noted. Temperature of the hot water should be measured.

- **Waste Heat Sources:** Most facilities have many sources of waste heat, providing possible opportunities for waste heat recovery to be used as the substantial or total source of needed hot water. Waste heat sources are air conditioners, air compressors, heaters and boilers, process cooling systems, ovens, furnaces, cookers, and many others. Temperature measurements for these waste heat sources are necessary to analyze them for replacing the operation of the existing water heaters.

- **Peak Equipment Loads:** The auditor should particularly look for any piece of electrically powered equipment that is used infrequently or whose use could be controlled and shifted to offpeak times. Examples of infrequently used equipment include trash compactors, fire sprinkler system pumps (testing), certain types of welders, drying ovens, and any type of back-up machine. Some production machines might be able to be scheduled for off-peak.

Water heating could be done off-peak, if a storage system is available and off-peak thermal storage can be accomplished for use in on-peak heating or cooling of buildings. Electrical measurements of voltages, currents, and wattages may be helpful. Any information which leads to a piece of equipment being used off-peak is valuable and could result in substantial savings on electric bills. The auditor should be especially alert for those infrequent on-peak uses that might help explain anomalies on the energy demand bills.

- **Other Energy-consuming Equipment:** Finally, an inventory of all other equipment that consumes a substantial amount of energy should be taken. Commercial facilities may have extensive computer and copying equipment, refrigeration and cooling equipment, cooking devices, printing equipment, water heaters, etc. Industrial facilities will have many highly specialized process and production operations and machines. Data on types, sizes, capacities, fuel use, electrical characteristics, age, and operating hours should be recorded for all of this equipment.

Preliminary Identification of ECOs: As the audit is being conducted, the auditor should take notes on potential ECOs that are evident. Identifying ECOs requires a good knowledge of the available energy efficiency technologies that can accomplish the same job with less energy and less cost. For example, overlighting indicates a potential lamp removal or lamp change ECO, and inefficient lamps indicate a potential lamp technology change. Motors with high use times are potential ECOs for high efficiency replacements. Notes on waste heat sources should indicate what other heating sources they might replace, as well as how far away they are from the end use point. Identifying any potential ECOs during the walk-through will make it easier later on to analyze the data and to determine the final ECO recommendations.

Post-Audit Analysis

Following the audit visit to the facility, the data collected should be examined, organized and reviewed for completeness. Any missing data should be obtained from the facility personnel or from revisiting the facility. The preliminary ECOs identified during the audit visit should now be reviewed, and the actual analysis of the equipment or operational change should be conducted. This involves determining the costs and the benefits of the potential ECO and making a judgment on the cost-effectiveness of that potential ECO.

Cost-effectiveness involves a judgment decision that is viewed differently by different people and different companies. Often, Simple Payback Period (SPP) is used to measure cost-effectiveness; most facilities want a SPP of two years or less. The SPP for an ECO is found by taking the initial cost and dividing it by the annual savings. This results in finding a period of time for the savings to repay the initial investment, without using the time value of money. One other common measure of cost-effectiveness is the discounted benefit-cost ratio. In this method, the annual savings are discounted when they occur in future years and are added together to find the present value of the annual savings over a specified period of time. The benefit-cost ratio is then calculated by dividing the present value of the savings by the initial cost. A ratio greater than one means that the investment will more than repay itself, even when the discounted future savings are taken into account.

Several ECO examples are given here in order to illustrate the relationship between the audit information obtained and the technology and operational changes recommended to save on energy bills.

Lighting ECO

First, an ECO technology is selected, such as replacing an existing 400 watt mercury vapor lamp with a 325 watt multi-vapor metal halide lamp when it burns out. The cost of the replacement lamp must be determined. Product catalogs can be used to get typical prices for the new lamp—about $10 more than the 400 watt mercury vapor lamp. The new lamp is a direct screw-in replacement, and no change is needed in the fixture or ballast. Labor cost is assumed to be the same to install either lamp. The benefits (cost savings) must be calculated next. The power savings is 400-325 = 75 watts. If the lamp operates for 4000 hours per year and electric energy costs $0.075/kWh, then the savings is (.075 kW)(4000 hr/year)($0.075/kWh) = $22.50/year. This gives an SPP = $10/$22.50/yr =.4 years, or about 5 months. This would be considered an extremely cost-effective ECO. (For illustration purposes, ballast wattage has been ignored, and average cost has been used to find the savings.)

Motor ECO

A ventilating fan at a fiberglass boat manufacturing company has a standard efficiency 5 hp motor that runs at full load two shifts a day, or 4160 hours per year. When this motor wears out, the company will have an ECO of using a high efficiency motor. A high efficiency 5 hp motor costs around $80 more to purchase than the standard efficiency motor. The standard motor is 83% efficient and the high efficiency model is 88.5% efficient. The cost savings is found by calculating (5 hp)(4160 hr/yr)(.746 kW/hp)[(1/.83) –(1/.885)]($.075/kWh) = (1162 kWh)*($0.075) = $87.15/year. The SPP = $80/$87.15/yr =.9 years, or about 11 months. This is also a very attractive ECO when evaluated by this economic measure.

The discounted benefit-cost ratio can be found, once a motor life is determined and a discount rate is selected. Companies generally have a corporate standard for the discount rate used in determining their measures used to make investment decisions. For a 10 year assumed life, and a 10% discount rate, the present worth factor is found as 6.144 (see Appendix IV). The benefit-cost ratio is found as B/C = ($87.15)(6.144)/$80 = 6.7. This is an extremely attractive benefit-cost ratio!

Peak Load Control ECO

A metals fabrication plant has a large shot-blast cleaner that is used to remove the rust from heavy steel blocks before they are machined and welded. The cleaner shoots out a stream of small metal balls (like shotgun pellets) to clean the metal blocks. A 150 hp motor provides the primary motive force for this cleaner. If turned on during the first shift, this machine requires a total electrical load of about 180 kW, which adds directly to the peak load billed by the electric utility. At $7.02/kW/month, this costs (180 kW)*($7.02/kW/month) = $1263.60/month. Discussions with line operating people resulted in the information that the need for the metal blocks was known well in advance and that the cleaning could easily be done on the evening shift before the blocks were needed. Based on this information, the recommended ECO is to restrict shot-blast cleaner use to the evening shift, saving the company $15,163.20 per year. Since there is no cost to implement this ECO, the SPP = O; that is, the payback is immediate.

The Energy Audit Report

The next step in the energy audit process is to prepare a report which details the final results and recommendations. The length and detail of this report will vary depending on the type of facility audited. A residential audit may result in a computer printout from the utility. An industrial audit is more likely to have a detailed explanation of the ECOs and benefit-cost analyses. The following discussion covers the more detailed audit reports.

The report should begin with an executive summary that provides the owners/managers of the audited facility with a brief synopsis of the total savings available and the highlights of each ECO. The report should then describe the facility that has been audited, providing information on the operation of the facility that relates to its energy costs. The energy bills should be presented, with tables and plots showing the costs and consumption. Following the energy cost analysis, the recommended ECOs should be presented, along with the calculations for the costs and benefits, as well as the cost-effectiveness criterion.

Regardless of the audience for the audit report, it should be written in a clear, concise and easy-to understand format and style. The executive summary should be tailored to non-technical personnel, and technical jargon should be minimized. A client who understands the report is more likely to implement the recommended ECOs. An outline for a complete energy audit report is shown below.

Energy Audit Report Format

Executive Summary

A brief summary of the recommendations and cost savings

Table of Contents

Introduction

Purpose of the energy audit

Need for a continuing energy cost control program

Facility Description

Product or service, and materials flow

Size, construction, facility layout, and hours of operation

Equipment list, with specifications

Energy Bill Analysis

Utility rate structures

Tables and graphs of energy consumptions and costs

Discussion of energy costs and energy bills

Energy conservation opportunities
- Listing of potential ECOs
- Cost and savings analysis
- Economic evaluation

Action Plan
- Recommended ECOs and an implementation schedule
- Designation of an energy monitor and ongoing program

Conclusion
- Additional comments not otherwise covered

The Energy Action Plan

The last step in the energy audit process is to recommend an action plan for the facility. Some companies will have an energy audit conducted by their electric utility or by an independent consulting firm and will then make changes to reduce their energy bills. They may not spend any further effort in the energy cost control area until several years in the future when another energy audit is conducted. In contrast to this is the company which establishes a permanent energy cost control program and assigns one person—or a team of people—to continually monitor and improve the energy efficiency and energy productivity of the company. Similar to a total quality management program where a company seeks to continually improve the quality of its products, services and operation, an energy cost control program seeks continual improvement in the amount of product produced for a given expenditure for energy.

The energy action plan lists the ECOs which should be implemented first and suggests an overall implementation schedule. Often, one or more of the recommended ECOs provides an immediate or very short payback period, so savings from those can be used to generate capital to pay for implementing the other ECOs. In addition, the action plan also suggests that a company designate one person as the energy monitor for the facility. This person can look at the monthly energy bills and see whether any unusual costs are occurring and can verify that the energy savings from ECOs is really being seen. Finally, this person can continue to look for other ways the company can save on energy costs and can be seen as evidence that the company is interested in a future program of energy cost control.

SPECIALIZED AUDIT TOOLS

Smoke Sources

Smoke is useful in determining airflow characteristics in buildings, air distribution systems, exhaust hoods and systems, cooling towers, and air intakes. There are several ways to produce smoke. Ideally, the smoke should be neutrally buoyant with the air mass around it so that no motion will be detected unless a force is applied. Cigarette and incense stick smoke, although inexpensive, do not meet this requirement.

Smoke generators using titanium tetrachloride ($TiCl_4$) provide an inexpensive and convenient way to produce and apply smoke. The smoke is a combination of hydrochloric acid (HCl) fumes and titanium oxides produced by the reaction of $TiCl_4$ and atmospheric water vapor. This smoke is both corrosive and toxic, so the use of a respirator mask utilizing activated carbon is strongly recommended. Commercial units typically use either glass or plastic cases. Glass has excellent longevity but is subject to breakage, since smoke generators are often used in difficult-to-reach areas. Most types of plastic containers will quickly degrade from the action of hydrochloric acid.

Small Teflon* squeeze bottles (i.e., 30 ml) with attached caps designed for laboratory reagent use resist degradation and are easy to use. The bottle should be stuffed with 2-3 real cotton balls, then filled with about 0.15 fluid ounces of liquid $TiCl_4$. Synthetic cotton balls typically disintegrate if used with titanium tetrachloride. This bottle should yield over a year of service with regular use. The neck will clog with debris but can be cleaned with a paper clip.

Some smoke generators are designed for short time use. These bottles are inexpensive and useful for a day of smoke generation but will quickly degrade. Smoke bombs are incendiary devices designed to emit a large volume of smoke over a short period of time. The smoke is available in various colors to provide good visibility. These are useful in determining airflow capabilities of exhaust air systems and large-scale ventilation systems. A crude smoke bomb can be constructed by placing a stick of elemental phosphorus in a metal pan and igniting it. A large volume of white smoke will be released. This is an inexpensive way of testing laboratory exhaust hoods, since many labs have phosphorus in stock.

More accurate results can be obtained by measuring the chemical composition of the airstream after injecting a known quantity of tracer gas such as sulphur hexafluoride into an area. The efficiency of an exhaust system can be determined by measuring the rate of tracer gas removal. Building infiltration/exfiltration rates can also be estimated with tracer gas.

Blower Door

The blower door is a device containing a fan, controller, several pressure gauges, and a frame which fits in the doorway of a building. It is used to study the pressurization and leakage rates of a building and its air distribution system under varying pressure conditions. The units currently available are designed for use in residences, although they can be used in small commercial buildings as well. The large quantities of ventilation air limit blower door use in large commercial and industrial buildings.

An air leakage/pressure curve can be developed for the building by measuring the fan flow rate necessary to achieve a pressure differential between the building interior and the ambient atmospheric pressure over a range of values. The natural air infiltration rate of the building under the prevailing pressure conditions can be estimated from the leakage/pressure curve and local air pressure data. Measurements made before and after sealing identified leaks can indicate the effectiveness of the work.

The blower door can help to locate the source of air leaks in the building by depressurizing to 30 Pascals and searching potential leakage areas with a smoke source. The air distribution system typically leaks on both the supply and return air sides. If the duct system is located outside the conditioned space (e.g., attic, under floor, etc.), supply leaks will depressurize the building and increase the air infiltration rate; return air leaks will pressurize the building, causing air to exfiltrate. A combination of supply and return air leaks is difficult to detect without sealing off the duct system at the registers and measuring the leakage rate of the building compared to that of the unsealed duct system. The difference between the two conditions is a measure of the leakage attributable to the air distribution system.

Airflow Measurement Devices

Two types of anemometers are available for measuring airflow: vane and hot-wire. The volume of air moving through an orifice can be determined by estimating the free area of the opening (e.g., supply air register, exhaust hood face, etc.) and multiplying by the air speed. This result is approximate, due to the difficulty in determining the average air speed and the free vent area. Regular calibrations are necessary to assure the accuracy of the instrument. The anemometer can also be used to optimize the face velocity of exhaust hoods by adjusting the door opening until the anemometer indicates the desired airspeed.

Airflow hoods also measure airflow. They contain an airspeed integrating manifold, which averages the velocity across the opening and reads out the airflow volume. The hoods are typically made of nylon fabric supported by an aluminum frame. The instrument is lightweight and easy to hold up against an air vent. The lip of the hood must fit snugly around the opening to assure that all the air volume is measured. Both supply and exhaust airflow can be measured. The result must be adjusted if test conditions fall outside the design range.

INDUSTRIAL AUDITS

Introduction

Industrial audits are some of the most complex and most interesting audits because of the tremendous variety of equipment found in these facilities. Much of the industrial equipment can be found during commercial audits too. Large chillers, boilers, ventilating fans, water heaters, coolers and freezers, and extensive lighting systems are often the same in most industrial operations as those found in large office buildings or shopping centers. Small cogeneration systems are often found in both commercial and industrial facilities.

The highly specialized equipment that is used in industrial processes is what differentiates these facilities from large commercial operations. The challenge for the auditor and energy management specialist is to learn how this complex—and often unique—industrial equipment operates and to come up with improvements to the processes and the equipment that can save energy and money. The sheer scope of the problem is so great that industrial firms often hire specialized consulting engineers to examine their processes and recommend operational and equipment changes that result in greater energy productivity.

Audit Services

A few electric and gas utilities are sufficiently large, and well staffed to offer industrial audits to their customers. These utilities have a trained staff of engineers and process specialists with extensive experience who can recommend operational changes or new equipment to reduce the energy costs in a particular production environment. Many gas and electric utilities, even if they do not offer audits, do offer financial incentives for facilities to install high efficiency lighting, motors, chillers, and other equipment. These incentives can make many ECOs very attractive.

Small and medium-sized industries that fall into the Manufacturing Sector—SIC 2000 to 3999—and are in the service area of one of the industrial assessment centers funded by the U.S. Department of Energy can receive free energy audits throughout this program. There are presently 26 IACs operating primarily in the eastern and mid-western areas of the U.S. These IACs are administered by Rutgers University, Piscataway, NJ. Companies that are interested in knowing if an IAC is located near them and whether they qualify for an IAC audit can call 215-387-2255 and ask for information on the industrial assessment center program. Information can also be found at www.iac.rutgers.edu.

Industrial Energy Rate Structures

Except for the smallest industries, facilities will be billed for energy services through a large commercial or industrial rate category. It is important to get this rate structure information for all sources of energy—electricity, gas, oil, coal, steam, etc. Gas, oil and coal are usually billed on a straight cost per unit basis, e.g. $0.90 per gallon of #2 fuel oil. Electricity and steam most often have complex rate structures with components for a fixed customer charge, a demand charge, and an energy charge. Gas, steam, and electric energy are often available with a time of day rate, or an interruptible rate that provides much cheaper energy service with the understanding that the customer may have his supply interrupted (stopped) for periods of several hours at a time. Advance notice of the interruption is almost always given, and the number of times a customer can be interrupted in a given period of time is limited.

Process and Technology Data Sources

For the industrial audit, it is critical to get in advance as much information as possible on the specialized process equipment so that study and research can be performed to understand the particular processes being used and what improvements in operation or technology are available. Data sources are extremely valuable here; auditors should maintain a library of information on processes and technology and should know where to find additional information from research organizations, government facilities, equipment suppliers, and other organizations.

EPRI/GRI

The Electric Power Research Institute (EPRI) and the Gas Research Institute (GRI) are both excellent sources of information on the latest technologies of using electric energy or gas. EPRI has a large number of on-going projects to show the cost-effectiveness of electro-technologies using new processes for heating, drying, cooling, etc. GRI also has a large number of projects underway to help promote the use of new cost-effective gas technologies for heating, drying, cooling, etc. Both of these organizations provide extensive documentation of their processes and technologies; they also have computer data bases to aid customer inquiries.

U.S. DOE Industrial Division

The U.S. Department of Energy has an Industrial Division that provides a rich source of information on new technologies and new processes. This division funds research into new processes and technologies, and it also funds many demonstration projects to help insure that promising improvements get implemented in appropriate industries. The Industrial Division of USDOE also maintains a wide network of contacts with government-related research laboratories, such as Oak Ridge National Laboratory, Brookhaven National Laboratory, Lawrence Berkeley National Laboratory, Sandia National Laboratory, and Battelle National Laboratory. These laboratories have many of their own research, development and demonstration programs for improved industrial and commercial technologies.

State Energy Offices

State energy offices are also good sources of information, as well as good contacts to see what kind of incentive programs might be available in the state. Many states offer programs of free boiler tune-ups, free air conditioning system checks, seminars on energy efficiency for various facilities, and other services. Most state energy offices have well-stocked energy libraries and are also tied into other state energy research organizations, national laboratories, and the USDOE.

Equipment Suppliers

Equipment suppliers provide additional sources for data on energy efficiency improvements to processes. Marketing new, cost-effective processes and technologies provides sales for the companies as well as helping industries to be more productive and more economically competitive. The energy auditor should compare the information from all of the sources described above.

CONDUCTING THE AUDIT

Safety Considerations

Safety is the primary consideration in any industrial audit. The possibility of injury from hot objects, hazardous materials, slippery surfaces, drive belts, and electric shocks is far greater than when conducting residential and commercial audits. Safety glasses, safety shoes, durable clothing, and possibly a safety hat and breathing mask might be needed during some audits. Gloves should be worn while making any electrical measurements and also while making any measurements around boilers, heaters, furnaces, steam lines, or other very hot pieces of equipment. In all cases, adequate attention to personal safety is a significant feature of any industrial audit.

Lighting

Lighting is not as great a percent of total industrial use as it is in the commercial sector on average, but lighting is still a big energy use and cost area for many industrial facilities. A complete inventory of all lighting should be taken during the audit visit. Hours of operation of lights are also necessary, since lights are commonly left on when they are not needed. Timers, energy management systems, and occupancy sensors are all valuable approaches to insuring that unneeded lights are not turned on. It is also important to look at the facility's outside lighting for parking and for storage areas.

During the lighting inventory, types of tasks being performed should also be noted, since light replacement with more efficient lamps often involves changing the color of the resultant light. For example, high pressure sodium lamps are much more efficient than mercury vapor lamps (or even metal halide lamps), but they produce a yellowish light that makes fine color distinction difficult. However, many assembly tasks can still be performed adequately under high pressure sodium lighting. These typically include metal fabrication, wood product fabrication, plastic extrusion, and many others.

Electric Motors

A common characteristic of many industries is their extensive use of electric motors. A complete inventory of all motors over 1 hp should be taken, as well as recording data on how long each motor operates during a day. For motors with substantial usage times, replacement with high-efficiency models is almost always cost effective. In addition, consideration should be given to replacement of standard drive belts with synchronous belts that transmit the motor energy more efficiently. For motors which are used infrequently, it may be possible to shift the use to off-peak times and achieve a kW demand reduction which would reduce energy cost.

HVAC Systems

An inventory of all space heaters and air conditioners should be taken. Btu per hour ratings and efficiencies of all units should be recorded, as well as usage patterns. Although many industries do not heat or air condition the production floor area, they almost always have office areas, cafeterias, and other areas that are normally heated and air conditioned. For these conditioned areas, the construction of the facility should be noted—how much insulation, what the walls and ceilings made of, how high the ceilings are. Adding additional insulation might be a cost effective ECO.

Production floors that are not air conditioned often have large numbers of ventilating fans that operate anywhere from one shift per day to 24 hours a day. Plants with high heat loads and plants in mild climate areas often leave these ventilating fans running all year long. These are good candidates for high efficiency motor replacements. Timers or an energy management system might be used to turn off these ventilating fans when the plant is shut down.

Boilers

All boilers should be checked for efficient operation using a stack gas combustion analyzer. Boiler specifications on Btu per hour ratings, pressures, and temperatures should be recorded. The boiler should be varied between low-fire, normal-fire, and high-fire, with combustion gas and temperature readings taken at each level. Boiler tune-up is one of the most common and most energy-saving operations available to many facilities. The auditor should check to see if any waste heat from the boiler is being recovered for use in a heat recuperator or for some other use, such as water heating. If not, this should be noted as a potential ECO.

Specialized Equipment

Most of the remaining equipment encountered during the industrial audit will be highly specialized process production equipment and machines. This equipment should all be examined, with operational data noted, as well as hours and periods of use. All heat sources should be considered carefully as to whether they could be replaced with sources using waste heat, or whether waste heat could serve as a provider of heat to another application. Operations where both heating and cooling occur periodically—such as a plastic extrusion machine—are good candidates for reclaiming waste heat, or in sharing heat from a machine needing cooling with another machine needing heat.

Air Compressors

Air compressors should be examined for size, operating pressures, and type (reciprocating or screw), and whether they use outside cool air for intake. Large air compressors are typically operated at night when much smaller units are sufficient. Also, screw-type air compressors use a large fraction of their rated power when they are idling, so control valves should be installed to prevent this loss. Efficiency is improved with intake air that is cool, so outside air should be used in most cases—except in extremely cold temperature areas.

The auditor should determine whether there are significant air leaks in air hoses, fittings, and machines. Air leaks are a major source of energy loss in many facilities and should be corrected by maintenance action. Finally, air compressors are a good source of waste heat. Nearly 90% of the energy used by an air compressor shows up as waste heat, so this is a large source of low temperature waste heat for heating input air to a heater or boiler, or for heating hot water for process use.

COMMERCIAL AUDITS

Introduction

Commercial audits span the range from very simple audits for small offices to very complex audits for multi-story office buildings or large shopping centers. Complex commercial audits are performed in substantially the same manner as industrial audits. The following discussion highlights those areas where commercial audits are likely to differ from industrial audits.

Commercial audits generally involve substantial consideration of the structural envelope features of the facility, as well as significant amounts of large or specialized equipment at the facility. Office buildings, shopping centers and malls all have complex building envelopes that should be examined and evaluated. Building materials, insulation levels, door and window construction, skylights, and many other envelope features must be considered in order to identify candidate ECOs.

Commercial facilities also have large capacity equipment, such as chillers, space heaters, water heaters, refrigerators, heaters, cookers, and office equipment like computers and copy machines. Small cogeneration systems are also commonly found in commercial facilities and institutions such as schools and hospitals. Much of the equipment in commercial facilities is the same type and size as that found in manufacturing or industrial facilities. Potential ECOs would look at more efficient equipment, use of waste heat, or operational changes to use less expensive energy.

Commercial Audit Services

Electric and gas utilities, and many engineering consulting firms, perform audits for commercial facilities. Some utilities offer free walk-through audits for commercial customers, as well as financial incentives for customers who change to more energy efficient equipment. Schools, hospitals, and some other government institutions can qualify for free audits under the ICP program described in the first part of this chapter. Whoever conducts the commercial audit must initiate the ICP process by collecting information on the energy rate structures, the equipment in use at the facility, and the operational procedures used there.

Commercial Energy Rate Structures

Small commercial customers are usually billed for energy on a per energy unit basis, while large commercial customers are billed under complex rate structures containing components related to energy, rate of energy use (power), time of day or season of year, power factor, and numerous other elements. One of the first steps in a commercial audit is to obtain the rate structures for all sources of energy and to analyze at least one to two year's worth of energy bills. This information should be put into a table and also plotted.

Conducting the Audit

A significant difference in industrial and commercial audits arises in the area of lighting. Lighting in commercial facilities is one of the largest energy costs—sometimes accounting for half or more of the entire electric bill. Lighting levels and lighting quality are extremely important to many commercial operations. Retail sales operations, in particular, want light levels that are far in excess of standard office values. Quality of light in terms of color is also a big concern in retail sales, so finding acceptable ECOs for reducing lighting costs is much more difficult for retail facilities than for office buildings. The challenge is to find new lighting technologies that allow high light levels and warm color while reducing the wattage required. New T8 and T10 fluorescent lamps, as well as metal halide lamp replacements for mercury vapor lamps, offer these features and usually represent cost-effective ECOs for retail sales and other facilities.

RESIDENTIAL AUDITS

Audits for large, multi-story apartment buildings can be very similar to commercial audits. (See section 3.6.) Audits of single-family residences, however, are generally fairly simple. For single-family structures, the energy audit focuses on the thermal envelope and the appliances, such as the heater, air conditioner, water heater, and "plug loads."

The residential auditor should start by obtaining past energy bills and analyzing them to determine any patterns or anomalies. During the audit visit, the structure is examined to determine the levels of insulation, the presence and condition of seals for windows and doors, and the integrity of the ducts. The space heater and/or air conditioner is inspected, along with the water heater. Equipment model numbers, age, size, and efficiencies are recorded. The post-audit analysis then evaluates potential ECOs, such as adding insulation, double-pane windows, and window shading or insulated doors. Changing to higher efficiency heaters, air conditioners, and water heaters is also considered. The auditor calculates costs, benefits, and simple payback periods and presents them to the owner or occupant. A simple audit report, often in the form of a computer printout, is given to the owner or occupant.

INDOOR AIR QUALITY

Introduction

Implementation of new energy-related standards and practices has contributed to an enhancement of indoor air quality. In fact, in many homes, businesses and factories, the quality of indoor air has been found to exceed the Environmental Protection Agency (EPA) standards for outdoor air! Thus, testing for indoor air quality problems is done in some energy audits both to prevent exacerbating any existing problems and to recommend ECOs that might improve air quality. Air quality standards for the industrial environment have been published by the American Council of Governmental Industrial Hygienists (ACGIH) in their booklet "Threshold Limit Values." No such standards currently exist for the residential and commercial environments, although the ACGIH standards are typically (perhaps inappropriately) used. The EPA has been working to develop residential and commercial standards for quite some time.

Symptoms of Air Quality Problems

Symptoms of poor indoor air quality include, but are not limited to: headaches; irritation of mucous membranes such as the nose, mouth, throat, lungs; tearing, redness and irritation of the eyes; numbness of the lips, mouth, and throat; mood swings; fatigue; allergies; coughing; nasal and throat discharge; and irritability. Chronic exposure to some compounds can lead to damage to internal organs such as the liver, kidney, lungs, and brain; even cancer and death have been associated with it.

Testing

Testing is required to determine if the air quality is acceptable. Many dangerous compounds, like carbon monoxide and methane without odorant added, are odorless and colorless. Some dangerous particulates such as asbestos fibers do not give any indication of a problem for up to twenty years after inhalation. Testing must be conducted in conjunction with pollution-producing processes to ensure capture of the contaminants. Such testing is usually performed by a Certified Industrial Hygienist (CIH).

Types of Pollutants

Airstreams have three types of contaminants: particulates like dust and asbestos; gases like carbon monoxide, ozone, carbon dioxide, volatile organic compounds, anhydrous ammonia, radon, outgassing from urea-formaldehyde insulation, low oxygen levels; and biologicals like mold, mildew, fungus, bacteria, and viruses.

POLLUTANT CONTROL MEASURES

Particulates

Particulates are controlled with adequate filtration near the source and in the air handling system. Mechanical filters are frequently used in return air streams, and baghouses are used for particulate capture. The coarse filters used in most residential air conditioners typically have filtration efficiencies below twenty percent. Mechanical filters called high efficiency particulate apparatus (HEPA) are capable of filtering particles as small as 0.3 microns at up to 99% efficiency. Electrostatic precipitators remove particulates by placing a positive charge on the walls of collection plates and allowing negatively charged particulates to attach to the surface. Periodic cleaning of the plates is necessary to maintain high filtration efficiency. Loose or friable asbestos fibers should be removed from the building or permanently encapsulated to prevent entry into the respirable airstream. While conducting an audit, it is important to determine exactly what type of insulation is in use before disturbing an area to make temperature measurements.

Problem Gases

Problem gases are typically removed by ventilating with outside air. Dilution with outside air is effective, but tempering the temperature and relative humidity of the outdoor air mass can be expensive in extreme conditions. Heat exchangers such as heat wheels, heat pipes, or other devices can accomplish this task with reduced energy use. Many gases can be removed from the airstream by using absorbent/adsorbent media such as activated carbon or zeolite. This strategy works well for spaces with limited ventilation or where contaminants are present in low concentrations. The media must be checked and periodically replaced to maintain effectiveness.

Radon gas—Ra 222—cannot be effectively filtered due to its short half life and the tendency for its Polonium daughters to plate out on surfaces. Low oxygen levels are a sign of inadequate outside ventilation air. A high level of carbon dioxide (e.g., 1000-10,000 ppm) is not a problem in itself, but levels above 1000 ppm indicate concentrated human or combustion activity or a lack of ventilation air. Carbon dioxide is useful as an indicator compound because it is easy and inexpensive to measure.

Microbiological Contaminants

Microbiological contaminants generally require particular conditions of temperature and relative humidity on a suitable substrate to grow. Mold and mildew are inhibited by relative humidity levels less than 50%. Air distribution systems often harbor colonies of microbial growth. Many people are allergic to microscopic dust mites. Cooling towers without properly adjusted automated chemical feed systems are an excellent breeding ground for all types of microbial growth.

Ventilation Rates

Recommended ventilation quantities for commercial and institutional buildings are published by the American Society of Heating, Refrigerating, and Air-Conditioning Engineers (ASHRAE) in standard 62.1-2004, "Ventilation for Acceptable Air Quality." These ventilation rates are for effective systems. Many existing systems fail in entraining the air mass efficiently. The density of the contaminants relative to air must be considered in locating the exhaust air intakes and ventilation supply air registers.

Liability

Liability related to indoor air problems appears to be a growing but uncertain issue, because few cases have made it through the court system. However, in retrospect, the asbestos and urea-formaldehyde pollution problems discovered in the last two decades suggest proceeding with caution and a proactive approach.

CONCLUSION

Energy audits are an important first step in the overall process of reducing energy costs for any building, company, or industry. A thorough audit identifies and analyzes the changes in equipment and operations that will result in cost-effective energy cost reduction. The energy auditor plays a key role in the successful conduction of an audit, as well as the implementation of the audit recommendations.

Bibliography

Instructions For Energy Auditors, Volumes I and II, U.S. Department of Energy, DOE/CS-0041/12&13, September, 1978. Available through National Technical Information Service, Springfield, VA.

Energy Conservation Guide for Industry and Commerce, National Bureau of Standards Handbook 115 and Supplement, 1978. Available through U.S. Government Printing Office, Washington, DC.

Guide to Energy Management, Fifth Edition, Capehart, B.L., Turner, W.C., and Kennedy, W.J., The Fairmont Press, Atlanta, GA, 2006.

Illuminating Engineering Society, *IES Lighting Handbook, Ninth Edition,* New York, NY, 2000.

Total Energy Management, A Handbook prepared by the National Electrical Contractors Association and the National Electrical Manufacturers Association, Washington, DC.

Handbook of Energy Audits, Thumann, Albert, and William J. Younger, Seventh Edition, The Fairmont Press, Atlanta, GA, 2006.

Industrial Energy Management and Utilization, Witte, Larry C., Schmidt, Philip S., and Brown, David R., Hemisphere Publishing Corporation, Washington, DC, 1988.

Threshold Limit Values for Chemical Substances and Physical Agents and Biological Exposure Indices, 1990-91 American Conference of Governmental Industrial Hygienists.

Ventilation for Acceptable Indoor Air Quality, ASHRAE 62.1-2004, American Society of Heating, Refrigerating and Air-Conditioning Engineers, Inc., 2004.

Facility Design and Planning Engineering Weather Data, Departments of the Air Force, the Army, and the Navy, 1978.

Handbook of Energy Engineering, Fifth Edition, Thumann, A., and Menta, D.P., The Fairmont Press, Atlanta, GA, 2004.

Energy Management Handbook, Sixth Edition, Turner, Wayne C. and Steve Doty, The Fairmont Press, Atlanta, GA 2006.

Encyclopedia of Energy Engineering and Technology, Barney L. Capehart, Editor, Taylor and Francis/CRC Publishers, New York, NY.

Appendix A
Economic Analysis

Dr. David Pratt

Editor's Note: Appendix A provides the foundations of economic analysis and the time value of money ("A dollar today is worth more than a dollar tomorrow.") For folks who are new to financing terminol-ogy, this appendix is a good reference. Near the end of this appendix there are 38 examples about how to do economic analyses of projects!

A.1 OBJECTIVE

The objective of this appendix is to present a coherent, consistent approach to economic analysis of capital investments (energy related or other). Adherence to the concepts and methods presented will lead to sound investment decisions with respect to time value of money principles. The appendix opens with material designed to elevate the importance of life cycle cost concepts in the economic analysis of projects. The next three sections provide foundational material necessary to fully develop time value of money concepts and techniques. These sections present general characteristics of capital investments, sources of funds for capital investment, and a brief summary of tax considerations, all of which are important for economic analysis. The next two sections introduce time value of money calculations and several approaches for calculating project measures of worth based on time value of money concepts. Next the measures of worth are applied to the process of making decisions when a set of potential projects are to be evaluated. The final concept and technique section of the appendix presents material to address several special problems that may be encountered in economic analysis. This material includes, among other things, discussions of inflation, non-annual compounding of interest, and sensitivity analysis. The appendix closes with a brief summary and a list of references which can provide additional depth in many of the areas covered in the appendix.

A.2 INTRODUCTION

Capital investment decisions arise in many circumstances. The circumstances range from evaluating business opportunities to personal retirement planning. Regardless of circumstances, the basic criterion for evaluating any investment decision is that the revenues (savings) generated by the investment must be greater than the costs incurred. The number of years over which the revenues accumulate and the comparative importance of future dollars (revenues or costs) relative to present dollars are important factors in making sound investment decisions. This consideration of costs over the entire life cycle of the investment gives rise to the name "life cycle cost analysis," which is commonly used to refer to the economic analysis approach presented in this appendix. An example of the importance of life cycle costs is shown in Figure A-1, which depicts the estimated costs of owning and operating an oil-fired furnace to heat a 2,000-square-foot house in the northeast United States. Of particular note is that the initial costs represent only 23% of the total costs incurred over the life of the furnace. The life cycle cost approach provides a significantly better evaluation of long-term implications of an investment than methods which focus on first cost or near-term results.

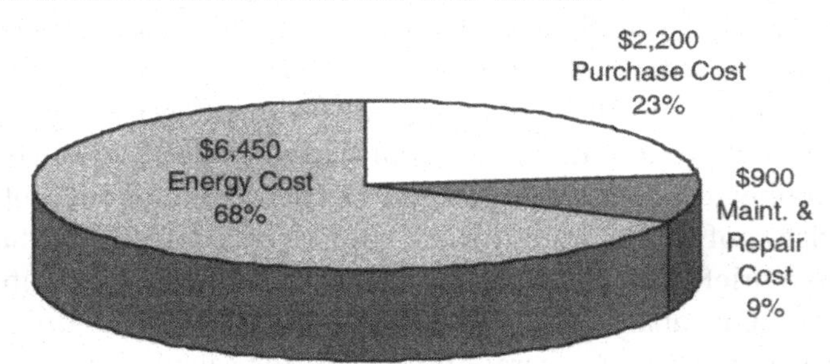

Figure A-1. 15-year life cycle costs of a heating system

Life cycle cost analysis methods can be applied to virtually any public or private business sector investment decision as well as to personal financial planning decisions. Energy related decisions, provide excellent examples for the application of this approach. Such decisions include: evaluation of alternative building designs that have different initial costs, operating and maintenance costs, and perhaps different lives; evaluation of investments to improve the thermal performance of an existing building (wall or roof insulation, window glazing); and evaluation of alternative heating, ventilating, or air conditioning systems. For federal buildings, Congress and the president have mandated, through legislation and executive order, energy conservation goals that must be met using cost-effective measures. The life cycle cost approach is mandated as the means of evaluating cost effectiveness.

A.3 GENERAL CHARACTERISTICS OF CAPITAL INVESTMENTS

A.3.1 Capital Investment Characteristics

When companies spend money, the outlay of cash can be broadly categorized into one of two classifications, expenses or capital investments. Expenses are generally those cash expenditures that are routine, on-going, and necessary for the ordinary operation of the business. Capital investments, on the other hand, are generally more strategic and have long-term effects. Decisions made regarding capital investments are usually made at higher levels within the organizational hierarchy and carry with them additional tax consequences as compared to expenses.

Three characteristics of capital investments are of concern when performing life cycle cost analysis. First, capital investments usually require a relatively large initial cost. "Relatively large" may mean several hundred dollars to a small company or many millions of dollars to a large company. The initial cost may occur as a single expenditure, such as purchasing a new heating system, or occur over a period of several years, such as designing and constructing a new building. It is not uncommon that the funds available for capital investments projects are limited. In other words, the sum of the initial costs of all the viable and attractive projects exceeds the total available funds. This creates a situation known as capital rationing that imposes special requirements on the investment analysis. This topic will be discussed in Section A.8.3.

The second important characteristic of a capital investment is that the benefits (revenues or savings) resulting from the initial cost occur in the future, normally over a period of years. The period between the initial cost and the last future cash flow is the life cycle or life of the investment. It is the fact that cash flows occur over the investment's life that requires the introduction of time value of money concepts to properly evaluate investments. If multiple investments are being evaluated and the lives of the investments are not equal, special consideration must be given to the issue of selecting an appropriate planning horizon for the analysis.

Planning horizon issues are introduced in Section A.8.5.

The last important characteristic of capital investments is that they are relatively irreversible. Frequently, after the initial investment has been made, terminating or significantly altering the nature of a capital investment has substantial (usually negative) cost consequences. This is one of the reasons that capital investment decisions are usually evaluated at higher levels of the organizational hierarchy than are operating expense decisions.

A.3.2 Capital Investment Cost Categories

In almost every case, the costs which occur over the life of a capital investment can be classified into one of the following categories:

- Initial Cost,
- Annual Expenses and Revenues,
- Periodic Replacement and Maintenance, or
- Salvage Value.

As a simplifying assumption, the cash flows which occur during a year are generally summed and regarded as a single end-of-year cash flow. While this approach does introduce some inaccuracy in the evaluation, it is generally not regarded as significantly relative to the level of estimation associated with projecting future cash flows.

Initial costs include all costs associated with preparing the investment for service. This includes purchase cost as well as installation and preparation costs. Initial costs are usually nonrecurring during the life of an investment. Annual expenses and revenues are the recurring costs and benefits generated throughout the life of the investment. Periodic replacement and maintenance costs are similar to annual expenses and revenues, except that they do not (or are not expected to) occur annually. The salvage (or residual) value of an investment is the revenue (or expense) attributed to disposing of the investment at the end of its useful life.

A.3.3 Cash Flow Diagrams

A convenient way to display the revenues (savings) and costs associated with an investment is a cash flow diagram. By using a cash flow diagram, the timing of the cash flows are more apparent, and the chances of properly applying time value of money concepts are increased. With practice, different cash flow patterns can be recognized and may suggest the most direct approach for analysis.

It is usually advantageous to determine the time frame over which the cash flows occur first. This establishes the horizontal scale of the cash flow diagram. This scale is divided into time periods that are frequently, but not always, years. Receipts and disbursements are then located on the time scale in accordance with the problem specifications. Individual outlays or receipts are indicated by drawing vertical lines appropriately placed along the time scale. The relative magnitudes can be suggested by the heights, but exact scaling generally does not enhance the meaningfulness of the diagram. Upward directed lines indicate cash inflow (revenues or savings) while downward directed lines indicate cash outflow (costs).

Figure A-2 illustrates a cash flow diagram. The cash flows depicted represent an economic evaluation of whether to choose a baseboard heating and window air conditioning system or a heat pump for a ranger's house in a national park [Fuller and Petersen, 1994]. The differential costs associated with the decision are:

- The heat pump costs (cash outflow) $1500 more than the baseboard system.
- The heat pump saves (cash inflow) $380 annually in electricity costs.
- The heat pump has a $50 higher annual maintenance cost (cash outflow).
- The heat pump has a $150 higher salvage value (cash inflow) at the end of 15 years.
- The heat pump requires $200 more in replacement maintenance (cash outflow) at the end of year 8.

Although cash flow diagrams are simply graphical representations of income and outlay, they should exhibit as much information as possible. During the analysis phase, it is useful to show the "minimum attractive rate of return" (an interest rate used to account for the time value of money within the problem) on the cash flow diagram, although this has been omitted in Figure A-2. The requirements for a good cash flow diagram are completeness, accuracy, and legibility. The measure of a successful diagram is that someone else can understand the problem fully from it.

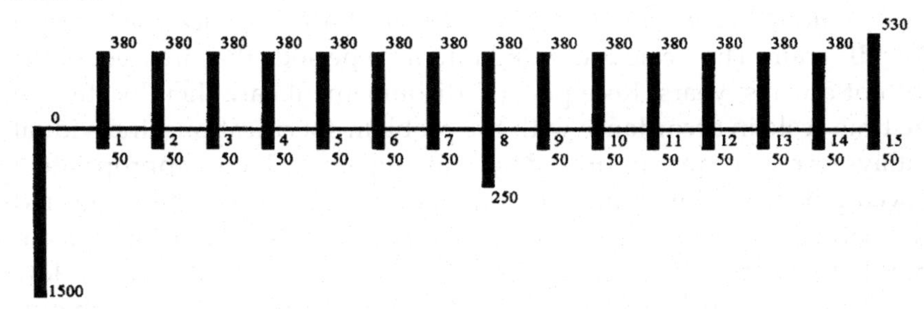

Figure A-2. Heat pump and baseboard system differential life cycle costs

A.4 SOURCES OF FUNDS

Capital investing requires a source of funds. For large companies multiple sources may be employed. The process of obtaining funds for capital investment is called financing. There are two broad sources of financial funding, debt financing and equity financing. Debt financing involves borrowing and utilizing money that is to be repaid at a later point in time. Interest is paid to the lending party for the privilege of using the money. Debt financing does not create an ownership position for the lender within the borrowing organization. The borrower is simply obligated to repay the borrowed funds, plus accrued interest according to a repayment schedule. Car loans and mortgage loans are two examples of this type of financing. The two primary sources of debt capital are loans and bonds. The cost of capital associated with debt financing is relatively easy to calculate, since interest rates and repayment schedules are usually clearly documented in the legal instruments controlling the financing arrangements. An added benefit to debt financing under current U.S. tax law (as of April 2000) is that the interest payments made by corporations on debt capital are tax deductible. This effectively lowers the cost of debt financing. For debt financing with deductible interest payments, the after-tax cost of capital is given by:

$$\text{Cost of Capital}_{\text{AFTERTAX}} =$$
$$\text{Cost of Capital}_{\text{BEFORETAX}} * (1 \text{ TaxRate}$$

where the tax rate is determined by applicable tax law.

The second broad source of funding is equity financing. Under equity financing the lender acquires an ownership (or equity) position within the borrower's organization. As a result of this ownership position, the lender has the right to participate in the financial success of the organization as a whole. The two primary sources of equity financing are stocks and retained earnings. The cost of capital associated with shares of stock is much debated within the financial community. A detailed presentation of the issues and approaches is beyond the scope of this appendix. Additional reference material can be found in Park and Sharp-Bette [1990]. One issue about which there is general agreement is that the cost of capital for stocks is higher than the cost of capital for debt financing. This is at least partially attributable to the fact that interest payments are tax deductible, while stock dividend payments are not.

If any subject is more widely debated in the financial community than the cost of capital for stocks, it is the cost of capital for retained earnings. Retained earnings are the accumulation of annual earnings surpluses that a company retains within the company's coffers rather than pays out to the stockholders as dividends. Although these earnings are held by the company, they truly belong to the stockholders. In essence the company is establishing the position that by retaining the earnings and investing them in capital projects, stockholders will achieve at least as high a return through future financial successes as they would have earned if the earnings had been paid out as dividends. Hence, one common approach to valuing the cost of capital for retained earnings is to apply the same cost of capital as for stock. This, therefore, leads to the same generally agreed result. The cost of capital for financing through retained earnings generally exceeds the cost of capital for debt financing.

In many cases the financing for a set of capital investments is obtained by packaging a combination of the above sources to achieve a desired level of available funds. When this approach is taken, the overall cost of capital is generally taken to be the weighted average cost of capital across all sources. The cost of each individual source's funds is weighted by the source's fraction of the total dollar amount available. By summing across all sources, a weighted average cost of capital is calculated, as shown in the following example:

Example 1

Determine the weighted average cost of capital for financing, which is composed of:

25% loans with a before tax cost of capital of 12%/yr and

75% retained earnings with a cost of capital of 10%/yr.

The company's effective tax rate is 34%.

$$\text{Cost of Capital}_{LOANS} = 12\% * (1 - 0.34) = 7.92\%$$

$$\text{Cost of Capital}_{RETAINEDEARNINGS} = 10\%$$

$$\text{Weighted Average Cost of Capital} = (0.25)*7.92\% + (0.75)*10.00\% = 9.48\%$$

A.5 TAX CONSIDERATIONS

A.5.1 After Tax Cash Flows

Taxes are a fact of life in both personal and business decision-making. Taxes occur in many forms and are primarily designed to generate revenues for governmental entities ranging from local authorities to the federal government. A few of the most common forms of taxes are income taxes, ad valorem taxes, sales taxes, and excise taxes. Cash flows used for economic analysis should always be adjusted for the combined impact of all relevant taxes. To do otherwise ignores the significant impact that taxes have on economic decision-making. Tax laws and regulations are complex and intricate. A detailed treatment of tax considerations as they apply to economic analysis is beyond the scope of this appendix and generally requires the assistance of a professional with specialized training in the subject. A high level summary of concepts and techniques that concentrate on federal income taxes is presented in the material which follows. The focus is on federal income taxes, since they impact most decisions and have relatively wide and general application.

The amount of federal taxes due are determined based on a tax rate multiplied by a taxable income. The rates (as of April 2000) are determined based on tables of rates published under the Omnibus Reconciliation Act of 1993 as shown in Table A-1. Depending on income range, the marginal tax rates vary from 15% of taxable income to 39% of taxable income. Taxable income is calculated by subtracting allowable deductions from gross income. Gross income is generated when a company sells its prod-

Table A-1. Federal tax rates based on the Omnibus Reconciliation Act of 1993

Taxable Income (TI)	Taxes Due	Marginal Tax Rate
$0 < TI ≤ $50,000	0.15*TI	0.15
$50,000 < TI ≤ $75,000	$7,500+0.25(TI-$50,000)	0.25
$75,000 < TI ≤ $100,000	$13,750+0.34(TI-$75,000)	0.34
$100,000 < TI ≤ $335,000	$22,250+0.39(TI-$100,000)	0.39
$335,000 < TI ≤ $10,000,000	$113,900+0.34(TI-$335,000)	0.34
$10,000,000 < TI ≤ $15,000,000	$3,400,000+0.35(TI-$10,000,000)	0.35
$15,000,000 < TI ≤ $18,333,333	$5,150,000+0.38(TI-$15,000,000)	0.38
$18,333,333 < TI	$6,416,667+0.35(TI-$18,333,333)	0.35

uct or service. Allowable deductions include salaries and wages, materials, interest payments, and depreciation, as well as other costs of doing business as detailed in the tax regulations.

The calculation of taxes owed and after tax cash flows (ATCF) requires knowledge of:

- Before Tax Cash Flows (BTCF), the net project cash flows before the consideration of taxes due, loan payments, and bond payments;
- Total loan payments attributable to the project, including a breakdown of principal and interest components of the payments;
- Total bond payments attributable to the project, including a breakdown of the redemption and interest components of the payments; and
- Depreciation allowances attributable to the project.

Given the availability of the above information, the procedure to determine the ATCF on a year-by-year basis proceeds by using the following calculation for each year:

- Taxable Income = BTCF – Loan Interest – Bond Interest – Depreciation

- Taxes = Taxable Income * Tax Rate

- ATCF = BTCF – Total Loan Payments – Total Bond Payments – Taxes

An important observation is that depreciation reduces taxable income (hence, taxes) but does not directly enter into the calculation of ATCF since it is not a true cash flow. It is not a true cash flow because no cash changes hands. Depreciation is an accounting concept designed to stimulate business by reducing taxes over the life of an asset. The next section provides additional information about depreciation.

A.5.2 Depreciation

Most assets used in the course of a business decrease in value over time. U.S. federal income tax law permits reasonable deductions from taxable income to allow for this. These deductions are called depreciation allowances. To be depreciable, an asset must meet three primary conditions: (1) it must be held by the business for the purpose of producing income, (2) it must wear out or be consumed in the course of its use, and (3) it must have a life longer than a year.

Many methods of depreciation have been allowed under U.S. tax law over the years. Among these methods are straight line, sum-of-the-years digits, declining balance, and the accelerated cost recovery system. Descriptions of these methods can be found in many references, including economic analysis text books [White, et al., 1998]. The method currently used for depreciation of assets placed in service after 1986 is the Modified Accelerated Cost Recovery System (MACRS). Determination of the allowable MACRS depreciation deduction for an asset is a function of (1) the asset's property class, (2) the asset's basis, and (3) the year within the asset's recovery period for which the deduction is calculated.

Eight property classes are defined for assets which are depreciable under MACRS. The property classes and several examples of property that fall into each class are shown in Table A-2. Professional tax guidance is recommended to determine the MACRS property class for a specific asset.

The basis of an asset is the cost of placing the asset in service. In most cases, the basis includes the purchase cost of the asset plus the costs

Table A-2. MACRS property classes

Property Class	Example Assets
3-Year Property	special handling devices for food special tools for motor vehicle manufacturing
5-Year Property	computers and office machines general purpose trucks
7-Year Property	office furniture most manufacturing machine tools
10-Year Property	tugs & water transport equipment petroleum refining assets
15-Year Property	fencing and landscaping cement manufacturing assets
20-Year Property	farm buildings utility transmission lines and poles
27.5-Year Residential Rental Property	rental houses and apartments
31.5-Year Nonresidential Real Property	business buildings

necessary to place the asset in service (e.g., installation charges).

Given an asset's property class and its depreciable basis, the depreciation allowance for each year of the asset's life can be determined from tabled values of MACRS percentages. The MACRS percentages specify the percentage of an asset's basis that are allowable as deductions during each year of an asset's recovery period. The MACRS percentages by recovery year (age of the asset) and property class are shown in Table A-3.

Table A-3. MACRS percentages by recovery year and property class

Recovery Year	3-Year Property	5-Year Property	7-Year Property	10-Year Property	15-Year Property	20-Year Property
1	33.33%	20.00%	14.29%	10.00%	5.00%	3.750%
2	44.45%	32.00%	24.49%	18.00%	9.50%	7.219%
3	14.81%	19.20%	17.49%	14.40%	8.55%	6.677%
4	7.41%	11.52%	12.49%	11.52%	7.70%	6.177%
5		11.52%	8.93%	9.22%	6.93%	5.713%
6		5.76%	8.92%	7.37%	6.23%	5.285%
7			8.93%	6.55%	5.90%	4.888%
8			4.46%	6.55%	5.90%	4.522%
9				6.56%	5.91%	4.462%
10				6.55%	5.90%	4.461%
11				3.28%	5.91%	4.462%
12					5.90%	4.461%
13					5.91%	4.462%
14					5.90%	4.461%
15					5.91%	4.462%
16					2.95%	4.461%
17						4.462%
18						4.461%
19						4.462%
20						4.461%
21						2.231%

Example 2

Determine depreciation allowances during each recovery year for a MACRS 5-year property with a basis of $10,000.

Year 1 deduction: $10,000 * 20.00% = $2,000
Year 2 deduction: $10,000 * 32.00% = $3,200
Year 3 deduction: $10,000 * 19.20% = $1,920
Year 4 deduction: $10,000 * 11.52% = $1,152
Year 5 deduction: $10,000 * 11.52% = $1,152
Year 6 deduction: $10,000 * 5.76% = $576

The sum of the deductions calculated in Example 2 is $10,000, which means that the asset is "fully depreciated" after six years. Though not shown here, tables similar to Table A-3 are available for the 27.5-year and 31.5-year property classes. Their usage is similar to that outlined above, except that depreciation is calculated monthly rather than annually.

A.6 TIME VALUE OF MONEY CONCEPTS

A.6.1 Introduction

Most people have an intuitive sense of the time value of money. Given a choice between $100 today and $100 one year from today, almost everyone would prefer the $100 today. Why is this the case? Two primary factors lead to this time preference associated with money; interest and inflation. Interest is the ability to earn a return on money which is loaned rather than consumed. By taking the $100 today and placing it in an interest bearing bank account (i.e., loaning it to the bank), one year from today an amount greater than $100 would be available for withdrawal. Thus, taking the $100 today and loaning it to earn interest, generates a sum greater than $100 one year from today and is thus preferred. The amount in excess of $100 that would be available depends upon the interest rate being paid by the bank. The next section develops the mathematics of the relationship between interest rates and the timing of cash flows.

The second factor which leads to the time preference associated with money is inflation. Inflation is a complex subject but in general can be described as a decrease in the purchasing power of money. The impact of inflation is that the "basket of goods" a consumer can buy today with $100 contains more than the "basket" the consumer could buy one year from today. This decrease in purchasing power is the result of inflation. The subject of inflation is addressed in Section A.9.4.

A.6.2 The Mathematics of Interest

The mathematics of interest must account for the amount and timing of cash flows. The basic formula for studying and understanding interest calculations is:

$$F_n = P + I_n$$

where: F_n = a future amount of money at the *end* of the nth year

P = a present amount of money at the beginning of the year which is n years prior to F_n

I_n = the amount of accumulated interest over n years

n = the number of years between P and F

The goal of studying the mathematics of interest is to develop a formula for F_n that is expressed only in terms of the present amount P, the annual interest rate i, and the number of years n. There are two major approaches for determining the value of I_n: simple interest and compound interest. Under simple interest, interest is earned (charged) only on the original amount loaned (borrowed). Under compound interest, interest is earned (charged) on the original amount loaned (borrowed) plus any interest accumulated from previous periods.

A.6.3 Simple Interest

For simple interest, interest is earned (charged) only on the original principal amount at the rate of i% per year (expressed as i%/yr). Table A-4 illustrates the annual calculation of simple interest. In Table A-4 and the formulas which follow, the interest rate i is to be expressed as a decimal amount (e.g., 8% interest is expressed as 0.08).

At the beginning of year 1 (end of year 0), P dollars (e.g., $100) are deposited in an account earning i%/yr (e.g., 8%/yr or 0.08) simple interest. Under simple compounding, during year 1 the P dollars ($100) earn

Table A-4. The mathematics of simple interest

Year (t)	Amount At Beginning Of Year	Interest Earned During Year	Amount At End Of Year (F_t)
0	-	-	P
1	P	Pi	P + Pi = P (1 + i)
2	P (1 + i)	Pi	P (1+ i) + Pi = P (1 + 2i)
3	P (1 + 2i)	Pi	P (1+ 2i) + Pi = P (1 + 3i)
n	P (1 + (n-1)i)	Pi	P (1+ (n-1)i) + Pi = P (1 + ni)

P*i dollars ($100*0.08 = $8) of interest. At the end of the year 1 the balance in the account is obtained by adding P dollars (the original principal, $100) plus P*i (the interest earned during year 1, $8) to obtain P+P*i ($100+$8 = $108). Through algebraic manipulation, the end of year 1 balance can be expressed mathematically as P*(1+i) dollars ($100*1.08 = $108).

The beginning of year 2 is the same point in time as the end of year 1 so the balance in the account is P*(1+i) dollars ($108). During year 2 the account again earns P*i dollars ($8) of interest, since under simple compounding interest is paid only on the original principal amount P ($100). Thus at the end of year 2, the balance in the account is obtained by adding P dollars (the original principal) plus P*i (the interest from year 1) plus P*i (the interest from year 2) to obtain P+P*i+P*i ($100+$8+$8 = $116). After some algebraic manipulation, this can conveniently be written mathematically as P*(1+2*i) dollars ($100*1.16 = $116).

Table A-4 extends the above logic to year 3 and then generalizes the approach for year n. If we return our attention to our original goal of developing a formula for F_n that is expressed only in terms of the present amount P, the annual interest rate i, and the number of years n, the above development and Table A-4 results can be summarized as follows:

For Simple Interest

$$F_n = P\ (1+n^*i)$$

Example 3

Determine the balance which will accumulate at the end of year 4 in an account which pays 10%/yr simple interest if a deposit of $500 is made today.

$F_n = P * (1 + n^*i)$

$F_4 = 500 * (1 + 4^*0.10)$

$F_4 = 500 * (1 + 0.40)$

$F_4 = 500 * (1.40)$

$F_4 = \$700$

A.6.4 Compound Interest

For compound interest, interest is earned (charged) on the original principal amount, plus any accumulated interest from previous years at the rate of i% per year (i%/yr). Table A-5 illustrates the annual calculation of compound interest. In the Table A-5 and the formulas which follow, i is expressed as a decimal amount (i.e., 8% interest is expressed as 0.08).

At the beginning of year 1 (end of year 0), P dollars (e.g., $100) are deposited in an account earning i%/yr (e.g., 8%/yr or 0.08) compound interest. Under compound interest, during year 1 the P dollars ($100) earn P*i dollars ($100*0.08 = $8) of interest. Notice that this the same as the amount earned under simple compounding. This result is expected since the interest earned in previous years is zero for year 1. At the end of the year 1 the balance in the account is obtain by adding P dollars (the original principal, $100) plus P*i (the interest earned during year 1, $8) to obtain P+P*i ($100+$8 = $108). Through algebraic manipulation, the end of year 1 balance can be expressed mathematically as P*(1+i) dollars ($100*1.08 = $108).

During year 2 and subsequent years, we begin to see the power (if you are a lender) or penalty (if you are a borrower) of compound interest over simple interest. The beginning of year 2 is the same point in time as the end of year 1 so the balance in the account is P*(1+i) dollars ($108). During year 2 the account earns i% interest on the original principal, P dollars ($100), *and* it earns i% interest on the accumulated interest from year 1, P*i dollars ($8). Thus the interest earned in year 2 is [P+P*i]*i dol-

Table A-5. The Mathematics of Compound Interest

Year (t)	Amount At Beginning Of Year	Interest Earned During Year	Amount At End Of Year (F_t)
0	-	-	P
1	P	Pi	P + Pi = P (1 + i)
2	P (1 + i)	P (1 + i) i	P (1+ i) + P (1 + i) i = P (1 + i) (1 + i) = $P (1+i)^2$
3	$P (1+i)^2$	$P (1+i)^2 i$	$P (1+ i)^2 + P (1 + i)^2 i$ $= P (1 + i)^2 (1 + i)$ $= P (1+i)^3$
n	$P (1+i)^{n-1}$	$P (1+i)^{n-1} i$	$P (1+ i)^{n-1} + P (1 + i)^{n-1} i$ $= P (1 + i)^{n-1} (1 + i)$ $= P (1+i)^n$

lars ([\$100+\$8]*0.08 = \$8.64). The balance at the end of year 2 is obtained by adding P dollars (the original principal) plus P*i (the interest from year 1) plus [P+P*i]*i (the interest from year 2) to obtain P+P*i+[P+P*i]*i dollars (\$100+\$8+\$8.64 = \$116.64). After some algebraic manipulation, this can be conveniently written mathematically as $P^*(1+i)^n$ dollars (\$100*1.082 = \$116.64).

Table A-5 extends the above logic to year 3 and then generalizes the approach for year n. If we return our attention to our original goal of developing a formula for F_n that is expressed only in terms of the present amount P, the annual interest rate i, and the number of years n, the above development and Table A-5 results can be summarized as follows:

For Compound Interest

$$F_n = P\,(1+i)^n$$

Example 4

Repeat Example 3 using compound interest rather than simple interest.

$F_n = P * (1 + i)^n$

$F_4 = 500 * (1 + 0.10)^4$

$F_4 = 500 * (1.10)^4$

$F_4 = 500 * (1.4641)$

$F_4 = \$732.05$

Notice that the balance available for withdrawal is higher under compound interest (\$732.05 > \$700.00). This is due to earning interest on principal plus interest rather than earning interest on just original principal. Since compound interest is by far more common in practice than simple interest, the remainder of this appendix is based on *compound interest* unless explicitly stated otherwise.

A.6.5 Single Sum Cash Flows

Time value of money problems involving compound interest are common. Because of this frequent need, tables of compound interest time value of money factors can be found in most books and reference manuals that deal with economic analysis. The factor $(1+i)^n$ is known as the *single sum, future worth factor,* or the *single payment, compound amount factor*. This factor is denoted (F | P,i,n), where F denotes a future amount, P denotes a present amount, i is an interest rate (expressed as a percentage amount), and n denotes a number of years. The factor (F | P,i,n) is read "to find F given P at i% for n years." Tables of values of (F | P,i,n) for selected values of i and n are provided in Appendix 4A. The tables of values in Appendix 4A are organized such that the annual interest rate (i) determines the appropriate page, the time value of money factor (F | P) determines the appropriate column, and the number of years (n) determines the appropriate row.

Example 5

Repeat Example 4 using the single sum, future worth factor.

$F_n = P * (1 + i)^n$
$F_n = P * (F | P,i,n)$
$F_4 = 500 * (F | P,10\%,4)$
$F_4 = 500 * (1.4641)$
$F_4 = 732.05$

The above formulas for compound interest allow us to solve for an unknown F, given P, i, and n. What if we want to determine P with known values of F, i, and n? We can derive this relationship from the compound interest formula above:

$$F_n = P\,(1+i)^n$$

Dividing both sides by $(1+i)^n$ yields

$$P = \frac{F_n}{(1+i)^n}$$

which can be rewritten as

$$P = F_n\,(1+i)^{-n}$$

The factor $(1+i)^{-n}$ is known as the single sum; present worth factor; or the single payment, present worth factor. This factor is denoted (P | F,i,n) and is read "to find P given F at i% for n years." Tables of (P | F,i,n) are provided in Appendix 4A.

Example 6

To accumulate $1000 five years from today in an account earning 8%/yr compound interest, how much must be deposited today?

$P = F_n * (1 + i)^{-n}$
P = F5 * (P | F,i,n)
P = 1000 * (P | F,8%,5)
P = 1000 * (0.6806)
P = 680.60

To verify your solution, try multiplying 680.60 * (F | P,8%,5). What would expect for a result? (Answer: $1000) If your still not convinced, try building a table like Table A-5 to calculate the year end balances each year for five years.

A.6.6 Series Cash Flows

Having considered the transformation of a single sum to a future worth when given a present amount and vice versa, let us generalize to a series of cash flows. The future worth of a series of cash flows is simply the sum of the future worths of each individual cash flow. Similarly, the present worth of a series of cash flows is the sum of the present worths of the individual cash flows.

Example 7

Determine the future worth (accumulated total) at the end of seven years in an account that earns 5%/yr if a $600 deposit is made today and a $1000 deposit is made at the end of year two?

for the $600 deposit, n = 7 (years between today and end of year 7)

for the $1000 deposit, n = 5 (years between end of year 2 and end of year 7)

F7 = 600 * (F | P,5%,7) + 1000 * (F | P,5%,5)

F7 = 600 * (1.4071) + 1000 * (1.2763)

F7 = 844.26 + 1276.30 = $2120.56

Example 8

Determine the amount that would have to be deposited today (present worth) in an account paying 6%/yr interest if you want to withdraw $500 four years from today and $600 eight years from today (leaving zero in the account after the $600 withdrawal).

For the $500 deposit n = 4, for the $600 deposit n = 8

P = 500 * (P | F,6%,4) + 600 * (P | F,6%,8)

P = 500 * (0.7921) + 600 * (0.6274)

P = 396.05 + 376.44 = $772.49

A.6.7 Uniform Series Cash Flows

A uniform series of cash flows exists when the cash flows are in a series, occur every year, and are all equal in value. Figure A-3 shows the cash flow diagram of a uniform series of withdrawals. The uniform series has length 4 and amount 2000. If we want to determine the amount of money that would have to be deposited today to support this series of withdrawals, starting one year from today, we could use the approach illustrated in Example 8 above to determine a present worth component for each individual cash flow. This approach would require us to sum the following series of factors (assuming the interest rate is 9% / yr):

P = 2000*(P | F,9%,1) + 2000*(P | F,9%,2) +
2000*(P | F,9%,3) + 2000*(P | F,9%,4)

After some algebraic manipulation, this expression can be restated as:

P = 2000*[(P | F,9%,1) + (P | F,9%,2) +
(P | F,9%,3) + (P | F,9%,4)]

P = 2000*[(0.9174) + (0.8417) + (0.7722) + (0.7084)]

P = 2000*[3.2397] = $6479.40

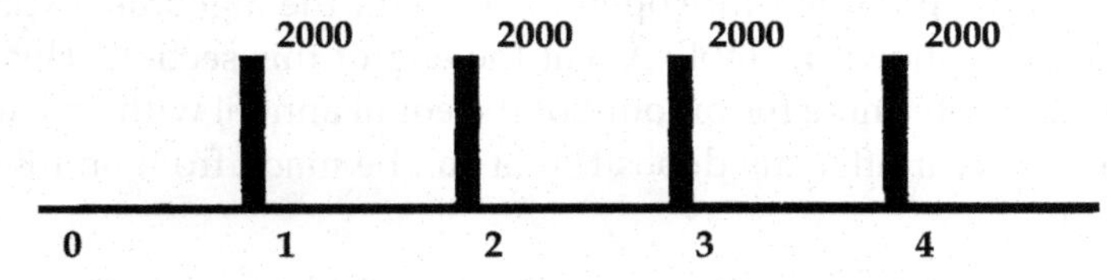

Figure A-3. Uniform series cash flow

Fortunately, uniform series occur frequently enough in practice to justify tabulating values to eliminate the need to repeatedly sum a series of (P | F,i,n) factors. To accommodate uniform series factors, we need to add a new symbol to our time value of money terminology in addition to the single sum symbols P and F. The symbol "A" is used to designate a uniform series of cash flows. When dealing with uniform series cash flows, the symbol A represents the amount of each annual cash flow and n represents the number of cash flows in the series. The factor (P | A,i,n) is known as the *uniform series, present worth factor* and is read "to find P given A at i% for n years." Tables of (P | A,i,n) are provided in Appendix 4A. An algebraic expression can also be derived for the (P | A,i,n) factor which expresses P in terms of A, i, and n. The derivation of this formula is omitted here, but the resulting expression is shown in the summary table (Table A-6) at the end of this section.

An important observation when using a (P | A,i,n) factor is that the "P" resulting from the calculation occurs one period prior to the first "A" cash flow. In our example the first withdrawal (the first "A") occurred one year after the deposit ("P"). Restating the example problem above using a (P | A,i,n) factor, it becomes:

P = A * (P | A,i,n)

P = 2000 * (P | A,9%,4)

P = 2000 * (3.2397) = $6479.40

This result is identical (as expected) to the result using the (P | F,i,n) factors. In both cases the interpretation of the result is: If we deposit $6479.40 in an account paying 9%/yr interest, we could make withdrawals of $2000 per year for four years, starting one year after the initial deposit, to deplete the account at the end of 4 years.

The reciprocal relationship between P and A is symbolized by the factor (A | P,i,n) and is called the *uniform series, capital recovery factor*. Tables of (A | P,i,n) are provided in Appendix 4A, and the algebraic expression for (A | P,i,n) is shown in Table A-6 at the end of this section. This factor enables us to determine the amount of the equal annual withdrawals "A" (starting one year after the deposit) that can be made from an initial deposit of "P."

Example 9

Determine the equal annual withdrawals that can be made for 8 years from an initial deposit of $9000 in an account that pays 12%/yr. The first withdrawal is to be made one year after the initial deposit.

A = P * (A | P,12%,8)

A = 9000 * (0.2013)

A = $1811.70

Factors are also available for the relationships between a future worth (accumulated amount) and a uniform series. The factor (F | A,i,n) is known as the *uniform series future worth* factor and is read "to find F given A at i% for n years." The reciprocal factor, (A | F,i,n), is known as the *uniform series sinking fund* factor and is read "to find A given F at i% for n years." An important observation when using an (F | A,i,n) factor or an (A | F,i,n) factor is that the "F" resulting from the calculation occurs at the same point in time as to the last "A" cash flow. The algebraic expressions for (A | F,i,n) and (F | A,i,n) are shown in Table 6 at the end of this section.

Example 10

If you deposit $2000 per year into an individual retirement account starting on your 24th birthday, how much will have accumulated in the account at the time of your deposit on your 65th birthday? The account pays 6% / yr.

n = 42 (birthdays between 24th and 65th, inclusive)

F = A * (F | A,6%,42)

F = 2000 * (175.9505) = $351,901

Example 11

If you want to be a millionaire on your 65th birthday, what equal annual deposits must be made in an account starting on your 24th birthday? The account pays 10% / yr.

n = 42 (birthdays between 24th and 65th, inclusive)

A = F * (A | F,10%,42)

A = 1000000 * (0.001860) = $1860

A.6.8 Gradient Series

A gradient series of cash flows occurs when the value of a given cash flow is greater than the value of the previous period's cash flow by a constant amount. The symbol used to represent the constant increment is G. The factor (P | G,i,n) is known as the *gradient series, present worth factor*.

Tables of (P | G,i,n) are provided in Appendix 4A. An algebraic expression can also be derived for the (P | G,i,n) factor, which expresses P in terms of G, i, and n. The derivation of this formula is omitted here, but the resulting expression is shown in the summary table (Table A-6) at the end of this section.

It is not uncommon to encounter a cash flow series that is the sum of a uniform series and a gradient series. Figure A-4 illustrates such a series. The uniform component of this series has a value of 1000 and the gradient series has a value of 500. By convention the first element of a gradient series has a zero value. Therefore, in Figure A-4, both the uniform series and the gradient series have length four (n = 4). Like the uniform series factor, the "P" calculated by a (P | G,i,n) factor is located one period before the first element of the series (which is the zero element for a gradient series).

Example 12

Assume you wish to make the series of withdrawals illustrated in Figure A-4 from an account which pays 15%/yr. How much money would you have to deposit today such that the account is depleted at the time of the last withdrawal?

This problem is best solved by recognizing that the cash flows are a combination of a uniform series of value 1000 and length 4 (starting at time = 1), plus a gradient series of size 500 and length 4 (starting at time = 1).

P = A * (P | A,15%,4) + G * (P | G,15%,4)
P = 1000 * (2.8550) + 500 * (3.7864)
P = 2855.00 + 1893.20 = $4748.20

Occasionally it is useful to convert a gradient series to an equivalent

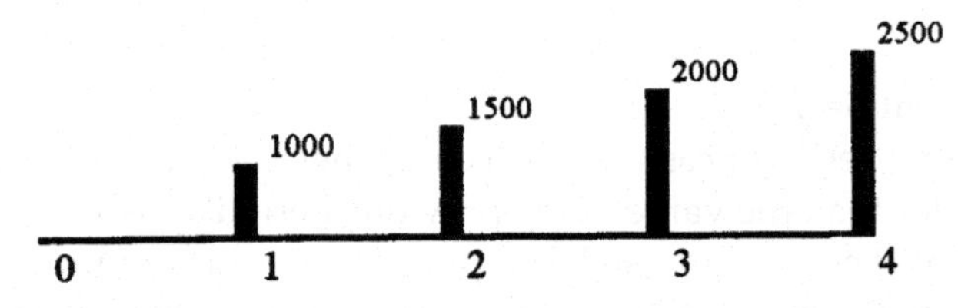

Figure A-4. Combined uniform series and gradient series cash flow

uniform series of the same length. Equivalence in this context means that the present value (P) calculated from the gradient series is numerically equal to the present value (P) calculated from the uniform series. One way to accomplish this task with the time value of money factors we have already considered is to convert the gradient series to a present value using a (P | G,i,n) factor and then convert this present value to a uniform series using an (A | P,i,n) factor. In other words:

$$A = [G * (P \mid G,i,n)] * (A \mid P,i,n)$$

An alternative approach is to use a factor known as the *gradient-to-uniform series conversion factor*, symbolized by (A | G,i,n). Tables of (A | G,i,n) are provided in Appendix 4A. An algebraic expression can also be derived for the (A | G,i,n) factor, which expresses A in terms of G, i, and n. The derivation of this formula is omitted here, but the resulting expression is shown in the summary table (Table A-6) at the end of this section.

A.6.9 Summary of Time Value of Money Factors

Table A-6 summarizes the time value of money factors introduced in this section. Time value of money factors are useful in economic analysis, because they provide a mechanism to accomplish two primary functions: (1) they allow us to replace a cash flow at one point in time with an equivalent cash flow (in a time value of money sense) at a different point in time and (2) they allow us to convert one cash flow pattern to another (e.g., convert a single sum of money to an equivalent cash flow series or convert a cash flow series to an equivalent single sum). The usefulness of these two functions when performing economic analysis of alternatives will become apparent in Sections A.7 and A.8, which follow.

A.6.10 The Concepts of Equivalence and Indifference

Up to this point the term "equivalence" has been used several times but never fully defined. It is appropriate at this point to formally define equivalence, as well as a related term, "indifference."

In economic analysis, "equivalence" means "the state of being equal in value." The concept is primarily applied to the comparison of two or more cash flow profiles. Specifically, two (or more) cash flow profiles are equivalent if their time value of money worths at a common point in time are equal.

Table A-6 Summary of discrete compounding time value of money factors

To Find	Given	Factor	Symbol	Name
P	F	$(1+i)^{-n}$	(P | F,i,n)	Single Payment, Present Worth Factor
F	P	$(1+i)^{n}$	(F | P,i,n)	Single Payment, Compound Amount Factor
P	A	$\frac{(1+i)^n-1}{i(1+i)^n}$	(P | A,i,n)	Uniform Series, Present Worth Factor
A	P	$\frac{i(1+i)^n}{(1+i)^n-1}$	(A | P,i,n)	Uniform Series, Capital Recovery Factor
F	A	$\frac{(1+i)^n-1}{i}$	(F | A,i,n)	Uniform Series, Compound Amount Factor
A	F	$\frac{i}{(1+i)^n-1}$	(A | F,i,n)	Uniform Series, Sinking Fund Factor
P	G	$\frac{1-(1+ni)(1+i)^{-n}}{i^2}$	(P | G,i,n)	Gradient Series, Present Worth Factor
A	G	$\frac{(1+i)^n-(1+ni)}{i[(1+i)^n-1]}$	(A | G,i,n)	Gradient Series, Uniform Series Factor

Question: Are the following two cash flows equivalent at 15%/yr?
Cash Flow 1: Receive $1,322.50 two years from today
Cash Flow 2: Receive $1,000.00 today

Analysis Approach 1: Compare worths at t = 0 (present worth).
PW(1) = 1,322.50*(P | F,15,2) = 1322.50*0.756147 = 1,000 PW(2) = 1,000
Answer: Cash Flow 1 and Cash Flow 2 are equivalent.

Analysis Approach 2: Compare worths at t = 2 (future worth).
FW(1) = 1,322.50
FW(2) = 1,000*(F | P,15,2) = 1,000*1.3225 = 1,322.50
Answer: Cash Flow 1 and Cash Flow 2 are equivalent.

Generally the comparison (hence the determination of equivalence) for the two cash flow series in this example would be made as present worths (t = 0) or future worths (t = 2), but the equivalence definition holds regardless of the point in time chosen. For example:

Analysis Approach 3: Compare worths at t = 1.
W1(1) = 1,322.50*(P | F,15,1)
= 1,322.50*0.869565 = 1,150.00
W1(2) = 1,000*(F | P,15,1) = 1,000*1.15 = 1,150.00
Answer: Cash Flow 1 and Cash Flow 2 are equivalent.

Thus, the selection of the point in time, t, at which to make the comparison is completely arbitrary. Clearly, however, some choices are more intuitively appealing than others (t = 0 and t = 2 in the above example).

In economic analysis, "indifference" means "to have no preference." The concept is primarily applied in the comparison of two or more cash flow profiles. Specifically, a potential investor is indifferent between two (or more) cash flow profiles if they are equivalent.

Question: Given the following two cash flows at 15%/yr, which do you prefer?
Cash Flow 1: Receive $1,322.50 two years from today
Cash Flow 2: Receive $1,000.00 today
Answer: Based on the equivalence calculations above, given these two choices, an investor is indifferent.

The concept of equivalence can be used to break a large, complex problem into a series of smaller, more manageable ones. This is done by taking advantage of the fact that, in calculating the economic worth of a cash flow profile, any part of the profile can be replaced by an equivalent representation at an arbitrary point in time without altering the worth of the profile.

Question: You are given a choice between (1) receiving P dollars today or (2) receiving the cash flow series illustrated in Figure A-5. What must the value of P be for you to be indifferent between the two choices if i = 12%/yr?

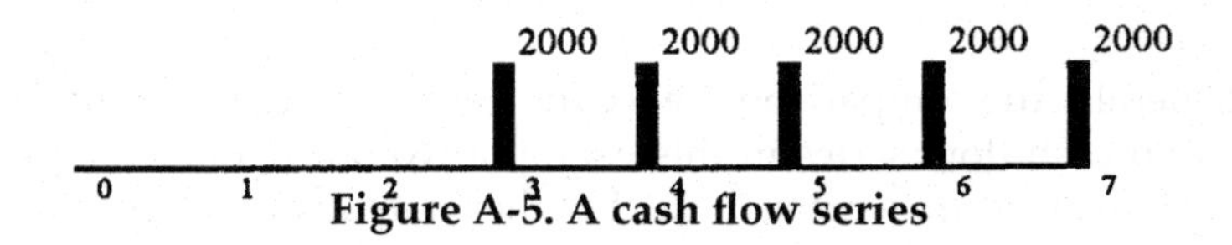

Figure A-5. A cash flow series

Analysis Approach: To be indifferent between the choices, P must have a value such that the two alternatives are equivalent at 12%/yr. If we select t = 0 as the common point in time upon which to base the analysis (present worth approach), then the analysis proceeds as follows:

PW(Alt 1) = P

Because P is already at t = 0 (today), no time value of money factors are involved.

PW(Alt 2)

Step 1— Replace the uniform series (t = 3 to 7) with an equivalent single sum, V_2, at t = 2 (one period before the first element of the series). V_2 = 2,000 * (P | A,12%,5) = 2,000 * 3.6048 = 7,209.60

Step 2— Replace the single sum V_2, with an equivalent value V_0 at t = 0. PW(Alt 2) = V_0 = V_2 * (P | F,12,2) = 7,209.60 * 0.7972 = 5,747.49

Answer: To be indifferent between the two alternatives, they must be equivalent at t = 0. To be equivalent, P must have a value of $5,747.49.

A.7 PROJECT MEASURES OF WORTH

A.7.1 Introduction

In this section measures of worth for investment projects are introduced. The measures are used to evaluate the attractiveness of a single investment opportunity. The measures to be presented are (1) present worth, (2) annual worth, (3) internal rate of return, (4) savings investment ratio, and (5) payback period. All but one of these measures of worth require an interest rate to calculate the worth of an investment. This interest rate is commonly referred to as the minimum attractive rate of return (MARR). There are many ways to determine a value of MARR for investment analysis, and no one way is proper for all applications. One principle is, however, generally accepted. MARR should always exceed the cost of capital as described in Section A.4, Sources of Funds, presented earlier in this appendix.

In all of the measures of worth below, the following conventions are used for defining cash flows. At any given point in time ($t = 0, 1, 2,\ldots, n$), there may exist both revenue (positive) cash flows, Rt, and cost (negative) cash flows, C_t. The net cash flow at t, A_t, is defined as $R_t - C_t$.

A.7.2 Present Worth

Consider again the cash flow series illustrated in Figure A-5. If you were given the opportunity to "buy" that cash flow series for $5,747.49, would you be interested in purchasing it? If you expected to earn a 12%/yr return on your money (MARR = 12%), based on the analysis in the previous section, your conclusion should be that you are indifferent between (1) retaining your $5,747.49 and (2) giving up your $5,747.49 in favor of the cash flow series. Figure A-6 illustrates the net cash flows of this second investment opportunity.

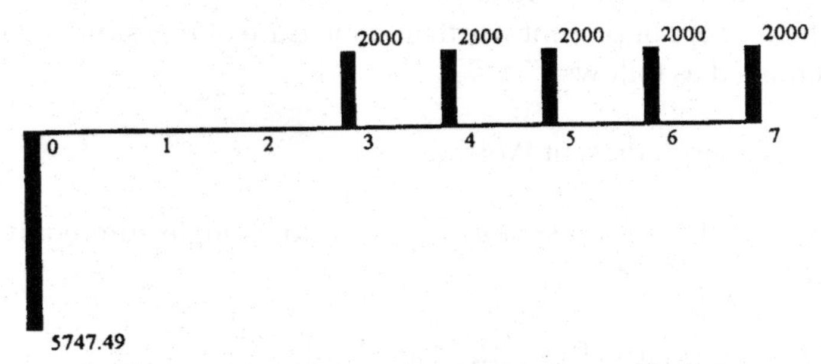

Figure A-6 An investment opportunity

What value would you expect if we calculated the present worth (equivalent value of all cash flows at t = 0) of Figure A-6? We must be careful with the signs (directions) of the cash flows in this analysis since some represent cash outflows (downward) and some represent cash inflows (upward).

PW = – 5747.49 + 2000*(P | A,12%,5)*(P | F,12%,2)

PW = – 5747.49 + 2000*(3.6048)*(0.7972)

PW = – 5747.49 + 5747.49 = $0.00

The value of zero for present worth indicates indifference regarding this investment opportunity. We would just as soon do nothing (i.e., retain our $5747.49) as invest in the opportunity.

What if the same returns (future cash inflows) were offered for a $5000 investment (t = 0 outflow)? Would this be more, or less attractive? Hopefully, after a little reflection, it is apparent that this would be a more attractive investment, because you are getting the same returns but paying less than the indifference amount for them. What happens if we calculate the present worth of this new opportunity?

PW = – 5000 + 2000*(P | A,12%,5)*(P | F,12%,2)

PW = – 5000 + 2000*(3.6048)*(0.7972)

PW = – 5000.00 + 5747.49 = $747.49

The positive value of present worth indicates an attractive investment. If we repeat the process with an initial cost greater than $5747.49, it should come as no surprise that the present worth will be negative, indicating an unattractive investment.

The concept of present worth as a measure of investment worth can be generalized as follows:

<u>Measure of Worth</u>: Present Worth

<u>Description</u>: All cash flows are converted to a single sum equivalent at time zero using i = MARR.

<u>Calculation Approach</u>: $PW = \sum_{t=1}^{n} A_t(P \mid F,i,t)$

Decision Rule: If PW ≥0, then the investment is attractive.

Example 13

Installing thermal windows on a small office building is estimated to cost $10,000. The windows are expected to last six years and have no salvage value at that time. The energy savings from the windows are expected to be $2525 each year for the first three years, and $3840 for each of the remaining three years. If MARR is 15%/yr and the present worth measure of worth is to be used, is this an attractive investment?

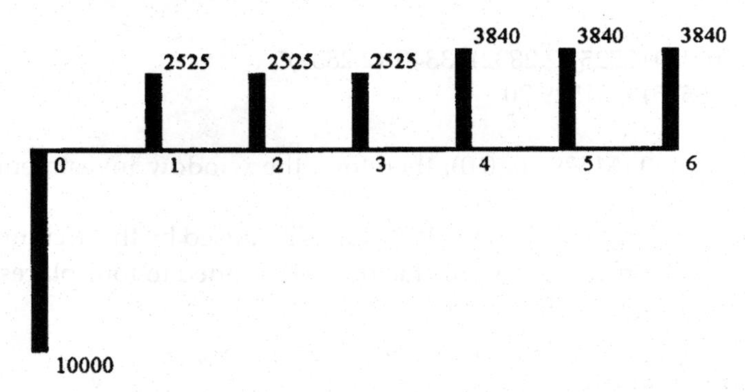

Figure A-7. Thermal windows investment

The cash flow diagram for the thermal windows is shown in Figure A-7.

PW =

−10000+2525*(P | F,15%,1)+2525*(P | F,15%,2)
+2525*(P | F,15%,3)+3840*(P | F,15%,4)+
3840*(P | F,15%,5)+
3840*(P | F,15%,6)

PW =

−10000+2525*(0.8696)+2525*(0.7561)
+2525*(0.6575)+
3840*(0.5718)+3840*(0.4972)+
3840*(0.4323)

PW =

−10000+2195.74+1909.15+1660.19+2195.71
+1909.25+1660.03

PW = $1530.07

Decision: PW≥0 ($1530.07≥0.0); therefore, the window investment is attractive.

An alternative (and simpler) approach to calculating PW is obtained by recognizing that the savings cash flows are two uniform series, one of value $2525 and length 3 starting at t = 1, and one of value $3840 and length 3 starting at t = 4.

PW = –10000+2525*(P | A,15%,3)+3840*
(P | A,15%,3)*(P | F,15%,3)

PW = –10000+2525*(2.2832)+3840*(2.2832)*
(0.6575) = $1529.70

Decision: PW≥0 ($1529.70>0.0); therefore, the window investment is attractive.

The slight difference in the PW values is caused by the accumulation of round off errors as the various factors are rounded to four places to the right of the decimal point.

A.7.3 Annual Worth

An alternative to present worth is annual worth. The annual worth measure converts all cash flows to an equivalent uniform annual series of cash flows over the investment life, using i = MARR. The annual worth measure is generally calculated by first calculating the present worth measure and then multiplying by the appropriate (A | P,i,n) factor. A thorough review of the tables in Appendix 4A or the equations in Table A-6 leads to the conclusion that for all values of i (i>0) and n (n>0), the value of (A | P,i,n) is greater than zero. Hence,

if PW>0, then AW>0;

if PW<0, then AW<0; and

if PW = 0, then AW = 0,

because the only difference between PW and AW is multiplication by a positive, non-zero value, namely (A | P,i,n). The decision rule for investment attractiveness for PW and AW are identical: positive values indicate an attractive investment; negative values indicate an unattractive investment; zero indicates indifference. Frequently the only reason for choosing between AW and PW as a measure of worth in an analysis is the preference of the decision maker.

The concept of annual worth as a measure of investment worth can be generalized as follows:

Measure of Worth: Annual Worth

Description: All cash flows are converted to an equivalent uniform annual series of cash flows over the planning horizon, using i = MARR.

Calculation Approach: AW = PW (A | P,i,n)

Decision Rule: If AW ≥0, then the investment is attractive.

Example 14

Reconsider the thermal window data of Example 13. If the annual worth measure of worth is to be used, is this an attractive investment?

AW = PW (A | P,15%,6)
AW = 1529.70 (0.2642) = $404.15/yr
Decision: AW ≥0 ($404.15>0.0); therefore, the window investment is attractive.

A.7.4 Internal Rate of Return

One of the problems associated with using the present worth or the annual worth measures of worth is that they depend upon knowing a value for MARR. As mentioned in the introduction to this section, the "proper" value for MARR is a much debated topic and tends to vary from company to company and decision-maker to decision-maker. If the value of MARR changes, the value of PW or AW must be recalculated to determine whether the attractiveness/unattractiveness of an investment has changed.

The internal rate of return (IRR) approach is designed to calculate a rate of return that is "internal" to the project. That is,

if IRR > MARR, the project is attractive;
if IRR < MARR, the project is unattractive; and
if IRR = MARR, the project is indifferent.

Thus, if MARR changes, no new calculations are required. We simply compare the calculated IRR for the project to the new value of MARR, and we have our decision.

The value of IRR is typically determined through a trial and error process. An expression for the present worth of an investment is written without specifying a value for i in the time value of money factors. Then various values of i are substituted until a value is found that sets the present worth (PW) equal to zero. The value of i found in this way is the IRR.

As appealing as the flexibility of this approach is, there are two major drawbacks. First, the iterations required to solve using the trial and error approach to solution can be time consuming. This factor is mitigated by the fact that most spreadsheets and financial calculators are pre-programmed to solve for an IRR value, given a cash flow series. The second (and more serious) drawback to the IRR approach is that some cash flow series have more than one value of IRR—i.e., more than one value of i sets the PW expression to zero. A detailed discussion of this multiple solution issue is beyond the scope of this appendix but can be found in White, et al. [1998], as well as most other economic analysis references. However, it can be shown that, if a cash flow series consists of an initial investment (negative cash flow at $t = 0$) followed by a series of future returns (positive or zero cash flows for all $t>0$) then a unique IRR exists. If these conditions are not satisfied, a unique IRR is not guaranteed and caution should be exercised in making decisions based on IRR.

The concept of internal rate of return as a measure of investment worth can be generalized as follows:

Measure of Worth: Internal Rate of Return

Description: An interest rate, IRR, is determined which yields a present worth of zero. IRR implicitly assumes the reinvestment of recovered funds at IRR.

Calculation Approach:

$$\text{Find IRR such that PW} = \sum_{t=1}^{n} A_t(P \mid F, IRR, t) = 0.$$

Important Note: Depending upon the cash flow series, multiple IRRs may exist! If the cash flow series consists of an initial investment (net negative cash flow) followed by a series of future returns (net non-negative cash flows), then a unique IRR exists.

<u>Decision Rule</u>: If IRR is unique and IRR ≥MARR, then the investment is attractive.

Example 15

Reconsider the thermal window data of Example 13. If the internal rate of return measure of worth is to be used, is this an attractive investment?

> First we note that the cash flow series has a single negative investment, followed by all positive returns; therefore, it has a unique value for IRR. For such a cash flow series it can also be shown that as i increases PW decreases.

From example 11, we know that for i = 15%:

PW = –10000+2525*(P | A,15%,3)+3840*(P | A,15%,3)*(P | F,15%,3)

PW = –10000+2525*(2.2832)+3840*(2.2832)*(0.6575) = $1529.70

Because PW>0, we must increase i to decrease PW toward zero for i = 18%:

PW = –10000+2525*(P | A,18%,3)+3840*(P | A,18%,3)*(P | F,18%,3)

PW = –10000+2525*(2.1743)+3840*(2.1743)*(0.6086) = $571.50

Since PW>0, we must increase i to decrease PW toward zero for i = 20%:

PW = –10000+2525*(P | A,20%,3)+3840*(P | A,20%,3)*(P | F,20%,3)

PW = –10000+2525*(2.1065)+3840*(2.1065)*(0.5787) = –$0.01

Although we could interpolate a value of i for which PW = 0 (rather than –0.01), for practical purposes PW = 0 at i = 20%; therefore, IRR = 20%.

Decision: IRR≥MARR (20%>15%); therefore, the window investment is attractive.

A.7.5 Saving Investment Ratio

Many companies are accustomed to working with benefit cost ratios. An investment measure of worth, which is consistent with the present worth measure and has the form of a benefit cost ratio, is the savings investment ratio (SIR). The SIR decision rule can be derived from the present worth decision rule as follows:

Starting with the PW decision rule

$$PW \geq 0$$

replacing PW with its calculation expression

$$\sum_{t=0}^{n} A_t(P\,|\,F,i,t) \geq 0$$

which, using the relationship $A_t = R_t - C_t$, can be restated

$$\sum_{t=0}^{n} (R_t - C_t)(P\,|\,F,i,t) \geq 0$$

which can be algebraically separated into

$$\sum_{t=0}^{n} R_t(P\,|\,F,i,t) - \sum_{t=0}^{n} C_t(P\,|\,F,i,t) \geq 0$$

adding the second term to both sides of the inequality

$$\sum_{t=0}^{n} R_t(P\,|\,F,i,t) \geq \sum_{t=0}^{n} C_t(P\,|\,F,i,t)$$

dividing both sides of the inequality
by the right side term

$$\frac{\sum_{t=0}^{n} R_t(P \mid F,i,t)}{\sum_{t=0}^{n} C_t(P \mid F,i,t)} \geq 1$$

which is the decision rule for SIR.

The SIR represents the ratio of the present worth of the revenues to the present worth of the costs. If this ratio exceeds one, the investment is attractive.

The concept of savings investment ratio as a measure of investment worth can be generalized as follows:

Measure of Worth: Savings Investment Ratio

Description: The ratio of the present worth of positive cash flows to the present worth of (the absolute value of) negative cash flows is formed using i = MARR.

Calculation Approach: $SIR = \frac{\sum_{t=0}^{n} R_t(P \mid F,i,t)}{\sum_{t=0}^{n} C_t(P \mid F,i,t)}$

Decision Rule: If SIR ≥1, then the investment is attractive.

Example 16

Reconsider the thermal window data of Example 13. If the savings investment ratio measure of worth is to be used, is this an attractive investment?

From example 13, we know that for i = 15%:

$$SIR = \frac{\sum_{t=0}^{n} R_t(P \mid F,i,t)}{\sum_{t=0}^{n} C_t(P \mid F,i,t)}$$

$$\text{SIR} = \frac{2525*(P \mid A, 15\%,3) + 3840*(P \mid A, 15\%,3)*(P \mid F, 15\%,3)}{10000}$$

$$\text{SIR} = \frac{11529.70}{10000.00} = 1.15297$$

Decision: SIR≥1.0 (1.15297>1.0); therefore, the window investment is attractive.

An important observation regarding the four measures of worth presented to this point (PW, AW, IRR, and SIR) is that they are all consistent and equivalent. In other words, an investment that is attractive under one measure of worth will be attractive under each of the other measures of worth. A review of the decisions determined in Examples 13 through 16 will confirm this observation. Because of their consistency, it is not necessary to calculate more than one measure of investment worth to determine the attractiveness of a project. The rationale for presenting multiple measures which are essentially identical for decision-making is that various individuals and companies may have a preference for one approach over another.

A.7.6 Payback Period

The payback period of an investment is generally taken to mean the number of years required to recover the initial investment through net project returns. The payback period is a popular measure of investment worth and appears in many forms in economic analysis literature and company procedure manuals. Unfortunately, all too frequently, payback period is used inappropriately and leads to decisions which focus exclusively on short term results and ignore time value of money concepts. After presenting a common form of payback period, these shortcomings will be discussed.

Measure of Worth: Payback Period

Description: The number of years required to recover the initial investment by accumulating net project returns is determined.

Calculation Approach:

PBP = the smaller *m* such that $\sum_{t=1}^{m} A_t \geq C_0$

Decision Rule: If PBP is less than or equal to a predetermined limit (often called a hurdle rate), then the investment is attractive.

Important Note: This form of payback period ignores the time value of money and ignores returns beyond the predetermined limit.

The fact that this approach ignores time value of money concepts is apparent by the fact that no time value of money factors are included in the determination of m. This implicitly assumes that the applicable interest rate to convert future amounts to present amounts is zero. This implies that people are indifferent between $100 today and $100 one year from today, which is an implication that is highly inconsistent with observable behavior.

The short-term focus of the payback period measure of worth can be illustrated using the cash flow diagrams of Figure A-8. Applying the PBP approach above yields a payback period for investment (a) of PBP = 2 (1200>1000 @ t = 2) and a payback period for investment (b) of PBP = 4 (1000300>1000) @ t = 4). If the decision hurdle rate is 3 years (a very common rate), then investment (a) is attractive but investment (b) is not. Hopefully, it is obvious that judging (b) unattractive is not good decision-making, since a $1,000,000 return four years after a $1,000 investment is attractive under almost any value of MARR. In point of fact, the IRR for (b) is 465%, so for any value of MARR less than 465%, investment (b) is attractive.

A.8 ECONOMIC ANALYSIS

A.8.1 Introduction

The general scenario for economic analysis is that a set of investment alternatives are available, and a decision must be made regarding which ones (if any) to accept and which ones (if any) to reject. If the analysis is deterministic, then an assumption is made that cash flow amounts, cash flow timing, and MARR are known with certainty. Frequently, although this assumption does not hold exactly, it is not considered restrictive in terms of potential investment decisions. If, however, the lack of certainty

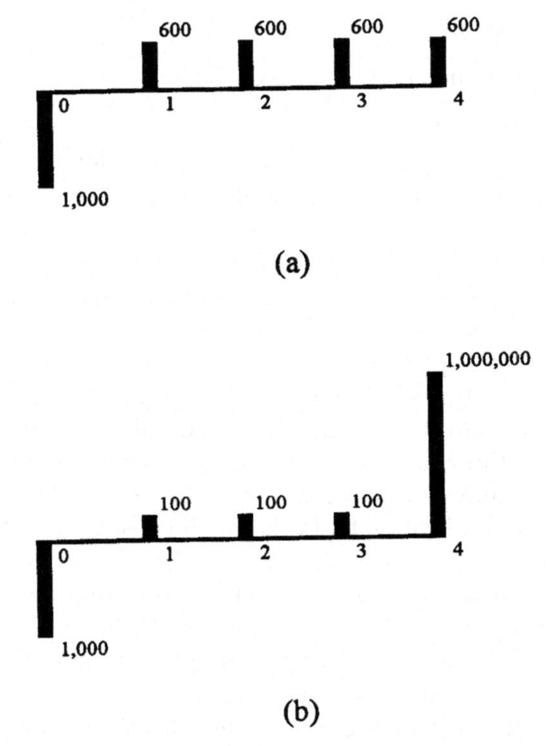

Figure A-8. Two investments evaluated using payback period

is a significant issue, then the analysis is stochastic and the assumptions of certainty are relaxed using probability distributions and statistical techniques to conduct the analysis. The remainder of this section deals with deterministic economic analysis, so the assumption of certainty will be assumed to hold. Stochastic techniques are introduced in Section A.9.5.

A.8.2 Deterministic Unconstrained Analysis

Deterministic economic analysis can be further classified into unconstrained deterministic analysis and constrained deterministic analysis. Under unconstrained analysis, all projects within the set available are assumed to be independent. The practical implication of this independence assumption is that an accept/reject decision can be made on each project without regard to the decisions made on other projects. In general this requires that (1) there are sufficient funds available to undertake all proposed projects, (2) there are no mutually exclusive projects, and (3) there are no contingent projects.

A funds restriction creates dependency, since before deciding on a project being evaluated, the evaluator would have to know what decisions had been made on other projects to determine whether sufficient funds were available to undertake the current project. Mutual exclusion creates dependency, since acceptance of one of the mutually exclusive projects precludes acceptance of the others. Contingency creates dependence, since prior to accepting a project, all projects on which it is contingent must be accepted.

If none of the above dependency situations are present and the projects are otherwise independent, then the evaluation of the set of projects is done by evaluating each individual project in turn and accepting the set of projects which were individually judged acceptable. This accept or reject judgment can be made using either the PW, AW, IRR, or SIR measure of worth. The unconstrained decision rules for each or these measures of worth are restated below for convenience:

Unconstrained PW Decision Rule: If PW ≥ 0, then the project is attractive.

Unconstrained AW Decision Rule: If AW ≥ 0, then the project is attractive.

Unconstrained IRR Decision Rule: If IRR is unique and IRR $\geq$MARR, then the project is attractive.

Unconstrained SIR Decision Rule: If SIR ≥ 1, then the project is attractive.

Example 17

Consider the set of four investment projects whose cash flow diagrams are illustrated in Figure A-9. If MARR is 12%/yr and the analysis is unconstrained, which projects should be accepted?

Using present worth as the measure of worth:

$PW_A = -1000+600*(P \mid A,12\%,4) = -1000+600(3.0373) = \$822.38 \Rightarrow$ Accept A

$PWB = -1300+800*(P \mid A,12\%,4) = -1300+800(3.0373) = \$1129.88 \Rightarrow$ Accept B

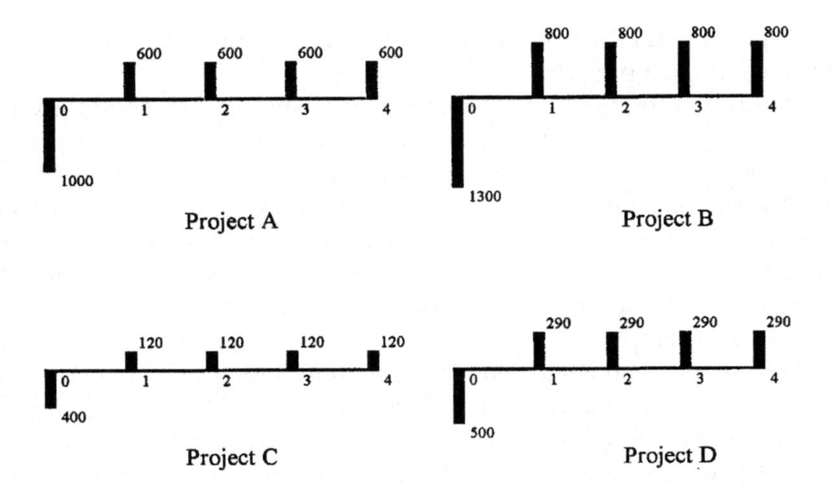

Figure A-9 Four investments projects

PWC = –400+120*(P | A,12%,4) = –400+120(3.0373) = –$35.52 ⇒ Reject C

PWD = –500+290*(P | A,12%,4) = –500+290(3.0373) = $380.83 ⇒ Accept D

Therefore,

Accept Projects A, B, and D and reject Project C

A.8.3 Deterministic Constrained Analysis

Constrained analysis is required any time a dependency relationship exists between any of the projects within the set to be analyzed. In general, dependency exists any time (1) there are insufficient funds available to undertake all proposed projects (commonly referred to as capital rationing), (2) there are mutually exclusive projects, or (3) there are contingent projects.

Several approaches have been proposed for selecting the best set of projects from a set of potential projects under constraints. Many of these approaches will select the optimal set of acceptable projects under some conditions (or will select a set that is near optimal). However, only a few approaches are guaranteed to select the optimal set of projects under all conditions. One of these approaches is presented below by way of a continuation of Example 17.

The first steps in the selection process are to specify the cash flow amounts and cash flow timings for each project in the potential project set. Additionally, a value of MARR to be used in the analysis must be specified. These issues have been addressed in previous sections, so further discussion will be omitted here. The next step is to form the set of all possible decision alternatives from the projects. A single decision alternative is a collection of zero, one, or more projects which could be accepted (all others not specified are to be rejected). As an illustration, the possible decision alternatives for the set of projects illustrated in Figure A-9 are listed in Table A-7. As a general rule, there will be 2^n possible decision alternatives generated from a set of n projects. Thus, for the projects of Figure A-9, there are $2^4 = 16$ possible decision alternatives. Since this set represents all possible decisions that could be made,

Table A-7. The decision alternatives from four projects

Accept A only Accept B only Accept C only Accept D only
Accept A and B only Accept A and C only Accept A and D only Accept B and C only Accept B and D only Accept C and D only
Accept A, B, and C only Accept A, B, and D only Accept A, C, and D only Accept B, C, and D only
Accept A, B, C, and D (frequently called the do everything alternative)
Accept none (frequently called the do nothing or null alternative)

one, and only one, will be selected as the best (optimal) decision. The set of decision alternatives developed in this way has the properties of being collectively exhaustive (all possible choices are listed) and mutually exclusive (only one will be selected).

The next step in the process is to eliminate decisions from the collectively exhaustive, mutually exclusive set that represent choices which would violate one (or more) of the constraints on the projects. For the projects of Figure A-9, assume the following two constraints exist:

Project B is contingent on Project C, and

A budget limit of \$1500 exists on capital expenditures at t = 0.

Based on these constraints the following decision alternatives must be removed from the collectively exhaustive, mutually exclusive set: any combination that includes B but not C (B only, A&B, B&D, A&B&D), any combination not already eliminated whose t = 0 costs exceed \$1500 (B&C, A&B&C, A&C&D, B&C&D, A&B&C&D). Thus, from the original set of 16 possible decision alternatives, 9 have been eliminated and need not be evaluated. These results are illustrated in Table A-8. It is frequently the case in practice that a significant percentage of the original collectively exhaustive, mutually exclusive set will be eliminated before measures of worth are calculated.

The next step is to create the cash flow series for the remaining (feasible) decision alternatives. This is a straight forward process and is accomplished by setting a decision alternative's annual cash flow equal to the sum of the annual cash flows (on a year by year basis) of all projects contained in the decision alternative. Table A-9 illustrates the results of this process for the feasible decision alternatives from Table A-8.

The next step is to calculate a measure of worth for each decision alternative. Any of the four consistent measures of worth presented previously (PW, AW, IRR, or SIR but NOT PBP) can be used. The measures are entirely consistent and will lead to the same decision alternative being selected. For illustrative purposes, PW will be calculated for the decision alternatives of Table A-9 assuming MARR = 12%.

Table A-8. The decision alternatives with constraints imposed

Alternative	Status
Accept A only	OK
Accept B only	infeasible, B contingent on C
Accept C only	OK
Accept D only	OK
Accept A and B only	infeasible, B contingent on C
Accept A and C only	OK
Accept A and D only	OK
Accept B and C only	infeasible, capital rationing
Accept B and D only	infeasible, B contingent on C
Accept C and D only	OK
Accept A, B, and C only	infeasible, capital rationing
Accept A, B, and D only	infeasible, B contingent on C
Accept A, C, and D only	infeasible, capital rationing
Accept B, C, and D only	infeasible, capital rationing
Do Everything	infeasible, capital rationing
null	OK

Table A-9. The decision alternatives cash flows

yr \ Alt	A only	C only	D only	A&C	A&D	C&D	null
0	-1000	-400	-500	-1400	-1500	-900	0
1	600	120	290	720	890	410	0
2	600	120	290	720	890	410	0
3	600	120	290	720	890	410	0
4	600	120	290	720	890	410	0

$PW_A =$ −1000 + 600*(P | A,12%,4) = −1000 + 600 (3.0373)
= $822.38

$PW_C =$ −400 + 120*(P | A,12%,4) = −400 + 120 (3.0373)
= −$35.52

$PW_D =$ −500 + 290*(P | A,12%,4) = −500 + 290 (3.0373)
= $380.83

$PW_{A\&C} =$ −1400 + 720*(P | A,12%,4) = −1400 + 720 (3.0373)
= $786.86

$PW_{A\&D} =$ −1500 + 890*(P | A,12%,4) = −1500 + 890 (3.0373)
= $1203.21

$PW_{C\&D} =$ −900 + 410*(P | A,12%,4) = −900 + 410 (3.0373)
= $345.31

$PW_{null} =$ −0 + 0*(P | A,12%,4) = −0 + 0 (3.0373) = $0.00

The decision rules for the various measures of worth under constrained analysis are list below:

Constrained PW Decision Rule: Accept the decision alternative with the highest PW.

Constrained AW Decision Rule: Accept the decision alternative with the highest AW.

Constrained IRR Decision Rule: Accept the decision alternative with the highest IRR.

Constrained SIR Decision Rule: Accept the decision alternative with the highest SIR.

For the example problem, the highest present worth ($1203.21) is associated with accepting projects A and D (rejecting all others). This decision is guaranteed to be optimal (i.e., no feasible combination of projects has a higher PW, AW, IRR, or SIR).

A.8.4 Some Interesting Observations Regarding Constrained Analysis

Several interesting observations can be made regarding the approach, measures of worth, and decisions associated with constrained analysis. Detailed development of these observations is omitted here but may be found in many engineering economic analysis texts [White, et al., 1998].

- The present worth of a decision alternative is the sum of the present worths of the projects contained within the alternative. (From above, $PW_{A\&D} = PW_A + PW_D$).

- The annual worth of a decision alternative is the sum of the annual worths of the projects contained within the alternative.

- The internal rate of return of a decision alternative is NOT the sum of internal rates of returns of the projects contained within the alternative. The IRR for the decision alternative must be calculated by the trial and error process of finding the value of i that sets the PW of the decision alternative to zero.

- The savings investment ratio of a decision alternative is NOT the sum of the savings investment ratios of the projects contained within the alternative. The SIR for the decision alternative must be calculated from the cash flows of the decision alternative.

- A common, but flawed, procedure for selecting the projects to accept from the set of potential projects involves ranking the projects (not decision alternatives) in preferred order based on a measure of worth calculated for the project (e.g., decreasing project PW) and then accepting projects as far down the list as funds allow. While this procedure will select the optimal set under some conditions (e.g., it works well if the initial investments of all projects are small relative to the capital budget limit), it is not guaranteed to select the optimal set under all conditions. The procedure outlined above will select the optimal set under all conditions.

- Table A-10 illustrates that the number of decision alternatives in the collectively exhaustive, mutually exclusive set can grow prohibitively large as the number of potential projects increases. The mitigating factor in this combinatorial growth problem is that in most practical situations a high percentage of the possible decision alternatives are infeasible and do not require evaluation.

Table A-10. The number of decision alternatives as a function of the number of projects

Number of Projects	Number of Decision Alternatives
1	2
2	4
3	8
4	16
5	32
6	64
7	128
8	256
0	512
10	1,024
15	32,768
20	1,048,576
25	33,554,432

A.8.5 The Planning Horizon Issue

When comparing projects, it is important to compare the costs and benefits over a common period of time. The intuitive sense of fairness here is based upon the recognition that most consumers expect an investment that generates savings over a longer period of time to cost more than an investment that generates savings over a shorter period of time. To facilitate a fair, comparable evaluation, a common period of time over which to conduct the evaluation is required. This period of time is referred to as the planning horizon. The planning horizon issue arises when at least one project has cash flows defined over a life which is greater than or less than the life of at least one other project. This situation did not occur in Example 17 of the previous section, since all projects had 4 year lives.

There are four common approaches to establishing a planning horizon for evaluating decision alternatives. These are (1) shortest life, (2) longest life, (3) least common multiple of lives, and (4) standard. The shortest life planning horizon is established by selecting the project with the shortest life and setting this life as the planning horizon. A significant issue in this approach is how to value the remaining cash flows for projects whose lives are truncated. The typical approach to this valuation is to estimate the value of the remaining cash flows as the salvage value (market value) of the investment at that point in its life.

Example 18

Determine the shortest life planning horizon for projects A, B, C with lives 3, 5, and 6 years, respectively.

> The shortest life planning horizon is 3 years based on Project A. A salvage value must be established at $t = 3$ for B's cash flows in years 4 and 5. A salvage value must be established at $t = 3$ for C's cash flows in years 4, 5, and 6.

The longest life planning horizon is established by selecting the project with the longest life and setting this life as the planning horizon. The significant issue in this approach is how to handle projects whose cash flows don't extend this long. The typical resolution for this problem is to assume that shorter projects are repeated consecutively (end-to-end) until one of the repetitions extends at least as far as the planning horizon. The assumption of project repeatability deserves careful consideration, since in some cases it is reasonable and in others it may be quite unreasonable. The reasonableness of the assumption is largely a function of the type of investment and the rate of innovation occurring within the investment's field. (For example, assuming repeatability of investments in high technology equipment is frequently ill advised, since the field is advancing rapidly.) If in repeating a project's cash flows, the last repetition's cash flows extend beyond the planning horizon, then the truncated cash flows (those that extend beyond the planning horizon) must be assigned a salvage value as above.

Example 19

Determine the longest life planning horizon for projects A, B, C with lives 3, 5, and 6 years, respectively.

The longest life planning horizon is 6 years based on Project C. Project A must be repeated twice, with the second repetition ending at year 6, so no termination of cash flows is required. Project B's second repetition extends to year 10; therefore, a salvage value at t = 6 must be established for B's repeated cash flows in years 7, 8, 9, and 10.

An approach that eliminates the truncation salvage value issue from the planning horizon question is the least common multiple approach. The least common multiple planning horizon is set by determining the smallest number of years at which repetitions of all projects would terminate simultaneously. The least common multiple for a set of numbers (lives) can be determined mathematically using algebra. Discussion of this approach is beyond the scope of this appendix. For a small number of projects, the value can be determined with trial and error by examining multiples of the longest life project.

Example 20

Determine the least common multiple planning horizon for projects A, B, C with lives 3, 5, and 6 years, respectively.

The least common multiple of 3, 5, and 6 is 30. This can be obtained by trial and error, starting with the longest project life (6) as follows:

1st trial: 6*1 = 6; 6 is a multiple of 3 but not 5; reject 6 and proceed

2nd trial: 6*2 = 12; 12 is a multiple of 3 but not 5; reject 12 and proceed

3rd trial: 6*3 = 18; 18 is a multiple of 3 but not 5; reject 18 and proceed

4th trial: 6*4 = 24; 24 is a multiple of 3 but not 5; reject 24 and proceed

5th trial: 6*5 = 30; 30 is a multiple of 3 and 5; accept 30 and stop

Under a 30-year planning horizon, A's cash flows are repeated 10 times, B's 6 times, and C's 5 times. No truncation is required.

The standard planning horizon approach uses a planning horizon that is independent of the projects being evaluated. Typically, this type of planning horizon is based on company policies or practices. The standard horizon may require repetition and/or truncation, depending upon the set of projects being evaluated.

Example 21

Determine the impact of a 5 year standard planning horizon on projects A, B, C with lives 3, 5, and 6 years, respectively.

With a 5-year planning horizon:

Project A must be repeated one time, with the second repetition truncated by one year.

Project B is a 5 year project and does not require repetition or truncation.

Project C must be truncated by one year.

There is no single accepted approach to resolving the planning horizon issue. Companies and individuals generally use one of the approaches outlined above. The decision of which to use in a particular analysis is generally a function of company practice, consideration of the reasonableness of the project repeatability assumption, and the availability of salvage value estimates at truncation points.

A.9 SPECIAL PROBLEMS

A.9.1 Introduction

The preceding sections of this appendix outline an approach for conducting deterministic economic analysis of investment opportunities. Adherence to the concepts and methods presented will lead to sound investment decisions with respect to time value of money principles. This section addresses several topics that are of special interest in some analysis situations.

A.9.2 Interpolating Interest Tables

All of the examples previously presented in this appendix conveniently used interest rates whose time value of money factors were tabulated in Appendix 4A. How does one proceed if non-tabulated time value of money factors are needed? There are two viable approaches, calculation of the exact values and interpolation. The best and theoretically correct approach is to calculate the exact values of needed factors, based on the formulas in Table A-6.

Example 22

Determine the exact value for (F | P,13%,7).

From Table A-6,

$(F \mid P,i,n) = (1+i)^n = (1+.13)^7 = 2.3526$

Interpolation is often used instead of calculation of exact values because, with practice, interpolated values can be calculated quickly. Interpolated values are not "exact," but for most practical problems they are "close enough," particularly if the range of interpolation is kept as narrow as possible. Interpolation of some factors, for instance (P | A,i,n), also tends to be less error prone than the exact calculation due of simpler mathematical operations.

Interpolation involves determining an unknown time value of money factor using two known values that bracket the value of interest. An assumption is made that the values of the time value of money factor vary linearly between the known values. Ratios are then used to estimate the unknown value. The example below illustrates the process.

Example 23

Determine an interpolated value for (F | P,13%,7).

The narrowest range of interest rates that bracket 13%, and for which time value of money factor tables are provided in Appendix 4A, is 12% to 15%.

The values necessary for this interpolation are

i values	(F«F,i%,7)
12%	2.2107
13%	(F | P,13%,7)
15%	2.6600

The interpolation proceeds by setting up ratios and solving for the unknown value, (F | P,13%,7), as follows:

$$\frac{\text{change between rows 2 \& 1 of left column}}{\text{change between rows 3 \& 1 of left column}} =$$

$$\frac{\text{change between rows 2 \& 1 of right column}}{\text{change between rows 3 \& 1 of right column}}$$

$$\frac{0.13 - 0.12}{0.15 - 0.12} = \frac{(F \mid P,13\%,7) - 2.2107}{2.6600 - 2{,}2107}$$

$$\frac{0.01}{0.03} = \frac{(F \mid P,13\%,7) - 2.2107}{0.4493}$$

$$0.1498 = (F \mid P,13\%,7) - 2.2107$$

$$(F \mid P,13\%,7) = 2.3605$$

The interpolated value for (F | P,13%,7), 2.3605, differs from the exact value, 2.3526, by 0.0079. This would imply a $7.90 difference in present worth for every thousand dollars of return at t = 7. The relative importance of this interpolation error can be judged only in the context of a specific problem.

A.9.3 Non-Annual Interest Compounding

Many practical economic analysis problems involve interest that is not compounded annually. It is common practice to express a non-annually compounded interest rate as follows:

12% per year compounded monthly or 12% / yr / mo.

When expressed in this form, 12% / yr / mo is known as the nominal annual interest rate. The techniques covered in this appendix up to this point can not be used directly to solve an economic analysis problem of this type because the interest period (per year) and compounding period (monthly) are not the same. Two approaches can be used to solve problems of this type. One approach involves determining a period interest rate; the other involves determining an effective interest rate.

To solve this type of problem using a period interest rate approach, we must define the period interest rate:

$$\text{Period Interest Rate} = \frac{\text{Nominal Annual Interest Rate}}{\text{Number of Interest Periods per Year}}$$

In our example,

$$\text{Period Interest Rate} = \frac{12\%/\text{yr}/\text{mo}}{12\ \text{mo}/\text{yr}} = 1\%/\text{mo}/\text{mo}$$

Because the interest period and the compounding period are now the same, the time value of money factors in Appendix 4A can be applied directly. Note, however, that the number of interest periods (n) must be adjusted to match the new frequency.

Example 24

$2,000 is invested in an account which pays 12% per year compounded monthly. What is the balance in the account after 3 years?

$$\text{Nominal Annual Interest Rate} = 12\%/\text{yr}/\text{mo}$$

$$\text{Period Interest Rate} = \frac{12\%/\text{yr}/\text{mo}}{12\ \text{mo}/\text{yr}} = 1\%/\text{mo}/\text{mo}$$

Number of Interest Periods = 3 years × 12 mo/yr = 36 interest periods (months)

$$F = P\ (F \mid P,i,n) = \$2{,}000\ (F \mid P,1,36) = \$2{,}000\ (1.4308) = \$2{,}861.60$$

Example 25

What are the monthly payments on a 5-year car loan of $12,500 at 6% per year compounded monthly?

$$\text{Nominal Annual Interest Rate} = 6\%/\text{yr}/\text{mo}$$

$$\text{Period Interest Rate} = \frac{6\%/\text{yr}/\text{mo}}{12\ \text{mo}/\text{yr}} = 0.5\%/\text{mo}/\text{mo}$$

Number of Interest Periods = 5 years × 12 mo/yr = 60 interest periods

$$A = P\ (A \mid P,i,n) = \$12{,}500\ (A \mid P,0.5,60) = \$12{,}500\ (0.0193) = \$241.25$$

To solve this type of problem using an effective interest rate approach, we must define the effective interest rate. The effective annual interest rate is the annualized interest rate that would yield results equivalent to the period interest rate as previously calculated. However, the effective annual interest rate approach should not be used if the cash flows are more frequent than annual (e.g., monthly). In general, the interest rate for time value of money factors should match the frequency of the cash flows. (For example, if the cash flows are monthly, use the period interest rate approach with monthly periods.)

As an example of the calculation of an effective interest rate, assume that the nominal interest rate is 12%/yr/qtr; therefore, the period interest rate is 3%/qtr/qtr. One dollar invested for 1 year at 3%/qtr/qtr would have a future worth as calculated:

$$F = P\ (F \mid P,i,n) = \$1\ (F \mid P,3,4) = \$1\ (1.03)^4$$
$$= \$1\ (1.1255) = \$1.1255$$

To get this same value in 1 year with an annual rate, the annual rate would have to be of 12.55%/yr/yr. This value is called the effective annual interest rate. The effective annual interest rate is given by $(1.03)^4 - 1 = 0.1255$ or 12.55%.

The general equation for the Effective Annual Interest Rate is:

$$\text{Effective Annual Interest Rate} = (1 + (r/m))^m - 1$$

where: r = nominal annual interest rate
m = number of interest periods per year

Example 26

What is the effective annual interest rate if the nominal rate is 12%/yr compounded monthly?

nominal annual interest rate = 12%/yr/mo

period interest rate = 1%/mo/mo

effective annual interest rate = $(1+0.12/12)^{12} - 1 = 0.1268$ or 12.68%

A.9.4 Economic Analysis Under Inflation

Inflation is characterized by a decrease in the purchasing power of money that is caused by an increase in general price levels of goods and services without an accompanying increase in value. Inflationary pressure is created when more dollars are put into an economy with no accompanying increase in goods and services. In other words, printing more money without an increase in economic output generates inflation. A complete treatment of inflation is beyond the scope of this appendix. A good summary can be found in Sullivan and Bontadelli [1980].

When consideration of inflation is introduced into economic analysis, future cash flows can be stated in terms of either constant-worth dollars or then-current dollars. Then-current cash flows are expressed in terms of the face amount of dollars (actual number of dollars) that will change hands when the cash flow occurs. Alternatively, constant-worth cash flows are expressed in terms of the purchasing power of dollars relative to a fixed point in time known as the base period.

Example 27

For the next 4 years, a family anticipates buying $1000 worth of groceries each year. If inflation is expected to be 3%/yr, what are the then-current cash flows required to purchase the groceries?

> To buy the groceries, the family will need to take the following face amount of dollars to the store. We will somewhat artificially assume that the family only shops once per year, buys the same set of items each year, and that the first trip to the store will be one year from today.
>
> Year 1: dollars required $1000.00*(1.03) = $1030.00
>
> Year 2: dollars required $1030.00*(1.03) = $1060.90
>
> Year 3: dollars required $1060.90*(1.03) = $1092.73
>
> Year 4: dollars required $1092.73*(1.03) = $1125.51

What are the constant-worth cash flows, if today's dollars are used as the base year?

> The constant worth dollars are inflation free dollars; therefore, the $1000 of groceries costs $1000 each year.

Year 1: $1000.00

Year 2: $1000.00

Year 3: $1000.00

Year 4: $1000.00

The key to proper economic analysis under inflation is to base the value of MARR on the types of cash flows. If the cash flows contain inflation, then the value of MARR should also be adjusted for inflation. Alternatively, if the cash flows do not contain inflation, then the value of MARR should be inflation-free. When MARR does not contain an adjustment for inflation, it is referred to as a real value for MARR. If it contains an inflation adjustment, it is referred to as a combined value for MARR. The relationship between inflation rate, the real value of MARR, and the combined value of MARR is given by:

$$1 + MARR_{COMBINED} = (1 + \text{inflation rate}) * (1 + MARR_{REAL})$$

Example 28
If the inflation rate is 3%/yr and the real value of MARR is 15%/yr, what is the combined value of MARR?

$$1 + MARR_{COMBINED} = (1 + \text{inflation rate}) * (1 + MARR_{REAL})$$

$$1 + MARR_{COMBINED} = (1 + 0.03) * (1 + 0.15)$$

$$1 + MARR_{COMBINED} = (1.03) * (1.15)$$

$$1 + MARR_{COMBINED} = 1.1845$$

$$MARR_{COMBINED} = 1.1845 - 1 = 0.1845 = 18.45\%$$

If the cash flows of a project are stated in terms of then-current dollars, the appropriate value of MARR is the combined value of MARR. Analysis done in this way is referred to as then current analysis. If the cash flows of a project are stated in terms of constant-worth dollars, the appropriate value of MARR is the real value of MARR. Analysis done in this way is referred to as then constant worth analysis.

Example 29

Using the cash flows of Examples 27 and interest rates of Example 28, determine the present worth of the grocery purchases using a constant worth analysis.

Constant worth analysis requires constant worth cash flows and the real value of MARR.

PW = 1000 * (P | A,15%,4)
= 1000 * (2.8550) = $2855.00

Example 30

Using the cash flows of Examples 27 and interest rates of Example 28, determine the present worth of the grocery purchases using a then current analysis.

Then current analysis requires then current cash flows and the combined value of MARR.

PW = 1030.00 * (P | F,18.45%,1) + 1060.90 * (P | F,18.45%,2) + 1092.73 * (P | F,18.45%,3) +1125.51 * (P | F,18.45%,4)

PW = 1030.00 * (0.8442) + 1060.90 * (0.7127) + 1092.73 * (0.6017) +1125.51 * (0.5080)

PW = 869.53 + 756.10 + 657.50 + 571.76 = 2854.89

The notable result of Examples 29 and 30 is that the present worths determined by the constant-worth approach ($2855.00) and the then-current approach ($2854.89) are equal. (The $0.11 difference is due to rounding.) This result is often unexpected but is mathematically sound. The important conclusion is that if care is taken to appropriately match the cash flows and value of MARR, the level of general price inflation is not a determining factor in the acceptability of projects. To make this important result hold, inflation must either (1) be included in both the cash flows, and MARR (the then-current approach) or (2) be included in neither the cash flows nor MARR (the constant-worth approach).

A.9.5 Sensitivity Analysis and Risk Analysis

Often times the certainty assumptions associated with deterministic analysis are questionable. These certainty assumptions include certain knowledge regarding amounts and timing of cash flows, as well as certain knowledge of MARR. Relaxing these assumptions requires the use of sensitivity analysis and risk analysis techniques.

Initial sensitivity analyses are usually conducted on the optimal decision alternative (or top two or three) on a single factor basis. Single factor sensitivity analysis involves holding all cost factors except one constant while varying the remaining cost factor through a range of percentage changes. The effect of cost factor changes on the measure of worth is observed, to determine whether the alternative remains attractive under the evaluated changes and to determine which cost factor effects the measure of worth the most.

Example 31

Conduct a sensitivity analysis of the optimal decision resulting from the constrained analysis of the data in Example 17. The sensitivity analysis should explore the sensitivity of present worth to changes in annual revenue over the range – 10% to +10%.

The PW of the optimal decision (Accept A & D only) was determined in Section A.8.3 to be:

$PW_{A\&D} = -1500 + 890^*(P \mid A,12\%,4) = -1500 + 890\ (3.0373) = \$1203.21$

If annual revenue decreases 10%, it becomes 890 – 0.10*890 = 801 and PW becomes

$PW_{A\&D} = -1500 + 801^*(P \mid A,12\%,4) = -1500 + 801\ (3.0373) = \$932.88$

If annual revenue increases 10%, it becomes 890 + 0.10*890 = 979 and PW becomes

$PW_{A\&D} = -1500 + 979^*(P \mid A,12\%,4) = -1500 + 979\ (3.0373) = \$1473.52$

The sensitivity of PW to changes in annual revenue over the range – 10% to +10% is +$540.64 (from $932.88 to $1473.52).

Example 32

Repeat Example 31, exploring the sensitivity of present worth to changes in initial cost over the range – 10% to +10%.

The PW of the optimal decision (Accept A & D only) was determined in Section A.8.3 to be:

$PW_{A\&D}$ = – 1500 + 890*(P | A,12%,4) = – 1500 + 890 (3.0373) = $1203.21

If initial cost decreases 10% it becomes 1500 – 0.10*1500 = 1350 and PW becomes

$PW_{A\&D}$ = – 1350 + 890*(P | A,12%,4) = – 1350 + 890 (3.0373) = $1353.20

If initial cost increases 10% it becomes 1500 + 0.10*1500 = 1650 and PW becomes

$PW_{A\&D}$ = – 1650 + 890*(P | A,12%,4) = – 1500 + 890 (3.0373) = $1053.20

The sensitivity of PW to changes in initial cost over the range – 10% to +10% is – $300.00 (from $1353.20 to $1053.20).

Example 33

Repeat Example 31 exploring the sensitivity of the present worth to changes in MARR over the range – 10% to +10%.

The PW of the optimal decision (Accept A & D only) was determined in Section A.8.3 to be:

$PW_{A\&D}$ = – 1500 + 890*(P | A,12%,4) = – 1500 + 890 (3.0373) = $1203.21

If MARR decreases 10% it becomes 12% – 0.10*12% = 10.8% and PW becomes

$PW_{A\&D}$ = – 1500 + 890*(P | A,10.8%,4) = – 1500 + 890 (3.1157) = $1272.97

If MARR increases 10% it becomes 12% + 0.10*12% = 13.2% and PW becomes

$PW_{A\&D}$ = – 1500 + 890*(P | A,13.2%,4) = – 1500 + 890 (2.9622) = $1136.36

The sensitivity of PW to changes in MARR over the range – 10% to +10% is – $136.61 (from $1272.97 to $1136.36).

The sensitivity data from Examples 31, 32, and 33 are summarized in Table A-11. A review of the table reveals that the decision alternative A&D remains attractive (PW ≥0) within the range of 10% changes in annual revenues, initial cost, and MARR. An appealing way to summarize single factor sensitivity data is to use a "spider" graph. A spider graph plots the PW values determined in the examples and connects them with lines, one line for each factor evaluated. Figure A-10 illustrates the spider graph for the data of Table A-11. On this graph, lines with large positive or negative slopes (angle relative to horizontal regardless of whether it is increasing or decreasing) indicate factors to which the present value measure of worth is sensitive. Figure A-10 shows that PW is least sensitive to changes in MARR (the MARR line is the most nearly horizontal) and most sensitive to changes in annual revenue. (The annual revenue line has the steepest slope.) Additional sensitivity could be explored in a similar manner.

When single factor sensitivity analysis is inadequate to assess the questions that surround the certainty assumptions of a deterministic analysis, risk analysis techniques can be employed. One approach to risk

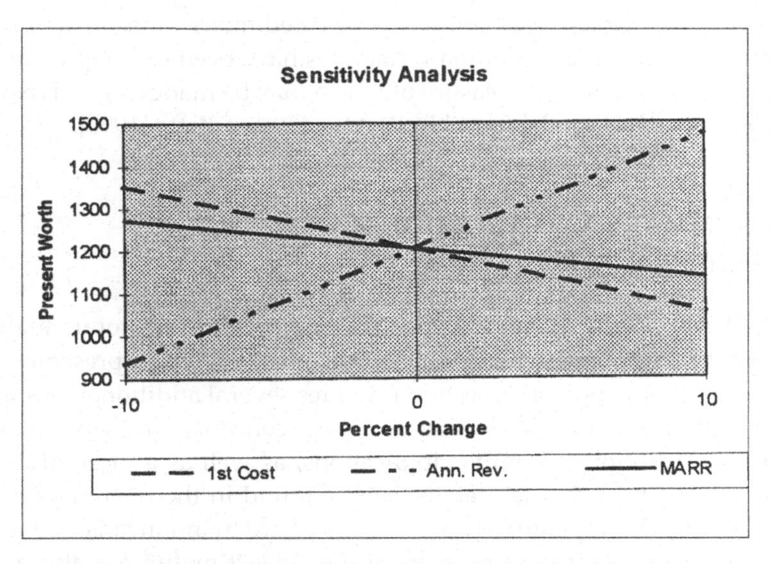

Figure A-10. Sensitivity analysis "spider" graph

Table A-11 Sensitivity analysis data table

Factor/ Percent Change	- 10%	Base	+ 10%
1st Cost	1353.20	1203.21	1053.20
Annual Revenue	932.88	1203.21	1473.52
MARR	1272.97	1203.21	1136.36

analysis is the application of probabilistic and statistical concepts to economic analysis. These techniques require information regarding the possible values that uncertain quantities may take on, as well as estimates of the probability that the various values will occur. A detailed treatment of this topic is beyond the scope of this appendix. A good discussion of this subject can be found in Park and Sharp-Bette [1990].

A second approach to risk analysis in economic analysis is through the use of simulation techniques and simulation software. Simulation involves using a computer simulation program to sample possible values for the uncertain quantities in an economic analysis and then calculating the measure of worth. This process is repeated many times using different samples each time. After many samples have been taken, probability statements regarding the measure of worth may be made. A good discussion of this subject can be found in Park and Sharp-Bette [1990].

A.10 SUMMARY AND ADDITIONAL EXAMPLE APPLICATIONS

In this appendix a coherent, consistent approach to economic analysis of capital investments (energy related or other) has been presented. To conclude the appendix, this section provides several additional examples to illustrate the use of time value of money concepts for energy related problems. Additional example applications, as well as a more in depth presentation of conceptual details, can be found in the references listed at the end of the appendix. These references are by no means exclusive; many other excellent presentations of the subject matter are also available. Adherence to the concepts and methods presented here and in the references will lead to sound investment decisions with respect to time value of money principles.

Example 34

In Section A.3.3 an example involving the evaluation of a baseboard heating and window air conditioner versus a heat pump was introduced to illustrate cash flow diagramming (Figure A-2). A summary of the differential costs is repeated here for convenience:

- The heat pump costs $1500 more than the baseboard system.
- The heat pump saves $380 annually in electricity costs.
- The heat pump has a $50 higher annual maintenance costs.
- The heat pump has a $150 higher salvage value at the end of 15 years.
- The heat pump requires $200 more in replacement maintenance at the end of year 8.

If MARR is 18%, is the additional investment in the heat pump attractive?

Using present worth as the measure of worth:

$$PW = -1500 + 380*(P \mid A,18\%,15) - 50*(P \mid A,18\%,15) + 150*(P \mid F,18\%,15) - 200*(P \mid F,18\%,8)$$

$$PW = -1500 + 380*(5.0916) - 50*(5.0916) + 150*(0.0835) - 200*(0.2660)$$

$$PW = -1500.00 + 1934.81 - 254.58 + 12.53 - 53.20 = \$139.56$$

<u>Decision</u>: PW≥0 ($139.56>0.0); therefore, the additional investment for the heat pump is attractive.

Example 35

A homeowner needs to decide whether to install R-11 or R-19 insulation in the attic of her home. The R-19 insulation costs $150 more to install and will save approximately 400 kWh per year. If the planning horizon is 20 years and electricity costs $0.08/kWh, is the additional investment attractive at MARR of 10%?

At $0.08/kWh, the annual savings are: 400 kWh * $0.08/kWh = $32.00

Using present worth as the measure of worth:

PW = – 150 + 32*(P I A,10%,20)

PW = – 150 + 32*(8.5136) = – 150 + 272.44 = $122.44

Decision: PW≥0 ($122.44>0.0); therefore, the R-19 insulation is attractive.

Example 36

The homeowner from Example 35 can install R-30 insulation in the attic of her home for $200 more than the R-19 insulation. The R-30 will save approximately 250 kWh per year over the R-19 insulation. Is the additional investment attractive?

Assuming the same MARR, electricity cost, and planning horizon, the additional annual savings are: 250 kWh * $0.08/kWh = $20.00

Using present worth as the measure of worth:

PW = – 200 + 20*(P I A,10%,20)

PW = – 200 + 20*(8.5136) = – 200 + 170.27 = – $29.73

Decision: PW<0 (-$29.73<0.0); therefore, the R-30 insulation is not attractive.

Example 37

An economizer costs $20,000 and will last 10 years. It will generate savings of $3,500 per year with maintenance costs of $500 per year. If M1ARR is 10%, is the economizer an attractive investment.

Using present worth as the measure of worth:

PW = – 20000 + 3500*(P | A,10%,10) – 500*(P | A,10%,10)

PW = – 20000 + 3500*(6.1446) – 500*(6.1446)

PW = – 20000.00 + 21506.10 – 3072.30 = – $1566.20

Decision: PW<0 (-$1566.20<0.0); therefore, the economizer is not attractive.

Example 38

If the economizer from Example 37 has a salvage value of $5000 at the end of 10 years, is the investment attractive?

Using present worth as the measure of worth:

PW = – 20000 + 3500*(P | A,10%,10)
– 500*(P | A,10%,10) + 5000*(P | F,10%,10)

PW = – 20000 + 3500*(6.1446) – 500*(6.1446) + 5000*(0.3855)

PW = – 20000.00 + 21506.10 – 3072.30 + 1927.50 = $361.30

Decision: PW≥0 ($361.30≥0.0); therefore, the economizer is now attractive.

A.11 References

Brown, R.J. and R.R. Yanuck, 1980, *Life Cycle Costing: A Practical Guide for Energy Managers*, The Fairmont Press, Inc., Atlanta, GA.

Fuller, S.K. and S.R. Petersen, 1994, NISTIR 5165: Life-Cycle Costing Workshop for Energy Conservation in Buildings: Student Manual, U.S. Department of Commerce, Office of Applied Economics, Gaithersburg, MD.

Fuller, S.K. and S.R. Petersen, 1995, NIST Handbook 135: Life-Cycle Costing Manual for the Federal Energy Management Program, National Technical Information Service, Springfield, VA.

Park, C.S. and G.P. Sharp-Bette, 1990, *Advanced Engineering Economics*, John Wiley & Sons, New York, NY.

Sullivan, W.G. and J.A. Bontadelli, 1980, "The Industrial Engineer and Inflation," *Industrial Engineering*, Vol. 12, No. 3, 24-33.

Thuesen, G.J. and W.J. Fabrycky, 1993, *Engineering Economy, 8th Edition*, Prentice Hall, Englewood Cliffs, NJ.

White, J.A., K.E. Case, D.B. Pratt, and M.H. Agee, 1998, *Principles of Engineering Economic Analysis, 3rd Edition*, John Wiley & Sons, New York, NY.

Time Value of Money Factors—Discrete Compounding
i = 1%

	Single Sums		Uniform Series				Gradient Series	
n	To Find F Given P (F|P,i%,n)	To Find P Given F (P|F,i%,n)	To Find F Given A (F|A,i%,n)	To Find A Given F (A|F,i%,n)	To Find P Given A (P|A,i%,n)	To Find A Given P (A|P,i%,n)	To Find P Given G (P|G,i%,n)	To Find A Given G (A|G,i%,n)
1	1.0100	0.9901	1.0000	1.0000	0.9901	1.0100	0.0000	0.0000
2	1.0201	0.9803	2.0100	0.4975	1.9704	0.5075	0.9803	0.4975
3	1.0303	0.9706	3.0301	0.3300	2.9410	0.3400	2.9215	0.9934
4	1.0406	0.9610	4.0604	0.2463	3.9020	0.2563	5.8044	1.4876
5	1.0510	0.9515	5.1010	0.1960	4.8534	0.2060	9.6103	1.9801
6	1.0615	0.9420	6.1520	0.1625	5.7955	0.1725	14.3205	2.4710
7	1.0721	0.9327	7.2135	0.1386	6.7282	0.1486	19.9168	2.9602
8	1.0829	0.9235	8.2857	0.1207	7.6517	0.1307	26.3812	3.4478
9	1.0937	0.9143	9.3685	0.1067	8.5660	0.1167	33.6959	3.9337
10	1.1046	0.9053	10.4622	0.0956	9.4713	0.1056	41.8435	4.4179
11	1.1157	0.8963	11.5668	0.0865	10.3676	0.0965	50.8067	4.9005
12	1.1268	0.8874	12.6825	0.0788	11.2551	0.0888	60.5687	5.3815
13	1.1381	0.8787	13.8093	0.0724	12.1337	0.0824	71.1126	5.8607
14	1.1495	0.8700	14.9474	0.0669	13.0037	0.0769	82.4221	6.3384
15	1.1610	0.8613	16.0969	0.0621	13.8651	0.0721	94.4810	6.8143
16	1.1726	0.8528	17.2579	0.0579	14.7179	0.0679	107.2734	7.2886
17	1.1843	0.8444	18.4304	0.0543	15.5623	0.0643	120.7834	7.7613
18	1.1961	0.8360	19.6147	0.0510	16.3983	0.0610	134.9957	8.2323
19	1.2081	0.8277	20.8109	0.0481	17.2260	0.0581	149.8950	8.7017
20	1.2202	0.8195	22.0190	0.0454	18.0456	0.0554	165.4664	9.1694
21	1.2324	0.8114	23.2392	0.0430	18.8570	0.0530	181.6950	9.6354
22	1.2447	0.8034	24.4716	0.0409	19.6604	0.0509	198.5663	10.0998
23	1.2572	0.7954	25.7163	0.0389	20.4558	0.0489	216.0660	10.5626
24	1.2697	0.7876	26.9735	0.0371	21.2434	0.0471	234.1800	11.0237
25	1.2824	0.7798	28.2432	0.0354	22.0232	0.0454	252.8945	11.4831
26	1.2953	0.7720	29.5256	0.0339	22.7952	0.0439	272.1957	11.9409
27	1.3082	0.7644	30.8209	0.0324	23.5596	0.0424	292.0702	12.3971
28	1.3213	0.7568	32.1291	0.0311	24.3164	0.0411	312.5047	12.8516
29	1.3345	0.7493	33.4504	0.0299	25.0658	0.0399	333.4863	13.3044
30	1.3478	0.7419	34.7849	0.0287	25.8077	0.0387	355.0021	13.7557
36	1.4308	0.6989	43.0769	0.0232	30.1075	0.0332	494.6207	16.4285
42	1.5188	0.6584	51.8790	0.0193	34.1581	0.0293	650.4514	19.0424
48	1.6122	0.6203	61.2226	0.0163	37.9740	0.0263	820.1460	21.5976
54	1.7114	0.5843	71.1410	0.0141	41.5687	0.0241	1.002E+03	24.0945
60	1.8167	0.5504	81.6697	0.0122	44.9550	0.0222	1.193E+03	26.5333
66	1.9285	0.5185	92.8460	0.0108	48.1452	0.0208	1.392E+03	28.9146
72	2.0471	0.4885	104.7099	9.550E-03	51.1504	0.0196	1.598E+03	31.2386
120	3.3004	0.3030	230.0387	4.347E-03	69.7005	0.0143	3.334E+03	47.8349
180	5.9958	0.1668	499.5802	2.002E-03	83.3217	0.0120	5.330E+03	63.9697
360	35.9496	0.0278	3.495E+03	2.861E-04	97.2183	0.0103	8.720E+03	89.6995

Time Value of Money Factors—Discrete Compounding
i = 2%

	Single Sums		Uniform Series				Gradient Series	
n	To Find F Given P (F|P,i%,n)	To Find P Given F (P|F,i%,n)	To Find F Given A (F|A,i%,n)	To Find A Given F (A|F,i%,n)	To Find P Given A (P|A,i%,n)	To Find A Given P (A|P,i%,n)	To Find P Given G (P|G,i%,n)	To Find A Given G (A|G,i%,n)
1	1.0200	0.9804	1.0000	1.0000	0.9804	1.0200	0.0000	0.0000
2	1.0404	0.9612	2.0200	0.4950	1.9416	0.5150	0.9612	0.4950
3	1.0612	0.9423	3.0604	0.3268	2.8839	0.3468	2.8458	0.9868
4	1.0824	0.9238	4.1216	0.2426	3.8077	0.2626	5.6173	1.4752
5	1.1041	0.9057	5.2040	0.1922	4.7135	0.2122	9.2403	1.9604
6	1.1262	0.8880	6.3081	0.1585	5.6014	0.1785	13.6801	2.4423
7	1.1487	0.8706	7.4343	0.1345	6.4720	0.1545	18.9035	2.9208
8	1.1717	0.8535	8.5830	0.1165	7.3255	0.1365	24.8779	3.3961
9	1.1951	0.8368	9.7546	0.1025	8.1622	0.1225	31.5720	3.8681
10	1.2190	0.8203	10.9497	0.0913	8.9826	0.1113	38.9551	4.3367
11	1.2434	0.8043	12.1687	0.0822	9.7868	0.1022	46.9977	4.8021
12	1.2682	0.7885	13.4121	0.0746	10.5753	0.0946	55.6712	5.2642
13	1.2936	0.7730	14.6803	0.0681	11.3484	0.0881	64.9475	5.7231
14	1.3195	0.7579	15.9739	0.0626	12.1062	0.0826	74.7999	6.1786
15	1.3459	0.7430	17.2934	0.0578	12.8493	0.0778	85.2021	6.6309
16	1.3728	0.7284	18.6393	0.0537	13.5777	0.0737	96.1288	7.0799
17	1.4002	0.7142	20.0121	0.0500	14.2919	0.0700	107.5554	7.5256
18	1.4282	0.7002	21.4123	0.0467	14.9920	0.0667	119.4581	7.9681
19	1.4568	0.6864	22.8406	0.0438	15.6785	0.0638	131.8139	8.4073
20	1.4859	0.6730	24.2974	0.0412	16.3514	0.0612	144.6003	8.8433
21	1.5157	0.6598	25.7833	0.0388	17.0112	0.0588	157.7959	9.2760
22	1.5460	0.6468	27.2990	0.0366	17.6580	0.0566	171.3795	9.7055
23	1.5769	0.6342	28.8450	0.0347	18.2922	0.0547	185.3309	10.1317
24	1.6084	0.6217	30.4219	0.0329	18.9139	0.0529	199.6305	10.5547
25	1.6406	0.6095	32.0303	0.0312	19.5235	0.0512	214.2592	10.9745
26	1.6734	0.5976	33.6709	0.0297	20.1210	0.0497	229.1987	11.3910
27	1.7069	0.5859	35.3443	0.0283	20.7069	0.0483	244.4311	11.8043
28	1.7410	0.5744	37.0512	0.0270	21.2813	0.0470	259.9392	12.2145
29	1.7758	0.5631	38.7922	0.0258	21.8444	0.0458	275.7064	12.6214
30	1.8114	0.5521	40.5681	0.0246	22.3965	0.0446	291.7164	13.0251
36	2.0399	0.4902	51.9944	0.0192	25.4888	0.0392	392.0405	15.3809
42	2.2972	0.4353	64.8622	0.0154	28.2348	0.0354	497.6010	17.6237
48	2.5871	0.3865	79.3535	0.0126	30.6731	0.0326	605.9657	19.7556
54	2.9135	0.3432	95.6731	0.0105	32.8383	0.0305	715.1815	21.7789
60	3.2810	0.3048	114.0515	8.768E-03	34.7609	0.0288	823.6975	23.6961
66	3.6950	0.2706	134.7487	7.421E-03	36.4681	0.0274	930.3000	25.5100
72	4.1611	0.2403	158.0570	6.327E-03	37.9841	0.0263	1.034E+03	27.2234
120	10.7652	0.0929	488.2582	2.048E-03	45.3554	0.0220	1.710E+03	37.7114
180	35.3208	0.0283	1.716E+03	5.827E-04	48.5844	0.0206	2.174E+03	44.7554
360	1.248E+03	8.016E-04	6.233E+04	1.604E-05	49.9599	0.0200	2.484E+03	49.7112

Time Value of Money Factors—Discrete Compounding
i = 3%

	Single Sums		Uniform Series				Gradient Series	
n	To Find F Given P (F|P,i%,n)	To Find P Given F (P|F,i%,n)	To Find F Given A (F|A,i%,n)	To Find A Given F (A|F,i%,n)	To Find P Given A (P|A,i%,n)	To Find A Given P (A|P,i%,n)	To Find P Given G (P|G,i%,n)	To Find A Given G (A|G,i%,n)
1	1.0300	0.9709	1.0000	1.0000	0.9709	1.0300	0.0000	0.0000
2	1.0609	0.9426	2.0300	0.4926	1.9135	0.5226	0.9426	0.4926
3	1.0927	0.9151	3.0909	0.3235	2.8286	0.3535	2.7729	0.9803
4	1.1255	0.8885	4.1836	0.2390	3.7171	0.2690	5.4383	1.4631
5	1.1593	0.8626	5.3091	0.1884	4.5797	0.2184	8.8888	1.9409
6	1.1941	0.8375	6.4684	0.1546	5.4172	0.1846	13.0762	2.4138
7	1.2299	0.8131	7.6625	0.1305	6.2303	0.1605	17.9547	2.8819
8	1.2668	0.7894	8.8923	0.1125	7.0197	0.1425	23.4806	3.3450
9	1.3048	0.7664	10.1591	0.0984	7.7861	0.1284	29.6119	3.8032
10	1.3439	0.7441	11.4639	0.0872	8.5302	0.1172	36.3088	4.2565
11	1.3842	0.7224	12.8078	0.0781	9.2526	0.1081	43.5330	4.7049
12	1.4258	0.7014	14.1920	0.0705	9.9540	0.1005	51.2482	5.1485
13	1.4685	0.6810	15.6178	0.0640	10.6350	0.0940	59.4196	5.5872
14	1.5126	0.6611	17.0863	0.0585	11.2961	0.0885	68.0141	6.0210
15	1.5580	0.6419	18.5989	0.0538	11.9379	0.0838	77.0002	6.4500
16	1.6047	0.6232	20.1569	0.0496	12.5611	0.0796	86.3477	6.8742
17	1.6528	0.6050	21.7616	0.0460	13.1661	0.0760	96.0280	7.2936
18	1.7024	0.5874	23.4144	0.0427	13.7535	0.0727	106.0137	7.7081
19	1.7535	0.5703	25.1169	0.0398	14.3238	0.0698	116.2788	8.1179
20	1.8061	0.5537	26.8704	0.0372	14.8775	0.0672	126.7987	8.5229
21	1.8603	0.5375	28.6765	0.0349	15.4150	0.0649	137.5496	8.9231
22	1.9161	0.5219	30.5368	0.0327	15.9369	0.0627	148.5094	9.3186
23	1.9736	0.5067	32.4529	0.0308	16.4436	0.0608	159.6566	9.7093
24	2.0328	0.4919	34.4265	0.0290	16.9355	0.0590	170.9711	10.0954
25	2.0938	0.4776	36.4593	0.0274	17.4131	0.0574	182.4336	10.4768
26	2.1566	0.4637	38.5530	0.0259	17.8768	0.0559	194.0260	10.8535
27	2.2213	0.4502	40.7096	0.0246	18.3270	0.0546	205.7309	11.2255
28	2.2879	0.4371	42.9309	0.0233	18.7641	0.0533	217.5320	11.5930
29	2.3566	0.4243	45.2189	0.0221	19.1885	0.0521	229.4137	11.9558
30	2.4273	0.4120	47.5754	0.0210	19.6004	0.0510	241.3613	12.3141
36	2.8983	0.3450	63.2759	0.0158	21.8323	0.0458	313.7028	14.3688
42	3.4607	0.2890	82.0232	0.0122	23.7014	0.0422	385.5024	16.2650
48	4.1323	0.2420	104.4084	9.578E-03	25.2667	0.0396	455.0255	18.0089
54	4.9341	0.2027	131.1375	7.626E-03	26.5777	0.0376	521.1157	19.6073
60	5.8916	0.1697	163.0534	6.133E-03	27.6756	0.0361	583.0526	21.0674
66	7.0349	0.1421	201.1627	4.971E-03	28.5950	0.0350	640.4407	22.3969
72	8.4000	0.1190	246.6672	4.054E-03	29.3651	0.0341	693.1226	23.6036
120	34.7110	0.0288	1.124E+03	8.899E-04	32.3730	0.0309	963.8635	29.7737
180	204.5034	4.890E-03	6.783E+03	1.474E-04	33.1703	0.0301	1.076E+03	32.4488
360	4.182E+04	2.391E-05	1.394E+06	7.173E-07	33.3325	0.0300	1.111E+03	33.3247

Time Value of Money Factors—Discrete Compounding
i = 4%

	Single Sums		Uniform Series				Gradient Series	
n	To Find F Given P (F|P,i%,n)	To Find P Given F (P|F,i%,n)	To Find F Given A (F|A,i%,n)	To Find A Given F (A|F,i%,n)	To Find P Given A (P|A,i%,n)	To Find A Given P (A|P,i%,n)	To Find P Given G (P|G,i%,n)	To Find A Given G (A|G,i%,n)
1	1.0400	0.9615	1.0000	1.0000	0.9615	1.0400	0.0000	0.0000
2	1.0816	0.9246	2.0400	0.4902	1.8861	0.5302	0.9246	0.4902
3	1.1249	0.8890	3.1216	0.3203	2.7751	0.3603	2.7025	0.9739
4	1.1699	0.8548	4.2465	0.2355	3.6299	0.2755	5.2670	1.4510
5	1.2167	0.8219	5.4163	0.1846	4.4518	0.2246	8.5547	1.9216
6	1.2653	0.7903	6.6330	0.1508	5.2421	0.1908	12.5062	2.3857
7	1.3159	0.7599	7.8983	0.1266	6.0021	0.1666	17.0657	2.8433
8	1.3686	0.7307	9.2142	0.1085	6.7327	0.1485	22.1806	3.2944
9	1.4233	0.7026	10.5828	0.0945	7.4353	0.1345	27.8013	3.7391
10	1.4802	0.6756	12.0061	0.0833	8.1109	0.1233	33.8814	4.1773
11	1.5395	0.6496	13.4864	0.0741	8.7605	0.1141	40.3772	4.6090
12	1.6010	0.6246	15.0258	0.0666	9.3851	0.1066	47.2477	5.0343
13	1.6651	0.6006	16.6268	0.0601	9.9856	0.1001	54.4546	5.4533
14	1.7317	0.5775	18.2919	0.0547	10.5631	0.0947	61.9618	5.8659
15	1.8009	0.5553	20.0236	0.0499	11.1184	0.0899	69.7355	6.2721
16	1.8730	0.5339	21.8245	0.0458	11.6523	0.0858	77.7441	6.6720
17	1.9479	0.5134	23.6975	0.0422	12.1657	0.0822	85.9581	7.0656
18	2.0258	0.4936	25.6454	0.0390	12.6593	0.0790	94.3498	7.4530
19	2.1068	0.4746	27.6712	0.0361	13.1339	0.0761	102.8933	7.8342
20	2.1911	0.4564	29.7781	0.0336	13.5903	0.0736	111.5647	8.2091
21	2.2788	0.4388	31.9692	0.0313	14.0292	0.0713	120.3414	8.5779
22	2.3699	0.4220	34.2480	0.0292	14.4511	0.0692	129.2024	8.9407
23	2.4647	0.4057	36.6179	0.0273	14.8568	0.0673	138.1284	9.2973
24	2.5633	0.3901	39.0826	0.0256	15.2470	0.0656	147.1012	9.6479
25	2.6658	0.3751	41.6459	0.0240	15.6221	0.0640	156.1040	9.9925
26	2.7725	0.3607	44.3117	0.0226	15.9828	0.0626	165.1212	10.3312
27	2.8834	0.3468	47.0842	0.0212	16.3296	0.0612	174.1385	10.6640
28	2.9987	0.3335	49.9676	0.0200	16.6631	0.0600	183.1424	10.9909
29	3.1187	0.3207	52.9663	0.0189	16.9837	0.0589	192.1206	11.3120
30	3.2434	0.3083	56.0849	0.0178	17.2920	0.0578	201.0618	11.6274
36	4.1039	0.2437	77.5983	0.0129	18.9083	0.0529	253.4052	13.4018
42	5.1928	0.1926	104.8196	9.540E-03	20.1856	0.0495	302.4370	14.9828
48	6.5705	0.1522	139.2632	7.181E-03	21.1951	0.0472	347.2446	16.3832
54	8.3138	0.1203	182.8454	5.469E-03	21.9930	0.0455	387.4436	17.6167
60	10.5196	0.0951	237.9907	4.202E-03	22.6235	0.0442	422.9966	18.6972
66	13.3107	0.0751	307.7671	3.249E-03	23.1218	0.0432	454.0847	19.6388
72	16.8423	0.0594	396.0566	2.525E-03	23.5156	0.0425	481.0170	20.4552
120	110.6626	9.036E-03	2.742E+03	3.648E-04	24.7741	0.0404	592.2428	23.9057
180	1.164E+03	8.590E-04	2.908E+04	3.439E-05	24.9785	0.0400	620.5976	24.8452
360	1.355E+06	7.379E-07	3.388E+07	2.952E-08	25.0000	0.0400	624.9929	24.9997

Time Value of Money Factors—Discrete Compounding
i = 5%

	Single Sums		Uniform Series				Gradient Series	
n	To Find F Given P (F|P,i%,n)	To Find P Given F (P|F,i%,n)	To Find F Given A (F|A,i%,n)	To Find A Given F (A|F,i%,n)	To Find P Given A (P|A,i%,n)	To Find A Given P (A|P,i%,n)	To Find P Given G (P|G,i%,n)	To Find A Given G (A|G,i%,n)
1	1.0500	0.9524	1.0000	1.0000	0.9524	1.0500	0.0000	0.0000
2	1.1025	0.9070	2.0500	0.4878	1.8594	0.5378	0.9070	0.4878
3	1.1576	0.8638	3.1525	0.3172	2.7232	0.3672	2.6347	0.9675
4	1.2155	0.8227	4.3101	0.2320	3.5460	0.2820	5.1028	1.4391
5	1.2763	0.7835	5.5256	0.1810	4.3295	0.2310	8.2369	1.9025
6	1.3401	0.7462	6.8019	0.1470	5.0757	0.1970	11.9680	2.3579
7	1.4071	0.7107	8.1420	0.1228	5.7864	0.1728	16.2321	2.8052
8	1.4775	0.6768	9.5491	0.1047	6.4632	0.1547	20.9700	3.2445
9	1.5513	0.6446	11.0266	0.0907	7.1078	0.1407	26.1268	3.6758
10	1.6289	0.6139	12.5779	0.0795	7.7217	0.1295	31.6520	4.0991
11	1.7103	0.5847	14.2068	0.0704	8.3064	0.1204	37.4988	4.5144
12	1.7959	0.5568	15.9171	0.0628	8.8633	0.1128	43.6241	4.9219
13	1.8856	0.5303	17.7130	0.0565	9.3936	0.1065	49.9879	5.3215
14	1.9799	0.5051	19.5986	0.0510	9.8986	0.1010	56.5538	5.7133
15	2.0789	0.4810	21.5786	0.0463	10.3797	0.0963	63.2880	6.0973
16	2.1829	0.4581	23.6575	0.0423	10.8378	0.0923	70.1597	6.4736
17	2.2920	0.4363	25.8404	0.0387	11.2741	0.0887	77.1405	6.8423
18	2.4066	0.4155	28.1324	0.0355	11.6896	0.0855	84.2043	7.2034
19	2.5270	0.3957	30.5390	0.0327	12.0853	0.0827	91.3275	7.5569
20	2.6533	0.3769	33.0660	0.0302	12.4622	0.0802	98.4884	7.9030
21	2.7860	0.3589	35.7193	0.0280	12.8212	0.0780	105.6673	8.2416
22	2.9253	0.3418	38.5052	0.0260	13.1630	0.0760	112.8461	8.5730
23	3.0715	0.3256	41.4305	0.0241	13.4886	0.0741	120.0087	8.8971
24	3.2251	0.3101	44.5020	0.0225	13.7986	0.0725	127.1402	9.2140
25	3.3864	0.2953	47.7271	0.0210	14.0939	0.0710	134.2275	9.5238
26	3.5557	0.2812	51.1135	0.0196	14.3752	0.0696	141.2585	9.8266
27	3.7335	0.2678	54.6691	0.0183	14.6430	0.0683	148.2226	10.1224
28	3.9201	0.2551	58.4026	0.0171	14.8981	0.0671	155.1101	10.4114
29	4.1161	0.2429	62.3227	0.0160	15.1411	0.0660	161.9126	10.6936
30	4.3219	0.2314	66.4388	0.0151	15.3725	0.0651	168.6226	10.9691
36	5.7918	0.1727	95.8363	0.0104	16.5469	0.0604	206.6237	12.4872
42	7.7616	0.1288	135.2318	7.395E-03	17.4232	0.0574	240.2389	13.7884
48	10.4013	0.0961	188.0254	5.318E-03	18.0772	0.0553	269.2467	14.8943
54	13.9387	0.0717	258.7739	3.864E-03	18.5651	0.0539	293.8208	15.8265
60	18.6792	0.0535	353.5837	2.828E-03	18.9293	0.0528	314.3432	16.6062
66	25.0319	0.0399	480.6379	2.081E-03	19.2010	0.0521	331.2877	17.2536
72	33.5451	0.0298	650.9027	1.536E-03	19.4038	0.0515	345.1485	17.7877
120	348.9120	2.866E-03	6.958E+03	1.437E-04	19.9427	0.0501	391.9751	19.6551
180	6.517E+03	1.534E-04	1.303E+05	7.673E-06	19.9969	0.0500	399.3863	19.9724
360	4.248E+07	2.354E-08	8.495E+08	1.177E-09	20.0000	0.0500	399.9998	20.0000

Time Value of Money Factors—Discrete Compounding
i = 6%

	Single Sums		Uniform Series				Gradient Series	
n	To Find F Given P (F|P,i%,n)	To Find P Given F (P|F,i%,n)	To Find F Given A (F|A,i%,n)	To Find A Given F (A|F,i%,n)	To Find P Given A (P|A,i%,n)	To Find A Given P (A|P,i%,n)	To Find P Given G (P|G,i%,n)	To Find A Given G (A|G,i%,n)
1	1.0600	0.9434	1.0000	1.0000	0.9434	1.0600	0.0000	0.0000
2	1.1236	0.8900	2.0600	0.4854	1.8334	0.5454	0.8900	0.4854
3	1.1910	0.8396	3.1836	0.3141	2.6730	0.3741	2.5692	0.9612
4	1.2625	0.7921	4.3746	0.2286	3.4651	0.2886	4.9455	1.4272
5	1.3382	0.7473	5.6371	0.1774	4.2124	0.2374	7.9345	1.8836
6	1.4185	0.7050	6.9753	0.1434	4.9173	0.2034	11.4594	2.3304
7	1.5036	0.6651	8.3938	0.1191	5.5824	0.1791	15.4497	2.7676
8	1.5938	0.6274	9.8975	0.1010	6.2098	0.1610	19.8416	3.1952
9	1.6895	0.5919	11.4913	0.0870	6.8017	0.1470	24.5768	3.6133
10	1.7908	0.5584	13.1808	0.0759	7.3601	0.1359	29.6023	4.0220
11	1.8983	0.5268	14.9716	0.0668	7.8869	0.1268	34.8702	4.4213
12	2.0122	0.4970	16.8699	0.0593	8.3838	0.1193	40.3369	4.8113
13	2.1329	0.4688	18.8821	0.0530	8.8527	0.1130	45.9629	5.1920
14	2.2609	0.4423	21.0151	0.0476	9.2950	0.1076	51.7128	5.5635
15	2.3966	0.4173	23.2760	0.0430	9.7122	0.1030	57.5546	5.9260
16	2.5404	0.3936	25.6725	0.0390	10.1059	0.0990	63.4592	6.2794
17	2.6928	0.3714	28.2129	0.0354	10.4773	0.0954	69.4011	6.6240
18	2.8543	0.3503	30.9057	0.0324	10.8276	0.0924	75.3569	6.9597
19	3.0256	0.3305	33.7600	0.0296	11.1581	0.0896	81.3062	7.2867
20	3.2071	0.3118	36.7856	0.0272	11.4699	0.0872	87.2304	7.6051
21	3.3996	0.2942	39.9927	0.0250	11.7641	0.0850	93.1136	7.9151
22	3.6035	0.2775	43.3923	0.0230	12.0416	0.0830	98.9412	8.2166
23	3.8197	0.2618	46.9958	0.0213	12.3034	0.0813	104.7007	8.5099
24	4.0489	0.2470	50.8156	0.0197	12.5504	0.0797	110.3812	8.7951
25	4.2919	0.2330	54.8645	0.0182	12.7834	0.0782	115.9732	9.0722
26	4.5494	0.2198	59.1564	0.0169	13.0032	0.0769	121.4684	9.3414
27	4.8223	0.2074	63.7058	0.0157	13.2105	0.0757	126.8600	9.6029
28	5.1117	0.1956	68.5281	0.0146	13.4062	0.0746	132.1420	9.8568
29	5.4184	0.1846	73.6398	0.0136	13.5907	0.0736	137.3096	10.1032
30	5.7435	0.1741	79.0582	0.0126	13.7648	0.0726	142.3588	10.3422
36	8.1473	0.1227	119.1209	8.395E-03	14.6210	0.0684	170.0387	11.6298
42	11.5570	0.0865	175.9505	5.683E-03	15.2245	0.0657	193.1732	12.6883
48	16.3939	0.0610	256.5645	3.898E-03	15.6500	0.0639	212.0351	13.5485
54	23.2550	0.0430	370.9170	2.696E-03	15.9500	0.0627	227.1316	14.2402
60	32.9877	0.0303	533.1282	1.876E-03	16.1614	0.0619	239.0428	14.7909
66	46.7937	0.0214	763.2278	1.310E-03	16.3105	0.0613	248.3341	15.2254
72	66.3777	0.0151	1.090E+03	9.177E-04	16.4156	0.0609	255.5146	15.5654
120	1.088E+03	9.190E-04	1.812E+04	5.519E-05	16.6514	0.0601	275.6846	16.5563
180	3.590E+04	2.786E-05	5.983E+05	1.672E-06	16.6662	0.0600	277.6865	16.6617
360	1.289E+09	7.760E-10	2.148E+10	4.656E-11	16.6667	0.0600	277.7778	16.6667

# Time Value of Money Factors—Discrete Compounding
i = 7%

	Single Sums		Uniform Series				Gradient Series	
n	To Find F Given P (F|P,i%,n)	To Find P Given F (P|F,i%,n)	To Find F Given A (F|A,i%,n)	To Find A Given F (A|F,i%,n)	To Find P Given A (P|A,i%,n)	To Find A Given P (A|P,i%,n)	To Find P Given G (P|G,i%,n)	To Find A Given G (A|G,i%,n)
1	1.0700	0.9346	1.0000	1.0000	0.9346	1.0700	0.0000	0.0000
2	1.1449	0.8734	2.0700	0.4831	1.8080	0.5531	0.8734	0.4831
3	1.2250	0.8163	3.2149	0.3111	2.6243	0.3811	2.5060	0.9549
4	1.3108	0.7629	4.4399	0.2252	3.3872	0.2952	4.7947	1.4155
5	1.4026	0.7130	5.7507	0.1739	4.1002	0.2439	7.6467	1.8650
6	1.5007	0.6663	7.1533	0.1398	4.7665	0.2098	10.9784	2.3032
7	1.6058	0.6227	8.6540	0.1156	5.3893	0.1856	14.7149	2.7304
8	1.7182	0.5820	10.2598	0.0975	5.9713	0.1675	18.7889	3.1465
9	1.8385	0.5439	11.9780	0.0835	6.5152	0.1535	23.1404	3.5517
10	1.9672	0.5083	13.8164	0.0724	7.0236	0.1424	27.7156	3.9461
11	2.1049	0.4751	15.7836	0.0634	7.4987	0.1334	32.4665	4.3296
12	2.2522	0.4440	17.8885	0.0559	7.9427	0.1259	37.3506	4.7025
13	2.4098	0.4150	20.1406	0.0497	8.3577	0.1197	42.3302	5.0648
14	2.5785	0.3878	22.5505	0.0443	8.7455	0.1143	47.3718	5.4167
15	2.7590	0.3624	25.1290	0.0398	9.1079	0.1098	52.4461	5.7583
16	2.9522	0.3387	27.8881	0.0359	9.4466	0.1059	57.5271	6.0897
17	3.1588	0.3166	30.8402	0.0324	9.7632	0.1024	62.5923	6.4110
18	3.3799	0.2959	33.9990	0.0294	10.0591	0.0994	67.6219	6.7225
19	3.6165	0.2765	37.3790	0.0268	10.3356	0.0968	72.5991	7.0242
20	3.8697	0.2584	40.9955	0.0244	10.5940	0.0944	77.5091	7.3163
21	4.1406	0.2415	44.8652	0.0223	10.8355	0.0923	82.3393	7.5990
22	4.4304	0.2257	49.0057	0.0204	11.0612	0.0904	87.0793	7.8725
23	4.7405	0.2109	53.4361	0.0187	11.2722	0.0887	91.7201	8.1369
24	5.0724	0.1971	58.1767	0.0172	11.4693	0.0872	96.2545	8.3923
25	5.4274	0.1842	63.2490	0.0158	11.6536	0.0858	100.6765	8.6391
26	5.8074	0.1722	68.6765	0.0146	11.8258	0.0846	104.9814	8.8773
27	6.2139	0.1609	74.4838	0.0134	11.9867	0.0834	109.1656	9.1072
28	6.6488	0.1504	80.6977	0.0124	12.1371	0.0824	113.2264	9.3289
29	7.1143	0.1406	87.3465	0.0114	12.2777	0.0814	117.1622	9.5427
30	7.6123	0.1314	94.4608	0.0106	12.4090	0.0806	120.9718	9.7487
36	11.4239	0.0875	148.9135	6.715E-03	13.0352	0.0767	141.1990	10.8321
42	17.1443	0.0583	230.6322	4.336E-03	13.4524	0.0743	157.1807	11.6842
48	25.7289	0.0389	353.2701	2.831E-03	13.7305	0.0728	169.4981	12.3447
54	38.6122	0.0259	537.3164	1.861E-03	13.9157	0.0719	178.8173	12.8500
60	57.9464	0.0173	813.5204	1.229E-03	14.0392	0.0712	185.7677	13.2321
66	86.9620	0.0115	1.228E+03	8.143E-04	14.1214	0.0708	190.8927	13.5179
72	130.5065	7.662E-03	1.850E+03	5.405E-04	14.1763	0.0705	194.6365	13.7298
120	3.358E+03	2.978E-04	4.795E+04	2.085E-05	14.2815	0.0700	203.5103	14.2500
180	1.946E+05	5.139E-06	2.780E+06	3.598E-07	14.2856	0.0700	204.0674	14.2848
360	3.786E+10	2.641E-11	5.408E+11	1.849E-12	14.2857	0.0700	204.0816	14.2857

Time Value of Money Factors—Discrete Compounding
i = 8%

	Single Sums		Uniform Series				Gradient Series	
n	To Find F Given P (F|P,i%,n)	To Find P Given F (P|F,i%,n)	To Find F Given A (F|A,i%,n)	To Find A Given F (A|F,i%,n)	To Find P Given A (P|A,i%,n)	To Find A Given P (A|P,i%,n)	To Find P Given G (P|G,i%,n)	To Find A Given G (A|G,i%,n)
1	1.0800	0.9259	1.0000	1.0000	0.9259	1.0800	0.0000	0.0000
2	1.1664	0.8573	2.0800	0.4808	1.7833	0.5608	0.8573	0.4808
3	1.2597	0.7938	3.2464	0.3080	2.5771	0.3880	2.4450	0.9487
4	1.3605	0.7350	4.5061	0.2219	3.3121	0.3019	4.6501	1.4040
5	1.4693	0.6806	5.8666	0.1705	3.9927	0.2505	7.3724	1.8465
6	1.5869	0.6302	7.3359	0.1363	4.6229	0.2163	10.5233	2.2763
7	1.7138	0.5835	8.9228	0.1121	5.2064	0.1921	14.0242	2.6937
8	1.8509	0.5403	10.6366	0.0940	5.7466	0.1740	17.8061	3.0985
9	1.9990	0.5002	12.4876	0.0801	6.2469	0.1601	21.8081	3.4910
10	2.1589	0.4632	14.4866	0.0690	6.7101	0.1490	25.9768	3.8713
11	2.3316	0.4289	16.6455	0.0601	7.1390	0.1401	30.2657	4.2395
12	2.5182	0.3971	18.9771	0.0527	7.5361	0.1327	34.6339	4.5957
13	2.7196	0.3677	21.4953	0.0465	7.9038	0.1265	39.0463	4.9402
14	2.9372	0.3405	24.2149	0.0413	8.2442	0.1213	43.4723	5.2731
15	3.1722	0.3152	27.1521	0.0368	8.5595	0.1168	47.8857	5.5945
16	3.4259	0.2919	30.3243	0.0330	8.8514	0.1130	52.2640	5.9046
17	3.7000	0.2703	33.7502	0.0296	9.1216	0.1096	56.5883	6.2037
18	3.9960	0.2502	37.4502	0.0267	9.3719	0.1067	60.8426	6.4920
19	4.3157	0.2317	41.4463	0.0241	9.6036	0.1041	65.0134	6.7697
20	4.6610	0.2145	45.7620	0.0219	9.8181	0.1019	69.0898	7.0369
21	5.0338	0.1987	50.4229	0.0198	10.0168	0.0998	73.0629	7.2940
22	5.4365	0.1839	55.4568	0.0180	10.2007	0.0980	76.9257	7.5412
23	5.8715	0.1703	60.8933	0.0164	10.3711	0.0964	80.6726	7.7786
24	6.3412	0.1577	66.7648	0.0150	10.5288	0.0950	84.2997	8.0066
25	6.8485	0.1460	73.1059	0.0137	10.6748	0.0937	87.8041	8.2254
26	7.3964	0.1352	79.9544	0.0125	10.8100	0.0925	91.1842	8.4352
27	7.9881	0.1252	87.3508	0.0114	10.9352	0.0914	94.4390	8.6363
28	8.6271	0.1159	95.3388	0.0105	11.0511	0.0905	97.5687	8.8289
29	9.3173	0.1073	103.9659	9.619E-03	11.1584	0.0896	100.5738	9.0133
30	10.0627	0.0994	113.2832	8.827E-03	11.2578	0.0888	103.4558	9.1897
36	15.9682	0.0626	187.1021	5.345E-03	11.7172	0.0853	118.2839	10.0949
42	25.3395	0.0395	304.2435	3.287E-03	12.0067	0.0833	129.3651	10.7744
48	40.2106	0.0249	490.1322	2.040E-03	12.1891	0.0820	137.4428	11.2758
54	63.8091	0.0157	785.1141	1.274E-03	12.3041	0.0813	143.2229	11.6403
60	101.2571	9.876E-03	1.253E+03	7.979E-04	12.3766	0.0808	147.3000	11.9015
66	160.6822	6.223E-03	1.996E+03	5.010E-04	12.4222	0.0805	150.1432	12.0867
72	254.9825	3.922E-03	3.175E+03	3.150E-04	12.4510	0.0803	152.1076	12.2165
120	1.025E+04	9.753E-05	1.281E+05	7.803E-06	12.4988	0.0800	156.0885	12.4883
180	1.038E+06	9.632E-07	1.298E+07	7.706E-08	12.5000	0.0800	156.2477	12.4998
360	1.078E+12	9.278E-13	1.347E+13	7.422E-14	12.5000	0.0800	156.2500	12.5000

# Time Value of Money Factors—Discrete Compounding
i = 9%

	Single Sums		Uniform Series				Gradient Series	
n	To Find F Given P (F|P,i%,n)	To Find P Given F (P|F,i%,n)	To Find F Given A (F|A,i%,n)	To Find A Given F (A|F,i%,n)	To Find P Given A (P|A,i%,n)	To Find A Given P (A|P,i%,n)	To Find P Given G (P|G,i%,n)	To Find A Given G (A|G,i%,n)
1	1.0900	0.9174	1.0000	1.0000	0.9174	1.0900	0.0000	0.0000
2	1.1881	0.8417	2.0900	0.4785	1.7591	0.5685	0.8417	0.4785
3	1.2950	0.7722	3.2781	0.3051	2.5313	0.3951	2.3860	0.9426
4	1.4116	0.7084	4.5731	0.2187	3.2397	0.3087	4.5113	1.3925
5	1.5386	0.6499	5.9847	0.1671	3.8897	0.2571	7.1110	1.8282
6	1.6771	0.5963	7.5233	0.1329	4.4859	0.2229	10.0924	2.2498
7	1.8280	0.5470	9.2004	0.1087	5.0330	0.1987	13.3746	2.6574
8	1.9926	0.5019	11.0285	0.0907	5.5348	0.1807	16.8877	3.0512
9	2.1719	0.4604	13.0210	0.0768	5.9952	0.1668	20.5711	3.4312
10	2.3674	0.4224	15.1929	0.0658	6.4177	0.1558	24.3728	3.7978
11	2.5804	0.3875	17.5603	0.0569	6.8052	0.1469	28.2481	4.1510
12	2.8127	0.3555	20.1407	0.0497	7.1607	0.1397	32.1590	4.4910
13	3.0658	0.3262	22.9534	0.0436	7.4869	0.1336	36.0731	4.8182
14	3.3417	0.2992	26.0192	0.0384	7.7862	0.1284	39.9633	5.1326
15	3.6425	0.2745	29.3609	0.0341	8.0607	0.1241	43.8069	5.4346
16	3.9703	0.2519	33.0034	0.0303	8.3126	0.1203	47.5849	5.7245
17	4.3276	0.2311	36.9737	0.0270	8.5436	0.1170	51.2821	6.0024
18	4.7171	0.2120	41.3013	0.0242	8.7556	0.1142	54.8860	6.2687
19	5.1417	0.1945	46.0185	0.0217	8.9501	0.1117	58.3868	6.5236
20	5.6044	0.1784	51.1601	0.0195	9.1285	0.1095	61.7770	6.7674
21	6.1088	0.1637	56.7645	0.0176	9.2922	0.1076	65.0509	7.0006
22	6.6586	0.1502	62.8733	0.0159	9.4424	0.1059	68.2048	7.2232
23	7.2579	0.1378	69.5319	0.0144	9.5802	0.1044	71.2359	7.4357
24	7.9111	0.1264	76.7898	0.0130	9.7066	0.1030	74.1433	7.6384
25	8.6231	0.1160	84.7009	0.0118	9.8226	0.1018	76.9265	7.8316
26	9.3992	0.1064	93.3240	0.0107	9.9290	0.1007	79.5863	8.0156
27	10.2451	0.0976	102.7231	9.735E-03	10.0266	0.0997	82.1241	8.1906
28	11.1671	0.0895	112.9682	8.852E-03	10.1161	0.0989	84.5419	8.3571
29	12.1722	0.0822	124.1354	8.056E-03	10.1983	0.0981	86.8422	8.5154
30	13.2677	0.0754	136.3075	7.336E-03	10.2737	0.0973	89.0280	8.6657
36	22.2512	0.0449	236.1247	4.235E-03	10.6118	0.0942	99.9319	9.4171
42	37.3175	0.0268	403.5281	2.478E-03	10.8134	0.0925	107.6432	9.9546
48	62.5852	0.0160	684.2804	1.461E-03	10.9336	0.0915	112.9625	10.3317
54	104.9617	9.527E-03	1.155E+03	8.657E-04	11.0053	0.0909	116.5642	10.5917
60	176.0313	5.681E-03	1.945E+03	5.142E-04	11.0480	0.0905	118.9683	10.7683
66	295.2221	3.387E-03	3.269E+03	3.059E-04	11.0735	0.0903	120.5546	10.8868
72	495.1170	2.020E-03	5.490E+03	1.821E-04	11.0887	0.0902	121.5917	10.9654
120	3.099E+04	3.227E-05	3.443E+05	2.905E-06	11.1108	0.0900	123.4098	11.1072
180	5.455E+06	1.833E-07	6.061E+07	1.650E-08	11.1111	0.0900	123.4564	11.1111
360	2.975E+13	3.361E-14	3.306E+14	3.025E-15	11.1111	0.0900	123.4568	11.1111

Time Value of Money Factors—Discrete Compounding
i = 10%

	Single Sums		Uniform Series				Gradient Series	
n	To Find F Given P (F|P,i%,n)	To Find P Given F (P|F,i%,n)	To Find F Given A (F|A,i%,n)	To Find A Given F (A|F,i%,n)	To Find P Given A (P|A,i%,n)	To Find A Given P (A|P,i%,n)	To Find P Given G (P|G,i%,n)	To Find A Given G (A|G,i%,n)
1	1.1000	0.9091	1.0000	1.0000	0.9091	1.1000	0.0000	0.0000
2	1.2100	0.8264	2.1000	0.4762	1.7355	0.5762	0.8264	0.4762
3	1.3310	0.7513	3.3100	0.3021	2.4869	0.4021	2.3291	0.9366
4	1.4641	0.6830	4.6410	0.2155	3.1699	0.3155	4.3781	1.3812
5	1.6105	0.6209	6.1051	0.1638	3.7908	0.2638	6.8618	1.8101
6	1.7716	0.5645	7.7156	0.1296	4.3553	0.2296	9.6842	2.2236
7	1.9487	0.5132	9.4872	0.1054	4.8684	0.2054	12.7631	2.6216
8	2.1436	0.4665	11.4359	0.0874	5.3349	0.1874	16.0287	3.0045
9	2.3579	0.4241	13.5795	0.0736	5.7590	0.1736	19.4215	3.3724
10	2.5937	0.3855	15.9374	0.0627	6.1446	0.1627	22.8913	3.7255
11	2.8531	0.3505	18.5312	0.0540	6.4951	0.1540	26.3963	4.0641
12	3.1384	0.3186	21.3843	0.0468	6.8137	0.1468	29.9012	4.3884
13	3.4523	0.2897	24.5227	0.0408	7.1034	0.1408	33.3772	4.6988
14	3.7975	0.2633	27.9750	0.0357	7.3667	0.1357	36.8005	4.9955
15	4.1772	0.2394	31.7725	0.0315	7.6061	0.1315	40.1520	5.2789
16	4.5950	0.2176	35.9497	0.0278	7.8237	0.1278	43.4164	5.5493
17	5.0545	0.1978	40.5447	0.0247	8.0216	0.1247	46.5819	5.8071
18	5.5599	0.1799	45.5992	0.0219	8.2014	0.1219	49.6395	6.0526
19	6.1159	0.1635	51.1591	0.0195	8.3649	0.1195	52.5827	6.2861
20	6.7275	0.1486	57.2750	0.0175	8.5136	0.1175	55.4069	6.5081
21	7.4002	0.1351	64.0025	0.0156	8.6487	0.1156	58.1095	6.7189
22	8.1403	0.1228	71.4027	0.0140	8.7715	0.1140	60.6893	6.9189
23	8.9543	0.1117	79.5430	0.0126	8.8832	0.1126	63.1462	7.1085
24	9.8497	0.1015	88.4973	0.0113	8.9847	0.1113	65.4813	7.2881
25	10.8347	0.0923	98.3471	0.0102	9.0770	0.1102	67.6964	7.4580
26	11.9182	0.0839	109.1818	9.159E-03	9.1609	0.1092	69.7940	7.6186
27	13.1100	0.0763	121.0999	8.258E-03	9.2372	0.1083	71.7773	7.7704
28	14.4210	0.0693	134.2099	7.451E-03	9.3066	0.1075	73.6495	7.9137
29	15.8631	0.0630	148.6309	6.728E-03	9.3696	0.1067	75.4146	8.0489
30	17.4494	0.0573	164.4940	6.079E-03	9.4269	0.1061	77.0766	8.1762
36	30.9127	0.0323	299.1268	3.343E-03	9.6765	0.1033	85.1194	8.7965
42	54.7637	0.0183	537.6370	1.860E-03	9.8174	0.1019	90.5047	9.2188
48	97.0172	0.0103	960.1723	1.041E-03	9.8969	0.1010	94.0217	9.5001
54	171.8719	5.818E-03	1.709E+03	5.852E-04	9.9418	0.1006	96.2763	9.6840
60	304.4816	3.284E-03	3.035E+03	3.295E-04	9.9672	0.1003	97.7010	9.8023
66	539.4078	1.854E-03	5.384E+03	1.857E-04	9.9815	0.1002	98.5910	9.8774
72	955.5938	1.046E-03	9.546E+03	1.048E-04	9.9895	0.1001	99.1419	9.9246
120	9.271E+04	1.079E-05	9.271E+05	1.079E-06	9.9999	0.1000	99.9860	9.9987
180	2.823E+07	3.543E-08	2.823E+08	3.543E-09	10.0000	0.1000	99.9999	10.0000
360	7.968E+14	1.255E-15	7.968E+15	1.255E-16	10.0000	0.1000	100.0000	10.0000

Time Value of Money Factors—Discrete Compounding
i = 12%

	Single Sums		Uniform Series				Gradient Series	
n	To Find F Given P (F|P,i%,n)	To Find P Given F (P|F,i%,n)	To Find F Given A (F|A,i%,n)	To Find A Given F (A|F,i%,n)	To Find P Given A (P|A,i%,n)	To Find A Given P (A|P,i%,n)	To Find P Given G (P|G,i%,n)	To Find A Given G (A|G,i%,n)
1	1.1200	0.8929	1.0000	1.0000	0.8929	1.1200	0.0000	0.0000
2	1.2544	0.7972	2.1200	0.4717	1.6901	0.5917	0.7972	0.4717
3	1.4049	0.7118	3.3744	0.2963	2.4018	0.4163	2.2208	0.9246
4	1.5735	0.6355	4.7793	0.2092	3.0373	0.3292	4.1273	1.3589
5	1.7623	0.5674	6.3528	0.1574	3.6048	0.2774	6.3970	1.7746
6	1.9738	0.5066	8.1152	0.1232	4.1114	0.2432	8.9302	2.1720
7	2.2107	0.4523	10.0890	0.0991	4.5638	0.2191	11.6443	2.5515
8	2.4760	0.4039	12.2997	0.0813	4.9676	0.2013	14.4714	2.9131
9	2.7731	0.3606	14.7757	0.0677	5.3282	0.1877	17.3563	3.2574
10	3.1058	0.3220	17.5487	0.0570	5.6502	0.1770	20.2541	3.5847
11	3.4785	0.2875	20.6546	0.0484	5.9377	0.1684	23.1288	3.8953
12	3.8960	0.2567	24.1331	0.0414	6.1944	0.1614	25.9523	4.1897
13	4.3635	0.2292	28.0291	0.0357	6.4235	0.1557	28.7024	4.4683
14	4.8871	0.2046	32.3926	0.0309	6.6282	0.1509	31.3624	4.7317
15	5.4736	0.1827	37.2797	0.0268	6.8109	0.1468	33.9202	4.9803
16	6.1304	0.1631	42.7533	0.0234	6.9740	0.1434	36.3670	5.2147
17	6.8660	0.1456	48.8837	0.0205	7.1196	0.1405	38.6973	5.4353
18	7.6900	0.1300	55.7497	0.0179	7.2497	0.1379	40.9080	5.6427
19	8.6128	0.1161	63.4397	0.0158	7.3658	0.1358	42.9979	5.8375
20	9.6463	0.1037	72.0524	0.0139	7.4694	0.1339	44.9676	6.0202
21	10.8038	0.0926	81.6987	0.0122	7.5620	0.1322	46.8188	6.1913
22	12.1003	0.0826	92.5026	0.0108	7.6446	0.1308	48.5543	6.3514
23	13.5523	0.0738	104.6029	9.560E-03	7.7184	0.1296	50.1776	6.5010
24	15.1786	0.0659	118.1552	8.463E-03	7.7843	0.1285	51.6929	6.6406
25	17.0001	0.0588	133.3339	7.500E-03	7.8431	0.1275	53.1046	6.7708
26	19.0401	0.0525	150.3339	6.652E-03	7.8957	0.1267	54.4177	6.8921
27	21.3249	0.0469	169.3740	5.904E-03	7.9426	0.1259	55.6369	7.0049
28	23.8839	0.0419	190.6989	5.244E-03	7.9844	0.1252	56.7674	7.1098
29	26.7499	0.0374	214.5828	4.660E-03	8.0218	0.1247	57.8141	7.2071
30	29.9599	0.0334	241.3327	4.144E-03	8.0552	0.1241	58.7821	7.2974
36	59.1356	0.0169	484.4631	2.064E-03	8.1924	0.1221	63.1970	7.7141
42	116.7231	8.567E-03	964.3595	1.037E-03	8.2619	0.1210	65.8509	7.9704
48	230.3908	4.340E-03	1.912E+03	5.231E-04	8.2972	0.1205	67.4068	8.1241
54	454.7505	2.199E-03	3.781E+03	2.645E-04	8.3150	0.1203	68.3022	8.2143
60	897.5969	1.114E-03	7.472E+03	1.338E-04	8.3240	0.1201	68.8100	8.2664
66	1.772E+03	5.644E-04	1.476E+04	6.777E-05	8.3286	0.1201	69.0948	8.2961
72	3.497E+03	2.860E-04	2.913E+04	3.432E-05	8.3310	0.1200	69.2530	8.3127
120	8.057E+05	1.241E-06	6.714E+06	1.489E-07	8.3333	0.1200	69.4431	8.3332
180	7.232E+08	1.383E-09	6.026E+09	1.659E-10	8.3333	0.1200	69.4444	8.3333
360	5.230E+17	1.912E-18	4.358E+18	2.295E-19	8.3333	0.1200	69.4444	8.3333

Time Value of Money Factors—Discrete Compounding
i = 15%

	Single Sums		Uniform Series				Gradient Series	
n	To Find F Given P (F|P,i%,n)	To Find P Given F (P|F,i%,n)	To Find F Given A (F|A,i%,n)	To Find A Given F (A|F,i%,n)	To Find P Given A (P|A,i%,n)	To Find A Given P (A|P,i%,n)	To Find P Given G (P|G,i%,n)	To Find A Given G (A|G,i%,n)
1	1.1500	0.8696	1.0000	1.0000	0.8696	1.1500	0.0000	0.0000
2	1.3225	0.7561	2.1500	0.4651	1.6257	0.6151	0.7561	0.4651
3	1.5209	0.6575	3.4725	0.2880	2.2832	0.4380	2.0712	0.9071
4	1.7490	0.5718	4.9934	0.2003	2.8550	0.3503	3.7864	1.3263
5	2.0114	0.4972	6.7424	0.1483	3.3522	0.2983	5.7751	1.7228
6	2.3131	0.4323	8.7537	0.1142	3.7845	0.2642	7.9368	2.0972
7	2.6600	0.3759	11.0668	0.0904	4.1604	0.2404	10.1924	2.4498
8	3.0590	0.3269	13.7268	0.0729	4.4873	0.2229	12.4807	2.7813
9	3.5179	0.2843	16.7858	0.0596	4.7716	0.2096	14.7548	3.0922
10	4.0456	0.2472	20.3037	0.0493	5.0188	0.1993	16.9795	3.3832
11	4.6524	0.2149	24.3493	0.0411	5.2337	0.1911	19.1289	3.6549
12	5.3503	0.1869	29.0017	0.0345	5.4206	0.1845	21.1849	3.9082
13	6.1528	0.1625	34.3519	0.0291	5.5831	0.1791	23.1352	4.1438
14	7.0757	0.1413	40.5047	0.0247	5.7245	0.1747	24.9725	4.3624
15	8.1371	0.1229	47.5804	0.0210	5.8474	0.1710	26.6930	4.5650
16	9.3576	0.1069	55.7175	0.0179	5.9542	0.1679	28.2960	4.7522
17	10.7613	0.0929	65.0751	0.0154	6.0472	0.1654	29.7828	4.9251
18	12.3755	0.0808	75.8364	0.0132	6.1280	0.1632	31.1565	5.0843
19	14.2318	0.0703	88.2118	0.0113	6.1982	0.1613	32.4213	5.2307
20	16.3665	0.0611	102.4436	9.761E-03	6.2593	0.1598	33.5822	5.3651
21	18.8215	0.0531	118.8101	8.417E-03	6.3125	0.1584	34.6448	5.4883
22	21.6447	0.0462	137.6316	7.266E-03	6.3587	0.1573	35.6150	5.6010
23	24.8915	0.0402	159.2764	6.278E-03	6.3988	0.1563	36.4988	5.7040
24	28.6252	0.0349	184.1678	5.430E-03	6.4338	0.1554	37.3023	5.7979
25	32.9190	0.0304	212.7930	4.699E-03	6.4641	0.1547	38.0314	5.8834
26	37.8568	0.0264	245.7120	4.070E-03	6.4906	0.1541	38.6918	5.9612
27	43.5353	0.0230	283.5688	3.526E-03	6.5135	0.1535	39.2890	6.0319
28	50.0656	0.0200	327.1041	3.057E-03	6.5335	0.1531	39.8283	6.0960
29	57.5755	0.0174	377.1697	2.651E-03	6.5509	0.1527	40.3146	6.1541
30	66.2118	0.0151	434.7451	2.300E-03	6.5660	0.1523	40.7526	6.2066
36	153.1519	6.529E-03	1.014E+03	9.859E-04	6.6231	0.1510	42.5872	6.4301
42	354.2495	2.823E-03	2.355E+03	4.246E-04	6.6478	0.1504	43.5286	6.5478
48	819.4007	1.220E-03	5.456E+03	1.833E-04	6.6585	0.1502	43.9997	6.6080
54	1.895E+03	5.276E-04	1.263E+04	7.918E-05	6.6631	0.1501	44.2311	6.6382
60	4.384E+03	2.281E-04	2.922E+04	3.422E-05	6.6651	0.1500	44.3431	6.6530
66	1.014E+04	9.861E-05	6.760E+04	1.479E-05	6.6660	0.1500	44.3967	6.6602
72	2.346E+04	4.263E-05	1.564E+05	6.395E-06	6.6664	0.1500	44.4221	6.6636
120	1.922E+07	5.203E-08	1.281E+08	7.805E-09	6.6667	0.1500	44.4444	6.6667
180	8.426E+10	1.187E-11	5.617E+11	1.780E-12	6.6667	0.1500	44.4444	6.6667
360	7.099E+21	1.409E-22	4.733E+22	2.113E-23	6.6667	0.1500	44.4444	6.6667

Time Value of Money Factors—Discrete Compounding
i = 18%

	Single Sums		Uniform Series				Gradient Series	
n	To Find F Given P (F|P,i%,n)	To Find P Given F (P|F,i%,n)	To Find F Given A (F|A,i%,n)	To Find A Given F (A|F,i%,n)	To Find P Given A (P|A,i%,n)	To Find A Given P (A|P,i%,n)	To Find P Given G (P|G,i%,n)	To Find A Given G (A|G,i%,n)
1	1.1800	0.8475	1.0000	1.0000	0.8475	1.1800	0.0000	0.0000
2	1.3924	0.7182	2.1800	0.4587	1.5656	0.6387	0.7182	0.4587
3	1.6430	0.6086	3.5724	0.2799	2.1743	0.4599	1.9354	0.8902
4	1.9388	0.5158	5.2154	0.1917	2.6901	0.3717	3.4828	1.2947
5	2.2878	0.4371	7.1542	0.1398	3.1272	0.3198	5.2312	1.6728
6	2.6996	0.3704	9.4420	0.1059	3.4976	0.2859	7.0834	2.0252
7	3.1855	0.3139	12.1415	0.0824	3.8115	0.2624	8.9670	2.3526
8	3.7589	0.2660	15.3270	0.0652	4.0776	0.2452	10.8292	2.6558
9	4.4355	0.2255	19.0859	0.0524	4.3030	0.2324	12.6329	2.9358
10	5.2338	0.1911	23.5213	0.0425	4.4941	0.2225	14.3525	3.1936
11	6.1759	0.1619	28.7551	0.0348	4.6560	0.2148	15.9716	3.4303
12	7.2876	0.1372	34.9311	0.0286	4.7932	0.2086	17.4811	3.6470
13	8.5994	0.1163	42.2187	0.0237	4.9095	0.2037	18.8765	3.8449
14	10.1472	0.0985	50.8180	0.0197	5.0081	0.1997	20.1576	4.0250
15	11.9737	0.0835	60.9653	0.0164	5.0916	0.1964	21.3269	4.1887
16	14.1290	0.0708	72.9390	0.0137	5.1624	0.1937	22.3885	4.3369
17	16.6722	0.0600	87.0680	0.0115	5.2223	0.1915	23.3482	4.4708
18	19.6733	0.0508	103.7403	9.639E-03	5.2732	0.1896	24.2123	4.5916
19	23.2144	0.0431	123.4135	8.103E-03	5.3162	0.1881	24.9877	4.7003
20	27.3930	0.0365	146.6280	6.820E-03	5.3527	0.1868	25.6813	4.7978
21	32.3238	0.0309	174.0210	5.746E-03	5.3837	0.1857	26.3000	4.8851
22	38.1421	0.0262	206.3448	4.846E-03	5.4099	0.1848	26.8506	4.9632
23	45.0076	0.0222	244.4868	4.090E-03	5.4321	0.1841	27.3394	5.0329
24	53.1090	0.0188	289.4945	3.454E-03	5.4509	0.1835	27.7725	5.0950
25	62.6686	0.0160	342.6035	2.919E-03	5.4669	0.1829	28.1555	5.1502
26	73.9490	0.0135	405.2721	2.467E-03	5.4804	0.1825	28.4935	5.1991
27	87.2598	0.0115	479.2211	2.087E-03	5.4919	0.1821	28.7915	5.2425
28	102.9666	9.712E-03	566.4809	1.765E-03	5.5016	0.1818	29.0537	5.2810
29	121.5005	8.230E-03	669.4475	1.494E-03	5.5098	0.1815	29.2842	5.3149
30	143.3706	6.975E-03	790.9480	1.264E-03	5.5168	0.1813	29.4864	5.3448
36	387.0368	2.584E-03	2.145E+03	4.663E-04	5.5412	0.1805	30.2677	5.4623
42	1.045E+03	9.571E-04	5.799E+03	1.724E-04	5.5502	0.1802	30.6113	5.5153
48	2.821E+03	3.545E-04	1.566E+04	6.384E-05	5.5536	0.1801	30.7587	5.5385
54	7.614E+03	1.313E-04	4.230E+04	2.364E-05	5.5548	0.1800	30.8207	5.5485
60	2.056E+04	4.865E-05	1.142E+05	8.757E-06	5.5553	0.1800	30.8465	5.5526
66	5.549E+04	1.802E-05	3.083E+05	3.244E-06	5.5555	0.1800	30.8570	5.5544
72	1.498E+05	6.676E-06	8.322E+05	1.202E-06	5.5555	0.1800	30.8613	5.5551
120	4.225E+08	2.367E-09	2.347E+09	4.260E-10	5.5556	0.1800	30.8642	5.5556
180	8.685E+12	1.151E-13	4.825E+13	2.073E-14	5.5556	0.1800	30.8642	5.5556
360	7.543E+25	1.326E-26	4.190E+26	2.386E-27	5.5556	0.1800	30.8642	5.5556

Time Value of Money Factors—Discrete Compounding
i = 20%

	Single Sums		Uniform Series				Gradient Series	
n	To Find F Given P (F|P,i%,n)	To Find P Given F (P|F,i%,n)	To Find F Given A (F|A,i%,n)	To Find A Given F (A|F,i%,n)	To Find P Given A (P|A,i%,n)	To Find A Given P (A|P,i%,n)	To Find P Given G (P|G,i%,n)	To Find A Given G (A|G,i%,n)
1	1.2000	0.8333	1.0000	1.0000	0.8333	1.2000	0.0000	0.0000
2	1.4400	0.6944	2.2000	0.4545	1.5278	0.6545	0.6944	0.4545
3	1.7280	0.5787	3.6400	0.2747	2.1065	0.4747	1.8519	0.8791
4	2.0736	0.4823	5.3680	0.1863	2.5887	0.3863	3.2986	1.2742
5	2.4883	0.4019	7.4416	0.1344	2.9906	0.3344	4.9061	1.6405
6	2.9860	0.3349	9.9299	0.1007	3.3255	0.3007	6.5806	1.9788
7	3.5832	0.2791	12.9159	0.0774	3.6046	0.2774	8.2551	2.2902
8	4.2998	0.2326	16.4991	0.0606	3.8372	0.2606	9.8831	2.5756
9	5.1598	0.1938	20.7989	0.0481	4.0310	0.2481	11.4335	2.8364
10	6.1917	0.1615	25.9587	0.0385	4.1925	0.2385	12.8871	3.0739
11	7.4301	0.1346	32.1504	0.0311	4.3271	0.2311	14.2330	3.2893
12	8.9161	0.1122	39.5805	0.0253	4.4392	0.2253	15.4667	3.4841
13	10.6993	0.0935	48.4966	0.0206	4.5327	0.2206	16.5883	3.6597
14	12.8392	0.0779	59.1959	0.0169	4.6106	0.2169	17.6008	3.8175
15	15.4070	0.0649	72.0351	0.0139	4.6755	0.2139	18.5095	3.9588
16	18.4884	0.0541	87.4421	0.0114	4.7296	0.2114	19.3208	4.0851
17	22.1861	0.0451	105.9306	9.440E-03	4.7746	0.2094	20.0419	4.1976
18	26.6233	0.0376	128.1167	7.805E-03	4.8122	0.2078	20.6805	4.2975
19	31.9480	0.0313	154.7400	6.462E-03	4.8435	0.2065	21.2439	4.3861
20	38.3376	0.0261	186.6880	5.357E-03	4.8696	0.2054	21.7395	4.4643
21	46.0051	0.0217	225.0256	4.444E-03	4.8913	0.2044	22.1742	4.5334
22	55.2061	0.0181	271.0307	3.690E-03	4.9094	0.2037	22.5546	4.5941
23	66.2474	0.0151	326.2369	3.065E-03	4.9245	0.2031	22.8867	4.6475
24	79.4968	0.0126	392.4842	2.548E-03	4.9371	0.2025	23.1760	4.6943
25	95.3962	0.0105	471.9811	2.119E-03	4.9476	0.2021	23.4276	4.7352
26	114.4755	8.735E-03	567.3773	1.762E-03	4.9563	0.2018	23.6460	4.7709
27	137.3706	7.280E-03	681.8528	1.467E-03	4.9636	0.2015	23.8353	4.8020
28	164.8447	6.066E-03	819.2233	1.221E-03	4.9697	0.2012	23.9991	4.8291
29	197.8136	5.055E-03	984.0680	1.016E-03	4.9747	0.2010	24.1406	4.8527
30	237.3763	4.213E-03	1.182E+03	8.461E-04	4.9789	0.2008	24.2628	4.8731
36	708.8019	1.411E-03	3.539E+03	2.826E-04	4.9929	0.2003	24.7108	4.9491
42	2.116E+03	4.725E-04	1.058E+04	9.454E-05	4.9976	0.2001	24.8890	4.9801
48	6.320E+03	1.582E-04	3.159E+04	3.165E-05	4.9992	0.2000	24.9581	4.9924
54	1.887E+04	5.299E-05	9.435E+04	1.060E-05	4.9997	0.2000	24.9844	4.9971
60	5.635E+04	1.775E-05	2.817E+05	3.549E-06	4.9999	0.2000	24.9942	4.9989
66	1.683E+05	5.943E-06	8.413E+05	1.189E-06	5.0000	0.2000	24.9979	4.9996
72	5.024E+05	1.990E-06	2.512E+06	3.981E-07	5.0000	0.2000	24.9992	4.9999
120	3.175E+09	3.150E-10	1.588E+10	6.299E-11	5.0000	0.2000	25.0000	5.0000
180	1.789E+14	5.590E-15	8.945E+14	1.118E-15	5.0000	0.2000	25.0000	5.0000
360	3.201E+28	3.124E-29	1.600E+29	6.249E-30	5.0000	0.2000	25.0000	5.0000

Time Value of Money Factors—Discrete Compounding
i = 25%

	Single Sums		Uniform Series				Gradient Series	
n	To Find F Given P (F|P,i%,n)	To Find P Given F (P|F,i%,n)	To Find F Given A (F|A,i%,n)	To Find A Given F (A|F,i%,n)	To Find P Given A (P|A,i%,n)	To Find A Given P (A|P,i%,n)	To Find P Given G (P|G,i%,n)	To Find A Given G (A|G,i%,n)
1	1.2500	0.8000	1.0000	1.0000	0.8000	1.2500	0.0000	0.0000
2	1.5625	0.6400	2.2500	0.4444	1.4400	0.6944	0.6400	0.4444
3	1.9531	0.5120	3.8125	0.2623	1.9520	0.5123	1.6640	0.8525
4	2.4414	0.4096	5.7656	0.1734	2.3616	0.4234	2.8928	1.2249
5	3.0518	0.3277	8.2070	0.1218	2.6893	0.3718	4.2035	1.5631
6	3.8147	0.2621	11.2588	0.0888	2.9514	0.3388	5.5142	1.8683
7	4.7684	0.2097	15.0735	0.0663	3.1611	0.3163	6.7725	2.1424
8	5.9605	0.1678	19.8419	0.0504	3.3289	0.3004	7.9469	2.3872
9	7.4506	0.1342	25.8023	0.0388	3.4631	0.2888	9.0207	2.6048
10	9.3132	0.1074	33.2529	0.0301	3.5705	0.2801	9.9870	2.7971
11	11.6415	0.0859	42.5661	0.0235	3.6564	0.2735	10.8460	2.9663
12	14.5519	0.0687	54.2077	0.0184	3.7251	0.2684	11.6020	3.1145
13	18.1899	0.0550	68.7596	0.0145	3.7801	0.2645	12.2617	3.2437
14	22.7374	0.0440	86.9495	0.0115	3.8241	0.2615	12.8334	3.3559
15	28.4217	0.0352	109.6868	9.117E-03	3.8593	0.2591	13.3260	3.4530
16	35.5271	0.0281	138.1085	7.241E-03	3.8874	0.2572	13.7482	3.5366
17	44.4089	0.0225	173.6357	5.759E-03	3.9099	0.2558	14.1085	3.6084
18	55.5112	0.0180	218.0446	4.586E-03	3.9279	0.2546	14.4147	3.6698
19	69.3889	0.0144	273.5558	3.656E-03	3.9424	0.2537	14.6741	3.7222
20	86.7362	0.0115	342.9447	2.916E-03	3.9539	0.2529	14.8932	3.7667
21	108.4202	9.223E-03	429.6809	2.327E-03	3.9631	0.2523	15.0777	3.8045
22	135.5253	7.379E-03	538.1011	1.858E-03	3.9705	0.2519	15.2326	3.8365
23	169.4066	5.903E-03	673.6264	1.485E-03	3.9764	0.2515	15.3625	3.8634
24	211.7582	4.722E-03	843.0329	1.186E-03	3.9811	0.2512	15.4711	3.8861
25	264.6978	3.778E-03	1.055E+03	9.481E-04	3.9849	0.2509	15.5618	3.9052
26	330.8722	3.022E-03	1.319E+03	7.579E-04	3.9879	0.2508	15.6373	3.9212
27	413.5903	2.418E-03	1.650E+03	6.059E-04	3.9903	0.2506	15.7002	3.9346
28	516.9879	1.934E-03	2.064E+03	4.845E-04	3.9923	0.2505	15.7524	3.9457
29	646.2349	1.547E-03	2.581E+03	3.875E-04	3.9938	0.2504	15.7957	3.9551
30	807.7936	1.238E-03	3.227E+03	3.099E-04	3.9950	0.2503	15.8316	3.9628
36	3.081E+03	3.245E-04	1.232E+04	8.116E-05	3.9987	0.2501	15.9481	3.9883
42	1.175E+04	8.507E-05	4.702E+04	2.127E-05	3.9997	0.2500	15.9843	3.9964
48	4.484E+04	2.230E-05	1.794E+05	5.575E-06	3.9999	0.2500	15.9954	3.9989
54	1.711E+05	5.846E-06	6.842E+05	1.462E-06	4.0000	0.2500	15.9986	3.9997
60	6.525E+05	1.532E-06	2.610E+06	3.831E-07	4.0000	0.2500	15.9996	3.9999
66	2.489E+06	4.017E-07	9.957E+06	1.004E-07	4.0000	0.2500	15.9999	4.0000
72	9.496E+06	1.053E-07	3.798E+07	2.633E-08	4.0000	0.2500	16.0000	4.0000
120	4.258E+11	2.349E-12	1.703E+12	5.871E-13	4.0000	0.2500	16.0000	4.0000
180	2.778E+17	3.599E-18	1.111E+18	8.998E-19	4.0000	0.2500	16.0000	4.0000
360	7.720E+34	1.295E-35	3.088E+35	3.238E-36	4.0000	0.2500	16.0000	4.0000

Time Value of Money Factors—Discrete Compounding i = 30%

	Single Sums		Uniform Series				Gradient Series	
	To Find F Given P	To Find P Given F	To Find F Given A	To Find A Given F	To Find P Given A	To Find A Given P	To Find P Given G	To Find A Given G
n	(F|P,i%,n)	(P|F,i%,n)	(F|A,i%,n)	(A|F,i%,n)	(P|A,i%,n)	(A|P,i%,n)	(P|G,i%,n)	(A|G,i%,n)
1	1.3000	0.7692	1.0000	1.0000	0.7692	1.3000	0.0000	0.0000
2	1.6900	0.5917	2.3000	0.4348	1.3609	0.7348	0.5917	0.4348
3	2.1970	0.4552	3.9900	0.2506	1.8161	0.5506	1.5020	0.8271
4	2.8561	0.3501	6.1870	0.1616	2.1662	0.4616	2.5524	1.1783
5	3.7129	0.2693	9.0431	0.1106	2.4356	0.4106	3.6297	1.4903
6	4.8268	0.2072	12.7560	0.0784	2.6427	0.3784	4.6656	1.7654
7	6.2749	0.1594	17.5828	0.0569	2.8021	0.3569	5.6218	2.0063
8	8.1573	0.1226	23.8577	0.0419	2.9247	0.3419	6.4800	2.2156
9	10.6045	0.0943	32.0150	0.0312	3.0190	0.3312	7.2343	2.3963
10	13.7858	0.0725	42.6195	0.0235	3.0915	0.3235	7.8872	2.5512
11	17.9216	0.0558	56.4053	0.0177	3.1473	0.3177	8.4452	2.6833
12	23.2981	0.0429	74.3270	0.0135	3.1903	0.3135	8.9173	2.7952
13	30.2875	0.0330	97.6250	0.0102	3.2233	0.3102	9.3135	2.8895
14	39.3738	0.0254	127.9125	7.818E-03	3.2487	0.3078	9.6437	2.9685
15	51.1859	0.0195	167.2863	5.978E-03	3.2682	0.3060	9.9172	3.0344
16	66.5417	0.0150	218.4722	4.577E-03	3.2832	0.3046	10.1426	3.0892
17	86.5042	0.0116	285.0139	3.509E-03	3.2948	0.3035	10.3276	3.1345
18	112.4554	8.892E-03	371.5180	2.692E-03	3.3037	0.3027	10.4788	3.1718
19	146.1920	6.840E-03	483.9734	2.066E-03	3.3105	0.3021	10.6019	3.2025
20	190.0496	5.262E-03	630.1655	1.587E-03	3.3158	0.3016	10.7019	3.2275
21	247.0645	4.048E-03	820.2151	1.219E-03	3.3198	0.3012	10.7828	3.2480
22	321.1839	3.113E-03	1.067E+03	9.370E-04	3.3230	0.3009	10.8482	3.2646
23	417.5391	2.395E-03	1.388E+03	7.202E-04	3.3254	0.3007	10.9009	3.2781
24	542.8008	1.842E-03	1.806E+03	5.537E-04	3.3272	0.3006	10.9433	3.2890
25	705.6410	1.417E-03	2.349E+03	4.257E-04	3.3286	0.3004	10.9773	3.2979
26	917.3333	1.090E-03	3.054E+03	3.274E-04	3.3297	0.3003	11.0045	3.3050
27	1.193E+03	8.386E-04	3.972E+03	2.518E-04	3.3305	0.3003	11.0263	3.3107
28	1.550E+03	6.450E-04	5.164E+03	1.936E-04	3.3312	0.3002	11.0437	3.3153
29	2.015E+03	4.962E-04	6.715E+03	1.489E-04	3.3317	0.3001	11.0576	3.3189
30	2.620E+03	3.817E-04	8.730E+03	1.145E-04	3.3321	0.3001	11.0687	3.3219
36	1.265E+04	7.908E-05	4.215E+04	2.372E-05	3.3331	0.3000	11.1007	3.3305
42	6.104E+04	1.638E-05	2.035E+05	4.915E-06	3.3333	0.3000	11.1086	3.3326
48	2.946E+05	3.394E-06	9.821E+05	1.018E-06	3.3333	0.3000	11.1105	3.3332
54	1.422E+06	7.032E-07	4.740E+06	2.110E-07	3.3333	0.3000	11.1110	3.3333
60	6.864E+06	1.457E-07	2.288E+07	4.370E-08	3.3333	0.3000	11.1111	3.3333
66	3.313E+07	3.018E-08	1.104E+08	9.054E-09	3.3333	0.3000	11.1111	3.3333
72	1.599E+08	6.253E-09	5.331E+08	1.876E-09	3.3333	0.3000	11.1111	3.3333
120	4.712E+13	2.122E-14	1.571E+14	6.367E-15	3.3333	0.3000	11.1111	3.3333
180	3.234E+20	3.092E-21	1.078E+21	9.275E-22	3.3333	0.3000	11.1111	3.3333
360	1.046E+41	9.559E-42	3.487E+41	2.868E-42	3.3333	0.3000	11.1111	3.3333

Appendix B

A non-profit corporation Efficiency Valuation Organization (EVO) (www.evo-world.org) has developed the International Performance Measurement and Verification Protocol (IPMVP), now in its fourth edition. In 2008 EVO expects to release the first edition of its new International Energy Efficiency Financing Protocol (IEEFP). The IEEFP will help financiers understand energy efficiency projects, specifically addressing:

- assessment of the quality of savings predictions and project plans;
- management of the installation and operational problems that can impede the actual savings;
- use of the verified energy savings stream to enhance the negligible collateral usually found in most efficiency projects;
- appropriate terms and conditions for lending agreements; and
- assessment of the validity of energy savings reports or claims.

The IPMVP was initially funded by the American government but is now funded by a wide range of interested parties and over 200 individuals. EVO also receives funding from its recent IPMVP training in Canada, China, Europe, South Africa, Taiwan and the United States. The work is supported by a large group of volunteers, many of who have been committed to the work for the past 11 years.

Development of the IEEFP has been supported by Asia-Pacific Economic Cooperation (APEC), and the British Global Opportunities Fund (GOF) in its first stages of development for Thailand and Mexico. It has engaged volunteer financiers from these two specific countries and around the world on EVO's IEEFP volunteer committee.

International Performance Measurement & Verification Protocol

Concepts and Options for Determining Energy and Water Savings Volume I

International Performance Measurement & Verification Protocol Committee

Revised March 2002
DOE/GO-102002-1554

Contents

Acknowledgments

PARTICIPATING ORGANIZATIONS

Brazil
- Institute Nacional De Eficiencia Energetica (INEE)
- Ministry of Mines and Energy

Bulgaria Bulgarian Foundation for Energy Efficiency (EnEffect)

Canada
- Canadian Association of Energy Service Companies (CAESCO)
- Natural Resources Canada (NRC)

China
- State Economic and Trade Commission
- Beijing Energy Efficiency Center (BECON)

Czech Republic Stredisko pro Efektivni Vyuzivani Energie (SEVEn7)

India Tata Energy Research Institute (TERI)

Japan Ministry of International Trade and Industry (MITI)

Korea	Korea Energy Management Corporation (KEMCO)
Mexico	• Comision Nacional Para El Ahorro De Energia (CONAE) • Fideicomiso De Apoyo Al Programa De Ahorro De Energia Del Sector Electrico (FIDE)
Poland	The Polish Foundation for Energy Efficiency (FEWE)
Russia	Center for Energy Efficiency (CENEf)
Sweden	Swedish National Board for Technical and Urban Development
Ukraine	Agency for Rational Energy Use and Ecology (ARENA—ECO)
United Kingdom	Association for the Conservation of Energy
United States	• American Society of Heating, Refrigerating and Air-Conditioning Engineers (ASHRAE) • Association of Energy Engineers (AEE) • Association of Energy Services Professionals (AESP) • Building Owners and Managers Association (BOMA) • National Association of Energy Service Companies (NAESCO) • National Association of State Energy Officials (NASEO) • National Realty Association • U.S. Department of Energy (DOE) • U.S. Environmental Protection Agency (EPA)

INTERNATIONAL PERFORMANCE MEASUREMENT AND VERIFICATION PROTOCOL COMMITTEES[1]

1. IPMVP Committee members—about 150 in all—are directly responsible for the content of the protocol. An additional approximately equal number of international experts served as advisors and reviewers and although there is not enough space to list them all here, their guidance and assistance is gratefully acknowledged. Committee members for the Adjustments, IEQ, New Buildings, Renewable Energy, and Water Committees are listed in Volumes II and III and on the IPMVP Web site.

EXECUTIVE COMMITTEE

1 Gregory Kats (Chair), Department of Energy, USA
2 Jim Halpern (Vice Chair), Measuring and Monitoring Services Inc., USA
3 John Armstrong, Hagler Bailly Services, USA
4 Flavio Conti, European Commission, Italy
5 Drury Crawley, US Department of Energy, USA
6 Dave Dayton, HEC Energy, USA
7 Adam Gula, Polish Foundation for Energy Efficiency, Poland
8 Shirley Hansen, Kiona International, USA
9 Leja Hattiangadi, TCE Consulting Engineers Limited, India
10 Maury Hepner, Enron Energy Services, USA
11 Chaan-min Lin, Hong Kong Productivity Council, Hong Kong
12 Arthur Rosenfeld, California Energy Commission, USA

TECHNICAL COMMITTEE

1 John Cowan (Co-chair and Technical Editor), Cowan Quality Buildings, Canada
2 Steve Kromer (Co-chair), Enron Energy Services, USA
3 David E. Claridge, Texas A & M University, USA
4 Ellen Franconi, Schiller Associates, USA
5 Jeff S. Haberl, Texas A & M University, USA
6 Maury Hepner, Enron Energy Services, USA
7 Satish Kumar, Lawrence Berkeley National Laboratory, USA
8 Eng Lock Lee, Supersymmetry Services Pvt. Ltd., Singapore
9 Mark Martinez, Southern California Edison, USA
10 David McGeown, NewEnergy, Inc., USA
11 Steve Schiller, Schiller Associates, USA

ADJUSTMENTS COMMITTEE

Co-chair: Hemant Joshi, Credit Rating Information Services of India Limited (CRISIL), India
Co-chair: Jayant Sathaye, Lawrence Berkeley National Laboratory, USA
Co-chair: Ed Vine, Lawrence Berkeley National Laboratory, USA

INDOOR ENVIRONMENTAL QUALITY COMMITTEE

Chair: Bill Fisk, Lawrence Berkeley National Laboratory, USA

NEW BUILDINGS COMMITTEE

Chair: Gordon Shymko, Tescor Pacific Energy Services, Inc., Canada

RENEWABLE ENERGY COMMITTEE

Co-chair: David Mills, University of New South Wales, Australia
Co-chair: Andy Walker, National Renewable Energy Laboratory, USA

WATER COMMITTEE

Co-chair: Tom Homer, Water Mangement Inc., USA
Co-chair: Warren Leibold, NYC Department of Environmental Protection, USA

TECHNICAL COORDINATOR

Satish Kumar, Lawrence Berkeley National Laboratory, USA Email: SKumar@lbl.gov, Phone: 202-646-7953

We would like to gratefully acknowledge the many organizations that made the IPMVP possible. In particular we would like to thank the Office of Building Technology, State and Community Programs in the U.S. Department of Energy's Office of Energy Efficiency and Renewable Energy, which has provided essential funding support to the IPMVP, including publication of this document.

The reprinting of Volume I and II of the IPMVP has been made possible through a generous grant from the U.S. Department of Energy's Federal Energy Management Program (FEMP). FEMP's gesture is greatly appreciated by the IPMVP.

DISCLAIMER

This Protocol serves as a framework to determine energy and water savings resulting from the implementation of an energy efficiency program. It is also intended to help monitor the performance of renewable energy systems and to enhance indoor environmental quality in buildings. The JPMVP does not create any legal rights or impose any legal obligations on any person or other legal entity. IPMVP has no legal authority or legal obligation to oversee, monitor or ensure compliance with provisions negotiated and included in contractual arrangements between third persons or third parties. It is the responsibility of the parties to a particular contract to reach agreement as to what, if *any, of this protocol is included in the contract and to ensure compliance.*

Preface

PURPOSE AND SCOPE

The International Performance Measurement and Verification Protocol (MVP) provides an overview of current best practice techniques available for verifying results of energy efficiency, water efficiency, and renewable energy projects. It may also be used by facility operators to assess and improve facility performance. Energy conservation measures[1] *(ECMS*[2]) covered herein include fuel saving measures, water efficiency measures, load shifting and energy reductions through installation or retrofit of equipment, and/or modification of operating procedures.

The IPMVP is not intended to prescribe contractual terms between buyers and sellers of efficiency services, although it provides guidance on some of these issues. Once other contractual issues are decided, this document can help in the selection of the *measurement & verification* (M&V) approach that best matches: i) project costs and savings magnitude, ii) technology-specific requirements, and iii) risk allocation between buyer and seller, i.e., which party is responsible for installed equipment performance and which party is responsible for achieving long term *energy savings.*

Two dimensions of ECM performance verification are addressed in this document:

- Savings determination technique using available data of suitable quality.
- Disclosure of data and analysis enabling one party to perform saving determinations while another verifies it.

1. Although there is some debate over the differences between the two terms energy conservation measure (ECM) and energy efficiency measure (EEM) they have been used interchangeably in this document.
2. The terms in italics are defined in Chapter 6.1

STRUCTURE OF IPMVP

Based on extensive user feedback, this version provides greater internal consistency, more precise definition of M&V Options, and treatment of additional issues, described below. Additional guidance is provided on how to adhere to the IPMVP. This edition of IPMVP is divided into three separate volumes:

Volume I Concepts and Options for Determining Savings
Volume II Indoor Environmental Quality (IEQ) Issues
Volume III Applications

Volume I defines basic terminology useful in the M&V field. It defines general procedures to achieve reliable and cost-effective determination of savings. Such definitions then can be customized for each project, with the help of other resources (see Chapter 1.4 and Chapter 6.2). *Verification* of savings is then done relative to the M&V Plan for the project. This volume is written for general application in measuring and verifying the performance of projects improving energy or water efficiency in buildings and industrial plants.

Volume I is largely drawn from the December 1997 edition of IPMVP. Apart from a general refocusing of the document for increased clarity, the definitions of Options A and B have been significantly modified in response to reactions received to earlier editions. These changes now include required field measurement of at least some variables under Option A, and all variables under Option B. Examples of each M&V Option have been added in Appendix A. Fortner sections on M&V for new buildings, residential and water efficiency have been moved to Volume III. The text has been updated and language tightened to achieve greater technical consistency and ease of use.

Volume II reviews indoor environmental quality issues as they may be influenced by an energy efficiency project. It focuses on measurement issues and project design and implementation practices associated with maintaining acceptable indoor conditions under an energy efficiency project, while advising on key related elements of M&V and *energy performance contracts.* Volume II is scheduled for publication concurrently with Volume I.

Volume III is planned for publication in early 2001, and reflects guidance and input of over 100 international experts. It will review application specific M&V issues. It is intended to address M&V specifics related to efficiency projects in industrial processes, new buildings, renewable energy, water efficiency, and emission trading. This volume is expected to be an area of continued development as more specific applications are defined.

NEW TOPICS

IPMVP 2000, in three volumes, introduces new topics of M&V for maintaining building indoor environmental quality (Volume II) and for renewable energy projects (Volume III), as summarized below.

Indoor Environmental Quality—Many building energy conservation measures have the potential to positively or negatively affect indoor pollutant concentrations, thermal comfort conditions, and lighting quality. These and other indoor environmental characteristics, which are collectively referred to as indoor environmental quality (IEQ), can influence the health, comfort, and productivity of building occupants. Even small changes in occupant health and productivity may be very significant financially, sometimes exceeding the financial benefits of energy conservation. Financial benefits resulting from improvements in IEQ can serve as a stimulus for energy efficiency investments. It is important that these IEQ considerations be explicitly recognized prior to selection and implementation of building energy efficiency measures. Volume II provides information that will help energy conservation professionals and building owners and managers maintain or improve IEQ when they implement building energy efficiency measures in non-industrial commercial and public buildings. This document also describes practical IEQ and ventilation measurements that can also help energy conservation professionals maintain or improve IEQ.

Volume II represents a consensus effort of approximately 25 committee members from 10 countries. The final work was then peer-reviewed by about 40 international experts whose comments and suggestions were incorporated in the final document.

Renewable Energy—The Renewable Energy section that will be part of Volume III provides a description of M&V Options for renewables within the IPMVP framework with examples and recommendations for specific applications. The term "renewable energy" refers to sources of energy that are regenerated by nature and sustainable in supply. Examples of renewable energy include solar, wind, biomass (sustainably harvested fuel crops, waste-to-energy, landfill gas), and geothermal energy. Strategies for M&V of renewables are important in designing, *commissioning,* serving as basis for financing payments, and providing ongoing diagnostics. In addition, good M&V can help reduce transaction costs by providing developers, investors, lenders, and customers with confidence regarding the value of projects and the allocation of risk. The section describes how the different M&V Options can be applied to renewable energy systems, and provides several examples.

Characteristics unique to renewable energy systems require M&V techniques distinct from those applied to energy efficiency projects. Renewable energy is generally capital-intensive and some sources, such as wind, rely on intermittent resources requiring special procedures to measure effects on the integrated energy system—including proper valuation of increased capacity and redundancy. Many of the benefits of renewables are external to conventional evaluation and accounting techniques. A sound protocol for measuring the performance and quantifying benefits unique to renewable energy systems can be a valuable part of recognizing real benefits of renewables that are often not part of current evaluation and accounting techniques.

The section represents a consensus effort of 65 committee members from 20 countries.

FUTURE WORK

The IPMVP is maintained with the sponsorship of the U.S. Department of Energy by a broad international coalition of facility owners/operators, financiers, contractors or *Energy Services Companies* (ESCOs) and other stakeholders. Continued international development and adoption of IPMVP will involve increasingly broad international participation and management of the document as well as its translation and adoption into a growing number of languages and application in a growing number of countries.

As a living document, every new version of IPMVP incorporates changes and improvements reflecting new research, improved methodologies and improved M&V data. At the same time, the protocol reflects a broad international consensus, and the vast majority of its work is accomplished by individual experts who volunteer their time to serve on committees. Please let us know how the IPMVP can be improved or expanded—it is updated and republished every two years.

Individuals interested in reviewing IPMVP progress and related

documents should visit www.ipmvp.org. The IPMVP web site contains new and/or modified content, interim revisions to the existing protocol and review drafts as they are prepared. Currently, the IPMVP web site has links to many of the organizations referenced herein, email archives containing the minutes of the conference calls and all the correspondence among members of the various committees, and contact information for many individuals associated with the protocol.

Chapter 1 Introduction

1.1 OVERVIEW

Energy efficiency offers the largest and most cost-effective opportunity for both industrialized and developing nations to limit the enormous financial, health and environmental costs associated with burning fossil fuels. Available, cost-effective investments in energy and water efficiency globally are estimated to be tens of billions of dollars per year. However, the actual investment level is far less, representing only a fraction of the existing, financially attractive opportunities for energy savings investments. In the interest of brevity, throughout this document the terms "energy" and "energy savings" represent both energy and water. Although there are differences between energy efficiency measures and water efficiency measures, they share many common attributes and are often part of the same project.

If all cost-effective efficiency investments were made public and commercial buildings in the U.S., for example, efficiency project spending would roughly triple, and within a decade would result in savings of $20 billion per year in energy and water costs, create over 100,000 permanent new jobs and significantly cut pollution. For developing countries with rapid economic growth and surging energy consumption, energy and water efficient design offers a very cost effective way to control the exploding costs of building power and water treatment plants, while limiting the expense of future energy imports and the widespread health and environmental damages and costs that result from burning fossil fuels.

These efficiency opportunities and their inherent benefits prompted the U.S. Department of Energy in early 1994 to begin working with industry to develop a consensus approach to measuring and verifying efficiency investments in order to overcome existing barriers to efficiency. The International Performance Measurement and Verification Protocol (IPMVP, or sometimes called the MVP) was first published in 1996, and contained methodologies that were compiled by the technical committee that comprised of hundreds of industry experts, initially from the United States, Canada and Mexico.

In 1996 and 1997, twenty national organizations from a dozen countries worked together to revise, extend and publish a new version of the IPMVP in December 1997. This second version has been widely adopted internationally, and has become the standard M&V documents in countries ranging from Brazil to Romania. According to Mykola Raptsun, former Deputy Chairman of State Committee of Ukraine Energy Conservation, now President of ARENA-ECO, the Ukrainian energy efficiency center:

The IPMVP has broad application for businessmen, energy managers, law makers and educators and could become the national standard document for M& V It has been important in helping the growth of the energy efficiency industry in Ukraine.

North America's energy service companies have adopted the IPMVP as the industry standard approach to measurement and verification (M&V). According to Steve Schiller, President of Schiller Associates, a leading energy efficiency consulting firm:

[In the United States], *referencing the International Performance Measurement and Verification Protocol (IPMVP) has become essentially a requirement associated with developing both individual energy efficiency performance contracting projects as well as performance contracting programs. Almost all performance-contracting firms now state that their work complies with the IPMVP. Thus, in a few short years the IPMVP has become the de-facto protocol for measurement and verification of performance contracts.*

Institutions such as the World Bank and International Finance Corporation (IFC) have found the Protocol beneficial and are incorporating it as a required part of new energy efficiency projects. According to Russell Sturm, Senior Projects Officer, Environmental Projects Unit, International Finance Corporation:

"In our work at the Environmental Projects Unit of the IFC we seek investments in the emerging ESCO markets of the developing and transition economies of the world. While these markets hold promise, the challenges on the road to commercial viability are formidable. IPMVP provides the founda-

tion necessary to build credibility for this emerging industry, helping us to establish a level of comfort among local players that is essential for broad-based acceptance in the marketplace. "

The IPMVP has been translated into Bulgarian, Chinese, Czech, Japanese, Korean, Polish, Portuguese, Romanian, Russian, Spanish and Ukrainian. The translated versions of the IPMVP in some of these languages are available through the website wwwipmvp.org.

As a result of strong and widespread interest, participation in developing this third edition has expanded to include a global network of professionals from around the world and includes national organizations from 16 countries and hundreds of individual experts from more than 25 nations. The work was drafted by volunteers serving on committees composed of leading international experts in their respective fields. Overall responsibility and direction is provided by the Executive Committee, composed of a dozen international experts who share a goal of strengthening and fostering the rapid growth of the energy and water efficiency industries. Our Financial Advisory Subcommittee has helped ensure that this document is valuable to the financial community in facilitating and enhancing efficiency investment financing.

1.2 WHY MEASURE AND VERIFY?

"You cannot manage what you do not measure"—Jack Welch, CEO of General Electric

When firms invest in energy efficiency, their executives naturally want to know how much they have saved and how long their savings will last. The determination of energy savings requires both accurate measurement and replicable methodology, known as a measurement and verification protocol.

The long-term success of energy and water management projects is often hampered by the inability of project partners to agree on an accurate, successful M&V Plan. This M&V Protocol discusses procedures that, when implemented, help buyers, sellers and financiers of energy and water projects to agree on an M&V Plan and quantify savings from Energy Conservation Measure (ECM) and Water Conservation Measure (WCM).

Simply put, the purpose of the IPMVP is to increase investment in energy efficiency and renewable energy. The IPMVP does so in at least six ways:

a) Increase energy savings

Accurate determination of savings gives facility owners and managers valuable feedback on the operation of their facility, allowing them to adjust facility management to deliver higher levels of energy savings, greater persistence of savings and reduced variability of savings. A growing body of data shows that better measurement and verification results in significantly higher levels of savings, greater persistence of savings over time and lower variability of savings (Kats et al. 1997 and 1999, Haberl et al. 1996). Logically this makes sense, since real time measurement at multiple measurement points provides a strong diagnostic tool for building managers that allows them to better understand, monitor and adjust energy systems to increase and maintain savings. This finding is consistent with the experience of the US Federal Energy Management Programs and reflects the very extensive long term *metering*[1] work done at the Texas A&M University Loan Star program (Claridge et al. 1996). Greater persistence and lower variability, in turn, can form the technical basis for rewarding energy efficiency projects which employ superior M&V techniques for determining energy savings.

b) Reduce cost of financing of projects

In early 1994, our financial advisors expressed concern that existing protocols (and those under development) created a patchwork of inconsistent and sometimes unreliable efficiency installation and measurement practices. This situation reduced reliability and performance of efficiency investments, increased project transaction costs, and prevented the development of new forms of lower cost financing. IPMVP is a response to this situation, providing guidance on risk management information helpful in structuring project financing contracts.

By providing greater and more reliable savings and a common approach to determining savings, widespread adoption of this Protocol has already made efficiency investments more reliable and profitable, and has fostered the development of new types of lower cost financing. By more clearly defining project M&V and defining generally accepted M&V methods, this Protocol provides lending institutions confidence in the credible assessment of savings and measurement of performance. This assessment and measurement then becomes the security which

1. The terms in italics are defined in Chapter 6.1

backs financing. If a sufficient level of confidence can be achieved, the door may be opened to "off-balance-sheet financing" where project debt does not appear on the credit line of the host facility—historically a major hurdle to energy efficiency project implementation.

The IPMVP is an important part of the credit equation for most lenders since it provides an established and independent mechanism to determine energy savings. For example, the US Department of Energy's Office of Energy Efficiency and Renewable Energy, in partnership with Virginia's Commonwealth Competition Council and New Jersey-based M/A Structured Finance Corp. has developed a pilot program for a $50 million pooled financing program for energy efficiency projects for K-12 schools and publicly owned colleges and universities. The goal of the program is to provide an off-balance sheet and procurement-friendly method of financing these projects for the public sector. The guidelines in the IPMVP have allowed participating financial institutions to lend on the basis of the energy savings, an important consideration in an off-balance sheet financing. The IPMVP provides the confidence and standardization to allow these institutions to fund upgrades based on future pooled energy savings, with borrowing "off-balance sheet" for the academic institutions.

c) Encourage better project engineering

Since good M&V practices are intimately related to good design of retrofit projects, IPMVP's direction on M&V practice encourages the good design of energy management projects. Good M&V design, and ongoing *monitoring* of performance will help in the creation of projects that work effectively for owners and users of the spaces or processes affected. Good energy management methods help reduce maintenance problems in facilities allowing them to run efficiently. Among the improvements that may be noted by complete engineering design of ECMs is an improvement in indoor air quality in occupied space.

d) Help demonstrate and capture the value of reduced emissions from energy efficiency and renewable energy investments.

Emissions reduced by efficiency projects include CO_2, the primary greenhouse gas (causing global warming), SO_2, NO_x and mercury. The failure to include the costs/benefits of these emissions has distorted price and market signals, and has resulted in a misallocation of energy investments and prevented a more rational and cost-effective energy investment strategy around the world. Determining the level of reduction of pollutants requires the ability to estimate with confidence actual energy savings.

The IPMVP provides a framework for calculating energy reductions before (baseline) and after the implementation of projects. The IPMVP can help achieve and document emissions reductions from projects that reduce energy consumption and help energy efficiency investments be recognized as an emission management strategy. Such profile will also help attract funding for energy efficiency projects through the sale of documented emission credits.

e) Increase public understanding of **energy management as a public policy tool**

By improving the credibility of energy management projects, M&V increases public acceptance of the related activities. Such public acceptance encourages investors to consider investing in energy efficiency projects or the emission credits they may create. By enhancing savings, good M&V practice also brings more attention to the public benefits provided by good energy management, such as improved community health, reduced environmental degradation, and increased employment.

f) Help national and industry organizations promote and achieve resource efficiency and environmental objectives

The IPMVP is being widely adopted by national and regional government agencies and by industry and trade organizations to help increase investment in energy efficiency and achieve environmental and health benefits. Chapter 1.4 provides examples of how the IPMVP is being used by a range of institutions in one country—the United States.

1.2.1 Role of Protocol

This Protocol:

- Provides energy efficiency project buyers, sellers and financiers a common set of terms to discuss key M&V project-related issues and establishes methods which can be used in energy performance contracts.

- Defines broad techniques for determining savings from both a "whole facility" and an individual technology.

- Applies to a variety of facilities including residential, commercial, institutional and industrial buildings, and industrial processes.

- Provides outline procedures which i) can be applied to similar projects throughout all geographic regions, and ii) are internationally accepted, impartial and reliable.

- Presents procedures, with varying levels of accuracy and cost, for measuring and/or verifying: i) baseline and project installation conditions, and ii) long-term energy savings.

- Provides a comprehensive approach to ensuring that building indoor environmental quality issues are addressed in all phases of ECM design, implementation and maintenance.

- Creates a living document that includes a set of methodologies and procedures that enable the document to evolve over time.

1.2.2 Audience for Protocol

The target audience for this Protocol includes:

— Facility Energy Managers
— Project Developers and/or Implementers
— ESCOs (Energy Service Companies)
— WASCOs (Water Service Companies)
— Non-Governmental Organizations (NGOs)
— Finance Firms
— Development Banks
— Consultants
— Government Policy Makers
— Utility Executives
— Environmental Managers
— Researchers

1.3 IPMVP ROLE IN INTERNATIONAL CLIMATE CHANGE MITIGATION

International efforts to reduce greenhouse gas emissions have also increased the need for standardized tools such as the IPMVP, to cost-effectively measure the economic and environmental benefits of energy efficiency projects. The vast majority of climate scientists have concluded that "the balance of evidence suggests that human activities are having a discernible influence on global climate" (IPCC, 1995). Responding to the mounting scientific call for action to reduce emissions of greenhouse gases (primarily those from fossil fuel use), the industrialized nations recently committed to binding emissions targets and timetables. The flexible, market mechanisms to reduce greenhouse gas emissions included in the 1997 Kyoto Protocol to the U.N. Framework Convention on Climate Change (FCCC) makes the need for an international consensus on M&V protocol more urgent.

Guidelines have recently been developed by the Lawrence Berkeley National Laboratory that addresses the monitoring, evaluation, reporting, verification, and certification of energy efficiency projects for climate change mitigation (Vine and Sathaye, 1999). The LBNL study determined that the IPMVP is the preferred international approach for monitoring and evaluating energy efficiency projects because of its international acceptance, because it covers many key issues in monitoring and evaluation and because it allows for flexibility.

The IPMVP Adjustment Committee will be working through 2000 to build on the leading intentional consensus approach to implementing, measuring and verifying efficiency investments to come up with agreed on estimates of future energy savings and emissions reductions. By achieving agreement on this issue, this committee will make a necessary and important contribution to establishing a framework on which international greenhouse gas trading can be built. For more information, contact Ed Vine (elvine@lbl.gov).

1.4 RELATIONSHIP TO U.S. PROGRAMS

The MVP is intended to include a framework approach that complements more detailed national, or regional energy efficiency guidelines in any country that it is used in. Following are the examples drawn from the US.

ASHRAE Guideline 14

IPMVP is complemented by the work of the American Society of Heating, Refrigeration and Air-Conditioning Engineers (ASHRAE) in the form of its draft Guideline 14P Measurement of Energy and Demand Savings. In contrast to the ASHRAE document, which focuses at a very technical level, the IPMVP establishes a general framework and terminology to assist buyers and sellers of M&V services. ASHRAE's Guideline 14 has completed its first public review and hence was available publicly for a period in the middle of 2000. The ASHRAE Guideline is expected to be fully available in 2001. It is advised that the reader use the ASHRAE or other relevant document, as well as others referenced herein, to help formulate a successful M&V Plan.

Federal Energy Management Program

The U.S. Department of Energy's Federal Energy Management Program (FEMP) was established, in part, to reduce energy costs to the U.S. Government from operating Federal facilities. FEMP assists Federal energy managers by identifying and procuring energy-saving projects.

The FEMP M&V Guideline follows the IPMVP, and provides guidance and methods for measuring and verifying the energy and cost savings associated with federal agency performance contracts. It is intended for federal energy managers, federal procurement officers, and contractors implementing performance contracts at federal facilities. Assistance is provided on choosing M&V methods that provide an appropriate level of accuracy for protection of the project investment. The FEMP M&V Guideline has two primary uses:

- It serves as a reference document for specifying M&V methods and procedures in delivery orders, requests for proposals (RFPs), and performance contracts.
- It is a resource for those developing project-specific M&V plans for federal performance contracting projects.

The first FEMP M&V Guideline was published in 1996, a new version has been published in 2000, Version 2.2, and contains the following updates to the 1996 version:

- A discussion of performance contracting responsibility issues and how they affect risk allocation.
- Quick M&V guidelines including procedural outlines, content checklists, and option summary tables.
- Measure-specific guidelines for assessing the most appropriate M&V Option for common measures.
- New M&V strategies and methods for cogeneration, new construc-

tion, operations and maintenance, renewable energy systems, and water conservation projects.

In addition to being a requirement for efficiency investments in U.S. Federal buildings, the FEMP Guideline provides a model for how to develop a specific application of the IPMVP. To secure a copy of the FEMP guideline, call 800-DOE-EREC.

State Performance Contracting Programs

Many states in the US have incorporated the IPMVP as an important part of a number of their energy efficiency programs and services for commercial, industrial and institutional customers. They use IPMVP as the basis of determining energy savings in energy performance contracting. IPMVP has been valuable in standardizing project performance metrics and has become an important component for facilitating wider acceptance of energy performance contracts that can reduce private sector transaction costs. IPMVP has helped cut transactions costs, improve project performance and has been important in securing low cost financing for our programs. Many states require that M&V Plans be developed for projects funded under the Standard Performance Contract Program. The New York State EnVest program, for example, is structured to be consistent with IPMVP and New York State Energy Research & Development Authority (NYSERDA) strongly recommends the use of IPMVP for institutional projects.

Other states which have incorporated IPMVP in state energy performance contracting and other energy efficiency programs are California, Colorado, Oregon, Texas, and Wisconsin.

Environmental Evaluation Initiatives in Buildings

The IPMVP is being integrated into the U.S. Green Building Council's (USGBC) Leadership in Energy and Environmental Design (LEED") Rating system, which is rapidly becoming the national green building design standard.

The USGBC-developed LEED" program provides a comprehensive green building rating system. In order to win a rating, a building must comply with several measures, including the IPMVP, for energy efficiency and water measures. Buildings are then rated on a range of environmental and life cycle issues to determine if the building achieves one of the LEED" performance levels. Applicants to LEED" will receive a point for complying with the IPMVP. For more information, please visit their website at wwwusgbc.org.

Chapter 2
The Importance of M&V in Financing Energy and Water Efficiency

2.1 FINANCING ENERGY AND WATER EFFICIENCY

The key to unlocking the enormous potential for energy and water efficiency worldwide is securing financing. Good measurement practices and verifiability are some of the important elements in providing the confidence needed to secure funding for projects. Securing financing requires confidence that energy efficiency investments will result in a savings stream sufficient to make debt payments. Measurement and verification practices allow project performance risks to be understood, managed, and allocated among the parties.

It is important that each M&V Plan clearly describe the tolerances associated with the measurement and savings determination methods. There can be significant variances in the tolerances within each of the measurement options presented in this protocol. Users are advised to understand the pros and cons of each option and the tolerance of the particular measurement method proposed. Each participant is then equipped to make an appropriate business decision about the risk and reward of an investment.

Energy and water efficiency projects meet a range of objectives, including upgrading equipment, improving performance, helping to achieve environmental compliance, or simply saving energy and money. All projects have one thing in common, an initial financial investment. The type of investment may be an internal allocation of funds (in-house project) or it may be a complex contractual agreement with an ESCO and/or third-party financier.

All types of financial investments have a common goal—making money or a "return" on investment. Rate of return is measured by various financial yardsticks such as simple payback, return on investment

(ROI) or internal rate of return (IRR). The expected rate of return is governed by the risk associated with the investment. Typically, the higher the project risk, the greater the return demanded. Risk takes a variety of forms in efficiency projects. Most risks can be measured; it is the accuracy of the measurement (tolerance) that is important. Many risks associated with investing in an energy or water efficiency project can be measured using tools common to the finance industry, such as internal rate of return or customer credit-worthiness. M&V, as defined in this Protocol, is primarily focused on risks that affect the measurement or determination of savings from energy or water efficiency programs. These risks are defined in the terms of the contracts between the participants.

This Protocol provides guidance on obtaining information needed to reduce and manage measurement uncertainties in order to structure project financing contracts. The value of ECM performance data can range from useful to absolutely critical, depending on the financing method and which party has accepted the contractual risk. For example, an ESCO typically will not be concerned about operating hours if the owner takes responsibility for equipment operation, though these risks should be highlighted and understood by the parties. Different investments require different measures of performance.

Accordingly, this Protocol provides four M&V Options to accommodate a variety of contractual arrangements.

Although this Protocol formalizes basic M&V language and techniques, it is not meant to prescribe an M&V Option for every type of ECM. Instead, this document offers Options available, provides guidance on which Option to choose and helps clarify the relationship of various M&V Options to the risks assumed by relevant parties, and thus places bounds on the financial risks of the deal.

2.2 DEFINITION AND ROLE OF PERFORMANCE CONTRACTS

When efficiency projects include a guarantee of performance, it is classified as a performance contract. It is important to recognize that there are two separate instruments in such transactions—the lending instrument and the guarantee. The lending takes place between the financier and the owner, or the ESCO. The guarantee is typically provided to the owner by the ESCO. Usually it guarantees the amount of energy that will be saved at some defined pricing level, and/or that energy savings will be sufficient to meet the financing payment obligations. However a guarantee may be as simple as a piece of equipment that is capable of operating at a stated level of efficiency ("rating of performance").

There are many types of financing in use with performance contracts. This Protocol does not recommend any particular approach, because the choice depends on many considerations beyond the scope of M&V disciplines. The availability of third-party financing in general, however, and the variety of applicable financial instruments, is growing. Those seeking financing of projects with measurable and verifiable savings should have no difficulty obtaining expert advice from more than one specialist, at least in the U.S.

Energy savings are commonly defined as reductions in energy use. Energy cost savings are reductions in the cost of energy and related O&M expenses, from a base cost established through a methodology set forth in an energy performance contract. (Energy saving activities may also reduce other costs such as pollution/health care costs through lowering of atmospheric emissions from boilers.) "Energy savings" and "energy cost savings" when defined in a performance contract are typically contractual terms.

Performance of equipment, both before and after a retrofit, can be measured with varying degrees of accuracy. Savings are often computed as energy cost avoidance and are the calculated difference between i) the measured performance and/or load of energy-using systems and ii) the amount of energy that the systems would have used in the absence of the ECM, such difference being multiplied by current unit prices for energy supplied. The *baseyear*[1] energy usage is defined using measured equipment performance data prior to the ECM coupled with assumptions about how that equipment would have operated in the *post-retrofit period*. Often, baseyear assumptions must incorporate expected and/or unforeseen changes that may alter the energy savings calculation. In these cases, the contract defines which party is responsible for the elements of the ECM that lead to energy savings and cost avoidance.

Broadly speaking, energy efficiency projects have two elements, performance and operation:

- performance of the project is related to its efficiency, defined with a metric such as improvements in lumens/watt or in tons of cooling per kW of demand.

1. The terms in italics are defined in Chapter 6.1

- operation of the project is related to its actual usage, defined as operation hours, ton hours etc.

Typically, an ESCO is responsible for the performance of any equipment or systems it installs. Depending on the energy performance contract, either the ESCO or the Owner may be responsible for the operation of the equipment. It is important to allow for changes in equipment operation that may result from factors outside either party's control, such as weather. Responsibility for maintenance may be assigned to either party or shared. Consider four categories of variables that account for all of the changes that might affect energy cost avoidance:

1 ESCO-controlled variables—retrofit performance

2 Owner-controlled variables—facility characteristics, operation

3 ESCO and/or owner controlled variable—maintenance

4 Variables that are outside of either party's control—weather, energy prices, natural disaster

The M&V Plan should clearly identify these variables for all ECMs before the project is implemented. The M&V process requires the skills of professionals familiar with measurement and collection techniques, data manipulation, interpretation, and technology performance. In some circumstances, it may be preferable that a third party be obtained by the owner to judge whether energy performance contract terms are appropriate and, later, are being applied correctly. In order to adequately understand the implications of various measurement strategies, the M&V professional should have a thorough understanding of the ECMs being installed and the services provided.

2.3 FINANCIAL RISK, MEASUREMENT

When creating financed energy efficiency project agreements, the parties enter into a contract defining and allocating risk among the parties. Generally, the lender will be looking for the most straight forward allocation of risks. In financing efficiency projects, most risks (beyond general creditworthiness of the parties) relate to one basic issue: will the project perform to expectation? Performance related risks that are scattered among several participants may make project financing more difficult. Usually, the lender wants the risk of performance to be between the ESCO and the owner only, acting as "Consultant" to the owner. It is difficult for a lender to assess creditworthiness if payments can be impacted by a variety of parties. In such cases the lender will price the financing to the creditworthiness of the lowest common denominator.

2.3.1 Ability to Pay

Debt service coverage, which is the ratio of the projected cash savings to repayment amount, is a critical measure of the project's financial viability. It serves as an indicator of the project's ability to be supported solely by the savings. When coverage falls below a certain level, (125% for example), the project will be subject to increased scrutiny by financiers. Most important to the calculation of coverage is the confidence with which savings are estimated and ultimately measured (or stipulated).

2.3.2 Construction Risks

Terms (risks) embodied in common construction contracts are also present in a financed energy efficiency project, if construction financing is used. (Often permanent financing is initiated after construction is finished and accepted.) Basic risks and questions include:

- Who is responsible for the design? Who builds what, by when?
- Who pays whom, how much and when?
- What cost overruns are likely, what contingencies are in the construction budget, and what recourse does the financier have in event of overruns?
- What is the maximum construction delay, what can cause it and how can it be cured.

Performance bonds cover the risk associated with the first item for both owners and lenders. Escrow and progress payment contracts cover risks associated with the second item. Lenders are interested in liquidated damage provisions and payment and performance bonds as a way to limit potential losses from construction delays.

2.3.3 Performance Risks

As discussed in Chapter 2.2, when an *energy savings performance contract* is used, capturing the effect of "change" is particularly important. For example, consider which party estimated the savings and which party carries the financial impact of. i) a change in operating hours, ii) a change in weather, iii) a degradation in chiller efficiency, iv) a change that requires compliance with new or existing standards, v) a partial facility closure, vi) an expansion to a third production shift, vii) quality of maintenance, etc. The financial impact of these changes can be either positive or negative. The contract must be clear who wins or loses. For example, an ESCO may not get credit for the savings created by actions of the owner. Similarly an ESCO should not be required to cover the higher costs incurred due to the owner's increased or decreased usage outside the parameters of the project; e.g., a new computer lab or, fewer shifts worked.

Energy savings estimates are usually based on an assumption that the facility will operate on a predicted schedule, or load profile. Changes to this schedule will affect project generated savings. Assignment of responsibility for these changes is a critical contract component. As well, these are all risks that need to be evaluated by each party in advance and accounted for using performance measurement as specified using an appropriate M&V method. Often these are examined in detail after implementation, when it is too late. For example, an executed contract may stipulate that the owner is responsible for the operating hours of a lighting system, and the ESCO is responsible for ensuring that the system power draw is correct. For this contract, Option A M&V method (as introduced in Chapter 3.4. 1) is appropriate. Cost avoidance is calculated using a stipulated value for operating hours and the measured change in the power draw of the lighting system.

Chapter 3
Basic Concepts and Methodology

3.1 INTRODUCTION

Energy or demand savings are determined by comparing measured energy use or demand before and after implementation of an energy savings program. In general:

Energy Savings = Baseyear Energy Use –
Post-Retrofit Energy Use ± Adjustments **Eq. 1**

The "Adjustments" term in this general equation brings energy use in the two time periods to the same set of conditions. Conditions commonly affecting energy use are weather, occupancy, plant throughput, and equipment operations required by these conditions. Adjustments may be positive or negative.

Adjustments are derived from identifiable physical facts. The adjustments are made either routinely such as for weather changes, or as necessary such as when a second shift is added, occupants are added to the space, or increased usage of electrical equipment in the building.

Adjustments are commonly made to restate baseyear energy use under post-retrofit conditions. Such adjustment process yields savings which are often described as "avoided energy use" of the post-retrofit period. The level of such savings are dependent on post-retrofit period operating conditions.

Adjustments may also be made to an agreed fixed set of conditions such as those of the baseyear or some other period. The level of savings computed in this situation is unaffected by post-retrofit period conditions, but reflects operation under a set of conditions which must be established in advance.

There are many other considerations and choices to make in determining savings. Chapter 3.4 describes four basic Options, any one of which may be adapted to a particular savings determination task. Chapter 4 gives guidance on common issues such as balancing costs and accuracy with the value of the energy savings program being evaluated. Chapter 5 reviews metering and instrumentation issues.

3.2 BASIC APPROACH

Proper savings determination is a necessary part of good design of the savings program itself. Therefore the basic approach in savings determination is closely linked with some elements of program design. The basic approach common to all good savings determination entails the following steps:

1 Select the IPMVP Option (see Chapter 3.4) that is consistent with

the intended scope of the project, and determine whether adjustment will be made to post-retrofit conditions or to some other set of conditions. (These fundamental decisions may be written into the terms of an energy performance contract.)

2 Gather relevant energy and operating data from the baseyear and record it in a way that can be accessed in the future.

3 Design the energy savings program. This design should include documentation of both the design intent and methods to be used for demonstrating achievement of the design intent.

4 Prepare a Measurement Plan, and a Verification Plan if necessary, (commonly together called an "M&V Plan"). The M&V Plan fundamentally defines the meaning of the word "savings" for each project. It will contain the results of steps 1 through 3 above, and will define the subsequent steps 5 through 8 (see Chapter 3.3).

5 Design, install and test any special measurement equipment needed under the M&V Plan.

6 After the energy savings program is implemented, inspect the installed equipment and revised operating procedures to ensure that they conform with the design intent defined in step 3. This process is commonly called "commissioning." ASHRAE defines good practice in commissioning most building modifications (ASHRAE 1996).

7 Gather energy and operating data from the post-retrofit period, consistent with that of the baseyear and as defined in the M&V Plan. The inspections needed for gathering these data should include periodic repetition of commissioning activities to ensure equipment is functioning as planned.

8 Compute and report savings in accordance with the M&V Plan.

Steps 7 and 8 are repeated periodically when a savings report is needed.

Savings are deemed to be statistically valid if the result of equation (1) is greater than the expected variances (noise) in the baseyear data. Chapter 4.2 discusses some methods of assessing this noise level. If noise is excessive, the unexplained random behavior of the facility is high and the resultant savings determination is unreliable. Where this criterion is not expected to be met, consideration should be given to using more independent variables in the model, or selecting an IPMVP Option that is less affected by unknown variables.

The balance of this document fleshes out some key details of this basic approach to determining savings.

Once a savings report has been prepared, a third party may verify that it complies with the M&V Plan, This third party should also verify that the M&V Plan itself is consistent with the objectives of the project.

3.3 M&V PLAN

The preparation of an M&V Plan is central to proper savings determination and the basis for verification. Advance planning ensures that all data needed for proper savings determination will be available after implementation of the energy savings program, within an acceptable budget.

Data from the baseyear and details of the ECMs may be lost over time. Therefore it is important to properly record them for future reference, should conditions change or ECMs fail. Documentation should be prepared in a fashion that is easily accessed by verifiers and other persons not involved in its development, since several years may pass before these data are needed.

An M&V Plan should include:

- A description Of the ECM and its intended result.

- Identification of the boundaries of the savings determination. The boundaries may be as narrow as the flow of energy through a pipe or wire, or as broad as the total energy use of one or many buildings. The nature of any energy effects beyond the boundaries should be described and their possible impacts estimated.

- Documentation of the facility's *baseyear conditions* and resultant *baseyear energy data.* In performance contracts, baseyear energy use and baseyear conditions may be defined by either the owner or the ESCO, providing the other party is given adequate opportunity to verify it. A preliminary energy audit used for establishing the objectives of a savings program or terms of an energy performance contract is typically not adequate for planning M&V activities. Usually a more comprehensive audit is required to gather the baseyear information relevant to M&V:

— energy consumption and demand profiles

— occupancy type, density and periods

— space conditions or plant throughput for each operating period and season. (For example in a building this would include light level and color, space temperature humidity and ventilation. An assessment of thermal comfort and/or indoor air quality (IAQ) may also prove useful in cases where the new system does not perform as well as the old inefficient system. See Volume II.)

— equipment inventory: nameplate data, location, condition. Photographs or videotapes are effective ways to record equipment condition.

— equipment operating practices (schedules and setpoint, actual temperature s/pressures)

— significant equipment problems or outages.

The extent of the information to be recorded is determined by the boundaries or scope of the savings determination. The baseyear documentation typically requires well documented audits, surveys, inspections and/or spot or short-term metering activities. Where whole building Option is employed (Chapter 3.4.3 or Chapter 3.4.4), all building equipment and conditions should be documented.

- Identification of any planned changes to conditions of the baseyear, such as night time temperatures.

- Identification of the post-retrofit period. This period may be as short as a one minute test following commissioning of an ECM, or as long as the time required to recover the investment cost of the ECM program.

- Establishment of the set of conditions to which all energy measurements will be adjusted. The conditions may be those of the post-retrofit period or some other set of fixed conditions. As discussed in the introductory remarks of Chapter 3, this choice determines whether reported savings are "avoided costs" or energy reductions under defined conditions.

- Documentation of the design intent of the ECM(s) and the commissioning procedures that will be used to verify successful implementation of each ECM.

- Specification of which Option from Chapter 3.4 will be used to determine savings.

- Specification of the exact data analysis procedures, algorithms and assumptions. For each mathematical model used, report all of its terms and the range of independent variables over which it is valid.

- Specification of the metering points, period(s) of metering, meter characteristics, meter reading and witnessing protocol, meter commissioning procedure, routine calibration process and method of dealing with lost data.

- For Option A, report the values to be used for any stipulated parameters. Show the overall significance of these parameters to the total expected saving and describe the uncertainty inherent in the stipulation.

- For Option D, report the name and version number of the simulation software to be used. Provide a paper and electronic copy of the input files, output files, and reference the weather files used for the simulation, noting which input parameters were measured and which assumed. Describe the process of obtaining any measured data. Report the accuracy with which the simulation results match the energy use data used for calibration.

- Specification of quality assurance procedures.

- Quantification of the expected accuracy associated with the measurement, data capture and analysis. Also describe qualitatively the expected impact of factors affecting the accuracy of results but which cannot be quantified.

- Specification of how results will be reported and documented. A sample of each report should be included.

- Specification of the data that will be available for another party to verify reported savings, if needed.

- Where the nature of future changes can be anticipated, methods for making the relevant non-routine *Baseline Adjustments*[1] should be defined.

- Definition of the budget and resource requirements for the savings determination, both initial setup costs and ongoing costs throughout the post-retrofit period.

When planning a savings measurement process, it is helpful to consider the nature of the facility's energy use pattern, and the ECM s impacts thereon. Consideration of the amount of variation in energy patterns and the change needing to be assessed will help to establish the amount of effort needed to determine savings. The following three examples show the range of scenarios that may arise.

- **ECM reduces a constant load without changing its operating hours.** Example: Lighting project where lamps and ballasts in an office building are changed, but the operating hours of the lights do not change.

- **ECM reduces operating hours while load is unchanged.** Example: Automatic controls shut down air handling equipment or lighting during unoccupied periods.

---

1. The terms in italics are defined in Chapter 6.1

- **ECM reduces both equipment load and operating hours.** Example: Resetting of temperature on hot water radiation system reduces overheating, thereby reducing boiler load and operating periods.

Generally, conditions of variable load or variable operating hours require more rigorous measurement and computation procedures.

It is important to realistically anticipate costs and effort associated with completing metering and data analysis activities. Time and budget requirements are often underestimated leading to incomplete data collection. It is better to complete a less accurate and less expensive savings determination than to have an incomplete or poorly done, yet theoretically more accurate determination that requires substantially more resources, experience and/or budget than available. Chapter 4.11 addresses cost/benefit tradeoffs.

Typical contents of four M&V Plans are outlined in the four examples shown in Appendix A.

3.4 METHODS

The Energy Use quantities in Equation 1 can be "measured" by one or more of the following techniques:

- Utility or fuel supplier invoices or meter readings.

- Special meters isolating a retrofit or portion of a facility from the rest of the facility. Measurements may be periodic for short intervals, or continuous throughout the post-retrofit period.

- Separate measurements of parameters used in computing energy use. For example, equipment operating parameters of electrical load and operating hours can be measured separately and factored together to compute the equipment's energy use.

- Computer simulation which is calibrated to some actual performance data for the system or facility being modeled, e.g., DOE-2 analysis for buildings.

- Agreed assumptions or stipulations of ECM parameters that are well known. The boundaries of the savings determination, the responsibilities of the parties involved in project implementation, and the significance of possible assumption error will determine where assumptions can reasonably replace actual measurement. For example, in an ECM involving the installation of more efficient light fixtures without changing lighting periods, savings can be determined by simply metering the lighting circuit power draw before and after retrofit while assuming the circuit operates for an agreed period of time. This example involves stipulation of operating periods, while equipment performance is measured.

The Adjustments term in equation (1) can be of two different types:

- **Routine** Adjustments for changes in parameters that can be expected to happen throughout the post-retrofit period and for which a relationship with energy use/demand can be identified. These changes are often seasonal or cyclical, such as weather or occupancy variations. This protocol defines four basic Options for deriving routine adjustments. Table B-1 summarizes the various Options.

- **Non-routine** Adjustments for changes in parameters which cannot be predicted and for which a significant impact on energy use/demand is expected. Non-routine adjustments should be based on known and agreed changes to the facility. Chapter 4.8 presents a general approach for handling non-routine adjustments, commonly called "baseline adjustments.

Options A and B focus on the performance of specific ECMs. They involve measuring the energy use of systems affected by each ECM separate from that of the rest of the facility. Option C assesses the energy savings at the whole facility level. Option D is based on simulations of the energy performance of equipment or whole facilities to enable determination of savings when baseyear or post-retrofit data are unreliable or unavailable.

An example of the use of each of the four Options is contained in Appendix A.

3.4.1 Option A: Partially Measured Retrofit Isolation

Option A involves isolation of the energy use of the equipment affected by an ECM from the energy use of the rest of the facility. Measurement equipment is used to isolate all relevant energy flows in the

Table A-1. Overview of M&V Options

M&V Option	How Savings Are Calculated	Typical Applications
A. Partially Measured Retrofit Isolation Savings are determined by partial field measurement of the energy use of the system(s) to which an ECM was applied, separate from the energy use of the rest of the facility. Measurements may be either short-term or continuous. Partial measurement means that some but not all parameter(s) may be stipulated, if the total impact of possible stipulation error(s) is not significant to the resultant savings. Careful review of ECM design and installation will ensure that stipulated values fairly represent the probable actual value. Stipulations should be shown in the M&V Plan along with analysis of the significance of the error they may introduce.	Engineering calculations using short term or continuous post-retrofit measurements and stipulations.	Lighting retrofit where power draw is measured periodically. Operating hours of the lights are assumed to be one half hour per day longer than store open hours.
B. Retrofit Isolation Savings are determined by field measurement of the energy use of the systems to which the ECM was applied, separate from the energy use of the rest of the facility. Short-term or continuous measurements are taken throughout the post-retrofit period.	Engineering calculations using short term or continuous measurements	Application of controls to vary the load on a constant speed pump using a variable speed drive. Electricity use is measured by a kWh meter installed on the electrical supply to the pump motor. In the baseyear this meter is in place for a week to verify constant loading. The meter is in place throughout the post-retrofit period to track variations in energy use.
C. Whole Facility Savings are determined by measuring energy use at the whole facility level. Short-term or continuous measurements are taken throughout the post-retrofit period.	Analysis of whole facility utility meter or sub-meter data using techniques from simple comparison to regression analysis.	Multifaceted energy management program affecting many systems in a building. Energy use is measured by the gas and electric utility meters for a twelve month baseyear period and throughout the post-retrofit period.
D. Calibrated Simulation Savings are determined through simulation of the energy use of components or the whole facility. Simulation routines must be demonstrated to adequately model actual energy performance measured in the facility. This option usually requires considerable skill in calibrated simulation.	Energy use simulation, calibrated with hourly or monthly utility billing data and/or end-use metering.	Multifaceted energy management program affecting many systems in a building but where no baseyear data are available. Post-retrofit period energy use is measured by the gas and electric utility meters. Baseyear energy use is determined by simulation using a model calibrated by the post-retrofit period utility data.

pre-retrofit and post-retrofit periods. Only partial measurement is used under Option A, with some parameter(s) being stipulated rather than measured. However such stipulation can only be made where it can be shown that the combined impact of the plausible errors from all such stipulations will not significantly affect overall reported savings.

3.4.1.1 Option A: Isolation Metering

Measurement equipment must be used to isolate the energy use of the equipment affected by the ECM from the energy use of the rest of the facility. The isolation metering should reflect the boundary between equipment which the ECM affects and that which it does not affect. For example, a lighting load reduction often has a related impact on HVAC system energy use, but the boundary for measurement may be defined to encompass only the lighting electricity. However if the boundary of the savings determination encompasses HVAC effects, measurement or stipulation will be required for both the lighting and HVAC energy flows.

Chapter 5 discusses metering issues.

3.4.1.2 Option A: Measurement vs. Stipulation

Some, but not all parameters of energy use may be stipulated under Option A. The decision of which parameters to measure and which to stipulate should consider the significance of the impact of all such stipulations on the overall reported savings. The stipulated values and analysis of their significance should be included in the M&V Plan (See Chapter 3.2).

Stipulation may be based on historical data, such as recorded operating hours from the baseyear. Wherever a parameter is not measured in the facility for the baseyear or post-retrofit period it should be treated as a stipulated value and the impact of possible error in the stipulation assessed relative to the expected savings.

Engineering estimates or mathematical modeling may be used to assess the significance of stipulation of any parameter in the reported savings. For example if a piece of equipment's operating hours are considered for stipulation, but may be between 2,100 and 2,300 hours per year, the estimated savings at 2,100 and 2,300 hours should be computed and the difference evaluated for its significance to the expected savings. The impact of all such possible stipulations should be totaled before determining whether sufficient measurement is in place.

The selection of factor(s) to measure may also be considered relative to the duties of a contractor undertaking some ECM performance risk. Where a factor is significant to assessing a contractor's performance, it should be measured, while other factors beyond the ESCO's control should be considered for stipulation.

3.4.1.3 Installation Verification

Since stipulation is allowed under this Option, great care is needed to review the engineering design and installation to ensure that the stipulations are realistic and achievable, i.e. the equipment truly has the potential to perform as assumed.

At defined intervals during the post-retrofit period the installation should be reinspected to verify continued existence of the equipment and its proper operation and maintenance. Such re-inspections will ensure continuation of the potential to generate predicted savings and validate stipulations. The frequency of these re-inspections can be determined by the likelihood of change. Such likelihood can be established through initial frequent inspections to establish the stability of equipment existence and performance. An example of a situation needing routine re-inspection is a lighting retrofit savings determination involving the sampling of the performance of fixtures and a count of the number of fixtures. In this case the continued existence of the fixtures and lamps is critical to the savings determination. Therefore periodic counts of the number of fixtures in place with all lamps burning would be appropriate. Similarly, where the performance of controls equipment is assumed but subject to being overridden, regular inspections or recordings of control settings are critical to limiting the uncertainty created by the stipulations.

3.4.1.4 Option A: Measurement Interval

Parameters may be continuously measured or periodically measured for short periods. The expected amount of variation in the parameter will govern the decision of whether to measure continuously or periodically.

Where a parameter is not expected to change it may be measured immediately after ECM installation and checked occasionally throughout the post-retrofit period. The frequency of this checking can be determined by beginning with frequent measurements to verify that the parameter is constant. Once proven constant, the frequency of measurement may be reduced.

If less than continuous measurement is used, the location of the measurement and the exact nature of the measurement device should be recorded in the M&V Plan, along with the procedure for calibrating the meter being used.

Where a parameter is expected to be constant, measurement intervals can be short and occasional. Lighting fixtures provide an example of constant power flow, assuming they have no dimming capability. However lighting operating periods may not be constant, for example outdoor lighting controlled by a photocell operates for shorter periods in seasons of long daylight than in seasons of short daylight. Where a parameter may change seasonally, such as this photocell case, measurements should be made under appropriate seasonal conditions.

Where a parameter may vary daily or hourly, as in most heating or cooling systems, continuous metering may be simplest. However for weather dependent loads, measurements may be taken over a long enough period to adequately characterize the load pattern (i.e., weekday/weekend and weather-dependent characteristics of the load) and repeated as necessary through the post-retrofit period. Examples of such day-type profiling can be found in Katipamula and Haberl (1991), Akbari et al. (1988), Hadley and Tomich (1986), Bou Saada and Haberl (1995a, 1995b) and Bou Saada et al. (1996).

3.4.1.5 Option A: Sampling

Where multiple versions of the same installation are included within the boundaries of a savings determination, statistically valid samples may be used as valid measurements of the total parameter. Such situation may arise, for example, where individual light fixtures are measured before and after retrofit to assess their power draw, while the total lighting power draw cannot be read at the electrical panel due to the presence of non-lighting loads on the same panel. Providing that a statistically significant sample of fixtures is measured before and after ECM installation, these data may be used as the 'measurement' of total lighting power draw. Appendix B discusses the statistical issues involved in sampling.

3.4.1.6 Option A: Uncertainty

Chapter 4.2 reviews the general issues surrounding uncertainty of savings determination. However, specific factors driving the uncertainty of Option A methods are:

- The magnitude of effects beyond the boundary of the retrofit isolation. For example, the significance of the mechanical cooling energy associated with a reduction in lighting power depends on the length of the mechanical cooling season and the number of hours of operation of the cooling equipment each day.

- The significance of the error introduced by possible variations between the stipulated and true values of parameters. This uncertainty is controlled through careful review of the ECM design, careful inspection of its implementation after installation and periodically thereafter.

- The variability in the measured parameters, if less than continuous measurement is employed. This uncertainty can be minimized through periodic measurements made frequently enough at the outset of the project to adequately characterize the variability.

- The degree to which the measured sample represents all components of an ECM

Savings uncertainty under Option A is generally inversely proportional to the complexity of the ECM and variability of operations in both the baseyear and post-retrofit period. Thus, the savings from a simple lighting retrofit may typically be more accurately determined with Option A than the savings from a chiller retrofit, since lighting stipulations may have less uncertainty.

3.4.1.7 Option A: Cost

Savings determinations under Option A can be less costly than under other Options, since the cost of deriving a stipulation may be less than the cost of making measurements. However in some situations where stipulation is the only possible route, the derivation of a good stipulation may require more cost than direct measurement. Cost of retrofit isolation should consider all elements: proper meter installation, commissioning and maintenance, proper stipulation analysis, and the ongoing cost to read and record data.

Portable meters may be used so that their costs can be shared with other objectives. However, meters which are permanently installed maybe useful in the facility to provide feedback to operating staff or automated control equipment for optimization of systems or billing of special users.

Savings determination cost is driven by the complexity of the ECM and the number of energy flows crossing the boundary of the ECM or retrofit isolation.

Cost is also driven by the frequency of measurement, whether continuous or periodic. Annual costs should be expected to be highest at the beginning of the post-retrofit period. At this stage in a project measurement processes are being refined, and closer monitoring of performance is needed to optimize ECM operation. Some projects may cease reporting savings after a defined "test" period, though metering maybe left in place for real time feedback to operating staff

The appropriate cost for each savings determination should be determined in proportion to the expected savings their potential variability.

3.4.1.8 Option A: Best Applications

Option A is best applied where:

- the performance of only the systems affected by the ECM is of concern, either due to the responsibilities assigned to the parties in a performance contract or due to the savings of the ECM being too small to be detected in the time available using Option C.
- interactive effects between ECMs or with other facility equipment can be measured or assumed to be not significant.
- isolation of the ECM from the rest of the facility and stipulation of key factors may avoid possibly difficult non-routine Baseline Adjustments for future changes to the facility.
- the independent variables that affect energy use are not complex and excessively difficult or expensive to monitor.
- submeters already exist to isolate energy use of systems.
- meters added for isolation purposes will be used for other purposes such as operational feedback or tenant billing.
- the uncertainty created by stipulations is acceptable.
- the continued effectiveness of the ECM can be assessed by routine visual inspection of stipulated parameters.
- stipulation of some parameters is less costly than measurement of them in Option B or simulation in Option D.

3.4.2 Option B: Retrofit Isolation

The savings determination techniques of Option B are identical to those of Option A except that no stipulations are allowed under Option B. In other words, full measurement is required.

Short term or continuous metering may be used under Option B. Continuous metering provides greater certainty in reported savings and more data about equipment operation. These data can be used to improve or optimize the operation of the equipment on a real-time basis, thereby improving the benefit of the retrofit itself. Results from several studies have shown five to fifteen percent annual energy savings can be achieved through careful use of continuous data logging (Claridge et al. 1994, 1996; Haberl et al. 1995).

Option B involves full measurement of the impact of the ECM. Therefore there is less need to verify the potential to perform than in Option A. The suggested installation verifications of Chapter 3.4.1.3 may be relaxed by eliminating ongoing re-inspections after the commissioning inspection.

The savings created by most types of ECMs can be determined with Option B. However, the degree of difficulty and costs associated with verification increases proportionately as metering complexity increases. Option B methods will generally be more difficult and costly than Option A. However Option B may produce less uncertain results where load and savings patterns are variable. Additional costs may be justifiable if a contractor is responsible for all aspects of ECM effectiveness.

ASHRAE's Guideline 14P is expected to provide technical details on a similar method (ASHRAE 2000).

3.4.2.1 Option B: Best Applications

Option B is best applied where:

- the performance of only the systems affected by the ECM is of concern, either due to the responsibilities assigned to the parties in a performance contract or due to the savings of the ECM being too small to be detected in the time available using Option C.

- interactive effects between ECMs or with other facility equipment can be measured or assumed to be immaterial.

- isolation of the ECM from the rest of the facility may avoid possibly difficult non-routine Baseline Adjustments for future changes to the facility.

- the independent variables that affect energy use are not complex and excessively difficult or expensive to monitor.
- submeters already exist to isolate energy use of systems.
- meters added for isolation purposes will be used for other purposes such as operational feedback or tenant billing.
- measurement of parameters is less costly than simulation in Option D.

3.4.3 Option C: Whole Building

Option C involves use of utility meters or whole building submeters to assess the energy performance of a total building. Option C assesses the impact of any type of ECM, but not individually if more than one is applied to an energy meter. This Option determines the collective savings of all ECMs applied to the part of the facility monitored by the energy meter. Also, since whole building meters are used, savings reported under Option C include the impact of any other changes made in facility energy use (positive or negative).

Option C may be used in cases where there is a high degree of interaction between installed ECMs or between ECMs and the rest of the building, or the isolation and measurement of individual ECM(s) is difficult or too costly.

This Option is intended for projects where savings are expected to be large enough to be discernible from the random or unexplained energy variations that are normally found at the level of the whole facility meter. The larger the saving, or the smaller the unexplained variations in the baseyear, the easier it will be to identify savings. Also the longer the period of savings analysis after ECM installation, the less significant is the impact of short term unexplained variations. Typically savings should be more than 10% of the baseyear energy use if they are to be separated from the noise in baseyear data.

Periodic inspections should be made of all equipment and operations in the facility after ECM installation. These inspections will identify changes from baseyear conditions or intended operations.

Accounting for changes (other than those caused by the ECMs) is the major challenge associated with Option C—particularly when savings are to be monitored for long periods. See also Chapter 4.8 on Baseline Adjustments.

ASHRAE's Guideline 14P is expected to provide technical details on a similar method (ASHRAE 2000).

3.4.3.1 Option C: Energy Data

Each energy flow into a building is measured separately by the utility or energy supplier. Where utility supply is only measured at a central point in a campus style facility, sub-meters are needed at each building or group of buildings on campus for which individual building performance is to be assessed.

Several meters may be used to measure the flow of one energy type into a building. To the extent any meter supplies energy use to a system that interacts with other energy systems directly or indirectly, it must be included in the whole building savings determinations. Meters serving non-interacting energy flows for which savings are not to be determined can be ignored, such as separately metered outdoor lighting circuits. If several different meters are read on separate days, then each meter having a unique billing period should be separately analyzed. The results can be combined after each individual analysis.

Savings should be determined separately for each meter or sub-meter serving a building so that performance changes can be assessed for separately metered parts of the facility. Where a meter measures a small fraction of one energy type's total use, it may be totaled with the larger meter(s) to reduce data management tasks. When electrical meters are so combined, it should be recognized that small consumption meters often do not have demand data associated with them so the totalized consumption data will no longer provide meaningful load factor information.

If energy data are missing from the post-retrofit period, a post-retrofit model can be created to fill in missing data. However the reported savings for the period should identify the report as "estimated."

Where changes to electric demand represent a significant amount of the calculated cost savings, the utility bill recorded demand may not be an adequate source of data due to the difficulties of deriving accurate models from single monthly demand readings. In this situation, the time of utility meter peaking must be known for each month so that the special demand recording equipment can be synchronized with the utility's resetting of the demand. Also the minimum time step for any demand recording meter should match the utility's demand time interval (see Chapter 5.2).

3.4.3.2 *Option C: Energy Invoices*

Energy data are often derived from utility meters, either through direct reading of the meter, or from utility invoices. Where utility bills are the source of energy use data, it should be recognized that a utility's needs for accuracy in meter reading may not be the same as that of savings determination. Utility bills can contain estimated data, especially for small accounts. Sometimes it cannot be determined from the bill itself that data come from an estimate rather than a meter reading. Unreported estimated meter readings create unknown errors for the month(s) of the estimate and the subsequent month when an actual reading is made. However the first bill with an actual reading after one or more estimates will correct the previous errors in energy quantities. When the fact of an estimate is shown on a utility bill, the associated savings report should reflect this fact.

Where electrical meter estimates are made, no valid data exist for electrical demand.

Energy may be supplied indirectly to a facility, through on-site storage facilities for oil, propane or coal. In such situations, information on the energy supplier shipment invoices is not representative of the facility's actual consumption during the period between shipments. Ideally a meter downstream of the storage facility should be used to measure energy use. However where there is no such meter, inventory level adjustments for each invoice period should be used to supplement the invoices.

3.4.3.3 *Option C: independent Variables*

Characteristics of a facility's use Or the environment which govern energy consumption are called independent variables. Common independent variables are weather and occupancy. Weather has many dimensions, but for whole building analysis weather is most often just outdoor temperature and possibly humidity depending upon the climate of the facility. Occupancy may be defined in many ways, such as: hotel room occupancy factor, office building core occupancy hours or maximum hours, number of occupied days (weekdays/weekends), or restaurant sales.

To the extent that independent variables have a cyclical nature to them, the significance of their impact on energy use can be assessed through mathematical modeling. Parameters found to have a significant effect in the baseyear period should be included in the routine adjustments when applying equation (1) for determining savings. Parameters having a less predictable but potentially significant effect should be measured and recorded in the baseyear conditions and post-retrofit periods so that non-routine baseline adjustments can be made if needed (see Chapter 4.8)

Independent variables should be measured and recorded at the same time as the energy meters. For example, weather data should be recorded daily so it can be totaled to correspond with the exact monthly energy metering period which may be different from the calendar month. Monthly mean temperature data for a non-calendar month would introduce unnecessary error into the model.

The number of independent variables to consider in the model of the baseyear data can be determined by regression analysis and other forms of mathematical modeling (Rabl 1988, Rabl and Rialhe 1992, ASHRAE 1997, Fels 1986, Ruch and Claridge 1991, Claridge et al. 1994).

3.4.3.4 Option C: Data Analysis and Models

The adjustment term of equation (1) under Option C is calculated by developing a valid model of each meter's baseyear energy use and/or demand. A model may be as simple as an ordered list of twelve actual baseyear monthly electrical demands without any adjustment factors. However they can often be a set of factors derived from regression analysis correlating energy use to one or more parameters such as *degree days,* metering period length, occupancy, and building operating mode (summer/winter). Models can also involve several sets of regression parameters each valid over a defined range of conditions such as ambient temperature, in the case of buildings, since buildings often use energy differently in different seasons.

Option C usually requires 12, 24, or 36 (i.e., one full year or multiple years) of continuous baseyear daily or monthly energy data, and continuous data during the post-retrofit period (Fels 1986) since models with more or less data (i.e., 13, 14, 15 or 9,10, 11 months) can cause the regression to have a statistical bias. Meter data can be hourly, daily or monthly whole-building data. Hourly data should be aggregated at least to the daily level to control the number of independent variables required to produce a reasonable model of the baseyear, without significant impact on the uncertainty in computed savings (Katipamula 1996, Kissock et al. 1992). Scatter found in daily data is often attributable to the weekly cycle of most facilities.

Many models appropriate for Option C are possible. To select the one most suited to the application, statistical evaluation indices should be considered, such as R^2 or *CV (RMSE)* (see Appendix B). Additional information concerning these selection procedures can be found in Reynolds and Fels (19 88), Kissock et al. (1992, 1994) and in the ASHRAE Handbook of Fundamentals (1997).

Statistical validity of the selected model should be assessed and demonstrated by reference to published statistical literature.

In certain types of facilities (such as schools) where there is a significant difference between the facility's energy use during the school year and summer break, separate *regression models* may need to be developed for different usage periods (Landman and Haberl 1996a; 1996b).

3.4.3.5 Option C: Computation of Routine Adjustments

The following steps are used to calculate the Adjustments term in Equation 1 for Option C.

1 Develop the appropriate model for the baseyear energy data and selected significant driving conditions (see Chapter 3.4.3.2 and Chapter 3.4.3.3).

2 Insert the post-retrofit period's independent variables (e.g. ambient temperature, metering period length) into the baseyear model from 1, above. This process derives the energy use that would have happened under post-retrofit conditions if the ECM had not been installed. (Note if some other set of conditions is selected for reporting savings (Chapter 3) the independent variables for this set of conditions would be used in place of the post-retrofit independent variables.)

3 Subtract the baseyear's energy use from the result of 2, above, for each month.

3.4.3.6 Option C: Cost

The cost of Option C methods depends on whether the energy

data come from utility bills or other special whole building meters. If such special whole building sub-meters were in place anyway there may be no extra cost, providing they are properly read, recorded and maintained. The primary cost elements in Option C are i) utility bill or data management and running of the model with each month's utility data, and ii) tracking and adjusting for conditions which change after the baseyear.

Option C is best applied where:

- the energy performance of the whole facility is to be assessed, not just the ECMs.
- there are many different types of ECMs in one building.
- the ECMs involve diffuse activities which cannot easily be isolated for the rest of the facility, such as operator training or wall and window upgrades.
- the savings are large enough to be separated from noise in the baseyear data during the time of monitoring.
- interactive effects between ECMs or with other facility equipment is substantial making isolation techniques of Options A and B excessively complex.
- Major future changes to the facility are not expected during the period of savings determination. A system of tracking key operating conditions can be established to facilitate possible future non-routine Baseline Adjustments.
- reasonable correlations can be found between energy use and other independent variables

3.4.4 Option D: Calibrated Simulation

Option D involves the use of computer simulation software to predict facility energy use for one or both of the energy use terms in Equation 1. Such *simulation model* must be "calibrated" so that it predicts an energy use and demand pattern that reasonably matches actual utility consumption and demand data from either the baseyear or a post-retrofit year.

Option D may be used to assess the performance of all ECMs in a facility, akin to Option C. However, different from Option C, multiple runs of the simulation tool in Option D allow estimates of the savings attributable to each ECM within a multiple ECM project.

Option D may also be used to assess just the performance of individual systems within a facility, akin to Options A and B. In this case, the system's energy use must be isolated from that of the rest of the facility by appropriate meters, as discussed in Chapter 3.4. 1. 1.

Option D is useful where:

- Baseyear energy data do not exist or are unavailable. Such situation may arise for a new facility containing particular energy efficiency measures needing to be assessed separately from the rest of the facility. It may also arise in a centrally metered campus of facilities where no individual facility meter exists in the baseyear period, but where individual meters will be available after ECM installation.

- Post-retrofit energy use data are unavailable or obscured by factors whose influence will be difficult to quantify. For example, such situation may arise where it would be too difficult to assess the impact of future facility usage changes that might significantly affect energy use. Industrial process changes or uncontrolled significant equipment additions often make the computation of future significant baseline adjustments so imprecise that the error in savings determination is excessive.

- The expected energy savings are not large enough to be separated from the facility's utility meter using Option C.

- It is desired to determine the savings associated with individual ECMs but Options A or B isolation and measurements are too difficult or costly.

If the post-retrofit energy use is predicted by the simulation software, the determined savings are actually maintained only if the simulated operating methods are maintained. Periodic inspections should be made of all equipment and operations in the facility after ECM installation (see Chapter 3.4.1.3). These inspections will identify changes from baseyear conditions and variances from modeled equipment performance.

The adjustments term in equation (1) is computed by running the simulation model under appropriate sets of conditions as needed to bring the two energy use terms to a common set of conditions.

Accurate computer modeling and calibration to measured data are the major challenges associated with Option D. To control the costs of this method while maintaining reasonable accuracy, the following points should be considered when using Option D:

1 Simulation analysis needs to be conducted by trained personnel who are experienced with the particular software and calibration techniques.
2 Input data should represent the best available information including as much as possible of actual performance data from key components in the facility.
3 The simulation needs to be adjusted ("calibrated") so its results match both the demand and consumption data from monthly utility bills within acceptable tolerances. The use of actual weather data may be necessary in cases where the actual weather data varies significantly from the average year weather data used in the simulation. Close agreement between predicted and actual annual total energy use is usually insufficient demonstration that the simulation adequately predicts the energy behavior of the facility.
4 Simulation analyses need to be well documented with paper and electronic copies of input and output files as well as the survey and metering/monitoring data used to define and calibrate the model. The particular version number of the software should be declared if it is publicly available so that any other party can fully review the many computations within the simulation.

ASHRAE's Guideline 14P is expected to provide technical details on a similar method (ASHRAE 2000).

3.4.4.1 Option D: Types of Simulation Programs

Information on the different types of building simulation models can be found in the ASHRAE Handbook (1997). DOE also maintains a current list of public domain and proprietary building energy simulation programs. This information can be obtained by accessing DOE's information server at www.eren.doe. gov/buildings/tools-directory.

Whole building simulation programs usually involve hourly calculation techniques. However techniques using ASHRAE's simplified energy analysis procedure may also be used if the building heat losses/gains, internal loads and HVAC systems are simple. ASHRAE's procedure features modified bin methods and simplified HVAC system models.

Many other types of special purpose programs may be used to simulate energy use and operating conditions of individual components or industrial processes. HVAC component models are available from ASHRAE in its $HVACO_2$ toolkit (Brandemuehl 1993), and for boiler/chiller equipment in the HVACO 1 toolkit (Bourdouxhe 1994a, 1994b, 1995). Simplified component air-side HVAC models are also available in a report by Knebel (1983). Equations for numerous other models have been identified as well (ASHRAE 1989, SEL 1996).

Any software used must be well documented and well understood by the user.

3.4.4.2 Option D: Calibration

Savings determined with Option D are based on one or more complex estimates of energy use. Therefore, the accuracy of the savings is completely dependent on how well the simulation models actual performance and how well calibrated it is to actual performance.

Calibration is achieved by verifying that the simulation model reasonably predicts the energy use of the facility by comparing model results to a set of calibration data. This calibration data should at a minimum be measured energy consumption and demand data, for the portion of the facility being simulated. Calibration of building simulations is usually done with 12 monthly utility bills. The calibration data set should be documented along with a description of its source(s).

Other operating data from the facility can be used as simulation input data as part of the calibration data set. These data might include operating characteristics and profiles of key variables such as use and occupancy, weather, known loads, equipment operating periods and efficiency. Some variables may be measured for short intervals, recorded for a day week or month, or extracted from existing operating logs. Accuracy of measurement equipment should be verified for critical measurements. If resources permit, actual building ventilation and infiltration should be measured since these quantities often vary widely from expectations. Snap-shot measurements will significantly improve simulation accuracy. Where resources are limited, on/off tests can be used to determine snap-shot end-use measurements of lighting, receptacle plug loads and motor control centers. These tests can be performed over a weekend using a data logger or EMCS to record whole-building electricity use, usually at one-minute intervals, and in some instances with inexpensive portable loggers that are synchronized to a common time stamp (Benton et al. 1996, Houeek et al. 1993, Soebarto 1996).

Following collection of as much calibration data as possible, the steps in calibrating the simulation are as shown below.

1 Assume other input parameters and document them.

2 Verify that the simulation predicts reasonable operating results such as space or process temperature/ humidity.

3 Compare simulated energy and demand results with metered data, on an hourly or monthly basis. Use actual weather data when conditions vary significantly from average year weather data. Assess patterns in the differences between simulation and calibration data. Bar charts, monthly percent difference time-series graphs and monthly x-y scatter plots give visual presentations which aid the identification of error patterns.

4 Revise assumed input data in step 1 and repeat steps 2 and 3 to bring predicted results reasonably close to actual energy use and demand. More actual operating data from the facility may also be needed to improve the calibration.

Buildings types which may not be easily simulated include those with:

- large atriums,
- a significant fraction of the space underground or ground coupled,
- unusual exterior shapes,
- complex shading configurations,
- a large number of distinct zones of temperature control.

Some building ECMs cannot be simulated without great difficulty, such as:

- addition of radiant barriers in an attic, and
- HVAC system changes not enabled by the fixed options within some whole-building hourly simulation programs.

The creation and calibration of a simulation can be time concerning. The use of monthly data for calibration is usually less costly than hourly calibration. Calibrations based on monthly utility data can achieve an approximate mean bias error (MBE) of ±20% compared to monthly energy use. Hourly calibrations can achieve ± 10% to ±20% CV (RMSE) of hourly energy use, or ± 1 % to ±5 % of the monthly utility bill.

3.4.4.3 Option D: Best Applications

Option D is best applied where:

- either baseyear or post-retrofit energy data unavailable or unreliable.
- there are too many ECMs to assess using Options A or B.
- the ECMs involve diffuse activities which cannot easily be isolated for the rest of the facility, such as operator training or wall and window upgrades.
- the impact of each ECM on its own is to be estimated within a multiple ECM project and the costs of Options A or B are excessive.
- interactive effects between ECMs or with other facility equipment is complex making isolation techniques of Options A and B excessively complex.
- major future changes to the facility are expected during the period of savings determination and no realistic means can be found to track or account for their energy impact.
- an experienced energy simulation professional is available and adequately funded for gathering suitable input data and calibrating the simulation model.
- the facility and the ECMs can be modeled by well documented simulation software, and reasonable calibration can be achieved

against actual metered energy and demand data.

3.5 ADHERENCE WITH IPMVP

The IPMVP is a framework of definitions and methods for assessing energy savings. The IPMVP framework was designed to allow users to develop an M&V plan for a specific project. The IPMVP was written to allow maximum flexibility in creating M&V plans that meet the needs of individual projects, but also adhere to the principles of accuracy, transparency and repeatability. In the case where users are required to demonstrate adherence, or wish to claim adherence with the IPMVP, the following issues should be addressed.

- The two parties should identify the organization/person responsible for M&V activities. This organization/person should be responsible for approving the site-specific M&V plan, and making sure that the M&V plan is followed for the duration of the contract.

- The M&V plan should clearly state which IPMVP Option (or combination of Options) and methods (linear regression, multiple regression, bin method etc.) will be used to determine the energy savings.

- The two parties should agree on a site-specific plan that specifies the metering/monitoring to be conducted. The plan should clearly state how the baseyear energy use and baseyear conditions are to be established including: what measurements are to be taken, how the data are to be used, what variables are to be stipulated and the basis for stipulation. The plan should provide information on the metering equipment, its calibration, the location of measurements, duration of the metering period, accuracy of the measurement process, etc.

- The M&V plan should specify the details of how calculations should be made by stating the variables (run-time hours, electrical consumption in a lighting fixture, kW/ton, etc.) that should be measured and any associated assumptions.

- The two parties should agree on how quality assurance should be maintained and replicability confirmed.

- The M&V plan should list the reports to be prepared, their contents and formats, and a stipulated time frame during which they should be furnished.

- All terminology should be consistent with IPMVP definitions.

Chapter 4

Common Issues[1]

4.1 FACTORS AFFECTING SAVINGS PERFORMANCE

Many factors affect the performance of equipment and achievement of savings. Depending upon the scope of the savings determination (its boundaries), the range of parameters of concern can be very focused (specific ECMs) or as wide as the whole facility.

Parameters that are predictable and measurable can be used for routine adjustments in Equation 1 of Chapter 3. Such adjustments reduce the variability in reported savings, or provide a greater degree of certainty in reported savings. Unpredictable parameters within the boundaries of a savings determination may require future non-routine Baseline Adjustments (e.g. future loss of tenants). Unmeasured parameters give rise to savings fluctuations for which no adjustment can be computed, only guessed (e.g. air infiltration rate).

Therefore, when planning an M&V process, consideration should be given to 1) predictability, 2) measurability and 3) likely impact of all plausible factors in each category below:

- Weather
- Occupancy level, schedule
- Installed equipment intensity, schedule
- Occupant or user demand for services (e.g. space temperature, plant throughput)

1. Common issues arising when using the options laid out in Chapter 3 are discussed in this Section. Measurement issues are in Chapter 5.

- Ability of the ECM as designed to achieve the intended savings
- ECM implementation effectiveness in meeting the design intent
- Occupant or operator cooperation in using ECM related equipment in accordance with direction
- Occupant or operator cooperation in using non-ECM related equipment in accordance with direction
- Equipment deterioration, both ECM related equipment and non-ECM related
- Equipment life, both ECM and non-ECM related

4.2 EVALUATING SAVINGS UNCERTAINTY

The effort undertaken in determining savings should focus on managing the uncertainty created in the determination process. ECMs with which the facility staff are familiar may require less effort than other, uncommon ECMs. The savings determination process itself introduces uncertainties through

- Instrumentation Error
- Modeling Error
- Sampling Error
- Planned and Unplanned assumptions

Methods of quantifying the first three errors are discussed in Appendix B. As used in this protocol, sampling error concerns do not refer to rigorous statistical procedures, but to the best practices as addressed in Appendix B. See also Reddy & Claridge (2000) that applies standard error analysis methods to the typical savings determination scenario.

The last category of error above, encompasses all the unquantifiable errors associated with stipulations, and the assumptions necessary for measurement and savings determination.

It is feasible to quantify many but not all dimensions of the uncertainty in savings determination. Therefore when planning an M&V process, consideration should be given to quantifying the quantifiable uncertainty factors and qualitatively assessing the unquantifiable. The objective is to consider all factors creating uncertainty, either qualitatively or quantitatively.

The accuracy of a savings estimate can be improved in two general ways. One is by reducing biases, by using better information or by using measured values in place of assumed or stipulated values. The second way is by reducing random errors, either by increasing the sample sizes, using a more efficient sample design or applying better measurement techniques. In most cases, improving accuracy by any of these means increases M&V cost. Such extra cost should be justified by the value of the improved information (see Chapter 4.11).

Quantified uncertainty should be expressed in a statistically meaningful way, namely declaring both accuracy and confidence levels. For example, "The quantifiable error is found, with 90% confidence, to be +20%." A statistical precision statement without a confidence level is meaningless since accuracy can sound very good if the confidence level is low.

The appropriate level of accuracy for any savings determination is established by the concerned parties. Appendix B discusses some issues in establishing a level of uncertainty.

For buildings, one or more full years of energy use and weather data should be used to construct regression models. Shorter periods introduce more uncertainty through not having data on all operating modes. The best predictors of both cooling and heating annual energy use are models from data sets with mean temperatures close to the annual mean temperature. The range of variation of daily temperature values in the data set seems to be of secondary importance. One month data sets in spring and fall, when the above condition applies, can be better predictors of annual energy use than five month data sets from winter and summer.

The required length of the metering or monitoring period depends on the type of ECM. If, for instance, the ECM affects a system that operated according to a well-defined schedule under a constant load, such as a constant-speed exhaust fan motor, the period required to determine annual savings could be quite short. In this case, short-term energy savings can be easily extrapolated to the entire year. However, if the project's energy use varies both across day and seasons, as with air-conditioning equipment, a much longer metering or monitoring period may be required to characterize the system. In this case, long-term data are used to determine annual energy savings.

If the energy consumption of the metered equipment or systems varies by more than ten percent from month to month, additional measurements must be taken at sufficient detail and over a long enough period of time to identify and document the source of the variances. Any major energy consumption variances due to seasonal production increases or periodic fluctuations in occupancy or use must also be tracked and recorded.

4.3 MINIMUM ENERGY STANDARDS

When a certain level of efficiency is required either by law or the owner's standard practice, savings may be based on the difference between the post-retrofit energy use and the minimum standard. In these situations, baseyear energy use may be set equal to or less than the applicable minimum energy standards. U.S. Department of Energy's Building Energy Standards and Guidelines Program (BSGP), available at www.eren.doe.gov/buildings/codes-standards/buildings, provides information about residential, commercial and Federal building codes.

4.4 MINIMUM OPERATING CONDITIONS

An energy efficiency program should not compromise the operations of the facility to which it is applied without the agreement of the facility users, whether building occupants or industrial process managers. Therefore the M&V Plan should record the agreed conditions that will be maintained (see Chapter 3.3).

Volume II of the IPMVP Concepts and Practices for Improved Indoor Environmental Quality suggests methods of monitoring indoor space conditions throughout an energy efficiency program.

4.5 ENERGY PRICES

Energy cost savings may be calculated by applying the price of each energy or demand unit to the determined savings. The price of energy should be the energy provider s rate schedule or an appropriate simplification thereof Appropriate simplifications use marginal prices which consider all aspects of billing affected by metered amounts, such as consumption charges, demand charges, transformer credits, power factor, demand ratchets, early payment discounts.

An example of the energy cost savings calculation is contained in Appendix A (Option D).

4.6 VERIFICATION BY A THIRD PARTY

Where the firm performing the energy savings determinations has more experience than the owner, the owner may seek assistance in reviewing savings reports. Such assistance should begin at the time of first review of the M&V Plan, to ensure that the design for the savings determination process will meet the owner's objectives. The review should continue with the routine savings reports and baseline adjustments. Full review of baseline adjustments requires good understanding of the facility and it operations. For this latter purpose, owner summaries of operating conditions will reduce the scope, work and cost of the third party verifier.

An energy performance contract requires that both parties believe the information on which the payments are based is valid and accurate. An experienced third party may be helpful to ensure agreement of measurement validity. Should conflicts arise over the course of the project payback period, this third party can help to resolve differences.

Third party savings verifiers are typically engineering consultants with experience and knowledge in verifying ECM savings, ECM technologies and, where relevant, energy performance contracting. Many are members of industry professional societies, though there is not yet any accreditation program for M&V professionals.

4.7 DATA FOR EMISSION TRADING

The IPMVP has already been recognized as valuable in some regions for verifying savings and securing financial benefits allowed under emissions trading programs, and is expected to be a part of an international trading regime. Application of this Protocol can provide increased confidence in the measurement of actual energy savings, and therefore provide greater confidence in determining associated reductions in emissions. It is becoming an important element in international greenhouse gas emission mitigation and trading programs because of the broad international participation in its development, and its growing adoption internationally.

Combined with the specific M&V Plan of each project, this Protocol enhances consistency of reporting and enables verification of energy savings. However to verify an emission credit this Protocol and the project's M&V Plan must be used in conjunction with the credit trading program's specific guidance on converting energy savings into equivalent emissions reductions.

Emission trading will be facilitated if the following energy reporting methods are considered when designing the savings determination process:

- Electrical savings should be split into peak period and off peak periods, and ozone season non-ozone season when NO_x or VOCs are involved. These periods will be defined by the relevant trading program.

- Reductions in purchases from the electrical grid should be divided into those due to load reduction and those due to increased self-generation at the facility.

- Savings should be separated into those that are 'surplus' or 'additional' to normal behavior and those that are simply 'business as usual' or needed to comply with existing regulations. These terms will be defined by the relevant trading program. For example, where equipment minimum efficiency standards limit the efficiency of new equipment on the market these standards may form the reference case for determining tradable credits derived from energy savings.

- Segregate energy savings at each site if a project spans a power pool's boundary line, or if emission quantities may be outside an air shed of concern.

- Segregate fuel savings by fuel or boiler type if different emission rates apply to each combustion device.

4.8 BASELINE ADJUSTMENTS (NON-ROUTINE)

Conditions which vary in a predictable fashion are normally included within the basic mathematical model used for routine adjustments, described in Chapter 3.4. Where unexpected or one-time changes occur they may require non-routine adjustments, normally called simply Baseline Adjustments.

Examples of situations often needing Baseline Adjustments are: i) changes in the amount of space being heated or air conditioned, ii) changes in the amount or use of equipment iii) changes in environmental conditions (lighting levels, set-point temperatures, etc.) for the sake of standards compliance, and iv) changes in occupancy, schedule or throughput.

Baseline Adjustments are not needed where:

- the variable is included in the mathematical model developed for the project

- changes affect a variable that was stipulated in the M&V Plan. For example if the number of ton-hours of cooling were stipulated for a chiller efficiency ECM, an increase in the cooling ton-hours will not affect the savings determined by the agreed simplified method, though actual savings will change.

- changes occur to equipment beyond the boundary of the savings determination. For example if the boundary includes only the lighting system, for a lighting retrofit, addition of personal computers to the space will not affect the savings determination.

Baseyear conditions need to be well documented in the M&V Plan so that proper adjustments can be made (see Chapter 3.3). It is also important to have a method of tracking and reporting changes to these conditions. This tracking of conditions may be performed by one or more of the facility owner, the agent determining savings, or a third party verifier. It should be established in the M&V Plain who will track and report each condition recorded for the baseyear and what, if any other aspects of facility operation will be monitored.

Where the nature of future changes can be anticipated, methods for making the relevant non-routine Baseline Adjustments should be included in the M&V Plan.

Non-routine Baseline Adjustments are determined from actual or assumed physical changes in equipment or operations. Sometimes it may be difficult to identify the impact of changes. If the facility's energy consumption record is used to identify such changes, the impact of the ECMs on the metered energy consumption must first be removed by Option B techniques.

4.9 WEATHER DATA

Where monthly energy measurements are used, weather data should be recorded daily and matched to the actual energy metering period.

For monthly or daily analysis, government published weather data should be treated as the most accurate and verifiable. However weather data from such source may not be available as quickly as site monitored weather data.

When analyzing the response of energy use to weather in mathematical modeling, daily mean temperature data or degree days may be used.

4.10 COST

The cost of determining savings depend on many factors such as:

- IPMVP Option selected
- ECM number, complexity and amount of interaction amongst them
- number of energy flows across the boundary drawn around the ECM to isolate it from the rest of the facility in Options A, B or D when applied to a system only
- level of detail and effort associated with establishing baseyear conditions needed for the Option selected
- amount and complexity of the measurement equipment (design, installation, maintenance, calibration, reading, removal)
- sample sizes used for metering representative equipment
- amount of engineering required to make and support the stipulations used in Option A or the calibrated simulations of Option D
- number and complexity of independent variables which are accounted for in mathematical models
- duration of metering and reporting activities
- accuracy requirements
- savings report requirements
- process of reviewing or verifying reported savings

- experience and professional qualifications of the people conducting the savings determination

Often these costs can be shared with other objectives such as real time control, operational feedback, or tenant sub-billing.

It is difficult to generalize about costs for the different IPMVP Options since each project will have its own unique set of constraints. However it should be an objective of M&V Planning to design the process to incur no more cost than needed to provide adequate certainty and verifiability in the reported savings, consistent with the overall budget for the ECMs. Typically however it would not be expected that average annual savings determination costs exceed more than about 10% of the average annual savings being assessed.

Table 2 highlights key cost governing factors unique to each Option, or not listed above.

Table 2: Unique Elements of M&V Costs

Option A	Number of measurement points Complexity of stipulation Frequency of post-retrofit inspection
Option B	Number of measurement points
Option C	Number of meters Number of independent variables needed to account for most of the variability in energy data.
Option D	Number and complexity of systems simulated. Number of field measurements needed to provide input data. Skill of professional simulator in achieving calibration

Commonly, since Option A involves stipulation, it will involve fewer measurement points and lower cost, providing stipulation and inspection costs do not dominate.

Since new measurement equipment is often involved in Options A or B, the cost of maintaining this equipment may make Option C a less costly endeavor for long monitoring periods. However, as mentioned above, the costs of extra meters for Options A or B may be shared with other objectives.

When multiple ECMs are installed at one site, it may be less costly to use the whole building methods of Options C or D than to isolate and measure multiple ECMs with Options A or B.

Though development and calibration of an Option D simulation model is often a time concerning process, it may have other uses such as for designing the ECMs themselves or designing a new facility.

Where a contractor (ESCO) is responsible for only certain aspects of project performance, other aspects may not have to be measured for contractual purposes, though the owner may still wish to measure all aspects for its own sake. In this situation, the costs of measurement may be shared between owner and contractor.

4.11 BALANCING UNCERTAINTY AND COST

The acceptable level of uncertainty required in a savings calculation is a function of the level of savings and the cost-effectiveness of decreasing uncertainty For example, suppose a project has an expected savings of $100,000 per year and that a basic M&V approach had an accuracy no better than ±25% with 90% confidence, or $25,000 per year. To improve the accuracy to within $ 10,000 it may be seen as reasonable to spend an extra $5,000 per year on M&V but not $30,000 per year. The quantity of savings at stake therefore places limits on the target expenditure for M&V

Further benefits of activities to reduce uncertainty may be the availability of better feedback to operations, enabling an enhancement of savings or other operational variables. The information may also be useful in assessing equipment sizing for planning plant expansions or replacement of equipment. It may also allow higher payments to be made under an energy performance contract based on measured vs. conservative stipulated values. Additional investments for improved accuracy should not exceed the expected increase in value. This issue is discussed in more detail by Goldberg (1996b).

Discussions and definitions of site-specific M&V plans should include consideration of accuracy requirements for M&V activities and the importance of relating M&V costs and accuracy to the value of ECM savings. However it should be recognized that not all uncertainties can be quantified (see Chapter 4.2). Therefore both quantitative and qualitative uncertainty statements must be considered when considering M&V

cost options for each project.

For a given savings determination model at a specific site, there will be an optimal savings determination plan. The method to identify that Plan includes iterative consideration of sensitivity of the savings uncertainty to each variable, estimating the cost of metering specified variables in the model and a criteria for valuing reduced uncertainty (e.g. risk-adjusting saving per a given formula).

Chapter 5
Measurement Issues

5.1 USING UTILITY METERS

Whole building energy measurements can utilize the same meters that the local power company uses to bill the owner if they are equipped or modified to provide an output that can be recorded by the facility's monitoring equipment. The "energy/pulse" constant of the pulse transmitter should be calibrated against a known reference such as similar data recorded by the power company s revenue meter.

5.2 ELECTRIC DEMAND

Electric demand measurement methods vary amongst utilities. The method used by any sub-meter or modeling routine should replicate the method the power company uses for the relevant billing meter. For example, if the local power company is calculating peak demand using a 15 minute "fixed window," then the recording equipment should be set to record data every 15 minutes. However if the power company uses a "sliding window" to record electric demand data, the data recorder should have sliding window recording capabilities. Such sliding window capability can be duplicated by recording data on one minute fixed window intervals and then recreating the sliding 15 minute window using post-processing software. Most often 15 minute fixed window measurements will represent sliding 15 minute data reasonably well. However, care should be taken to ensure that the facility does not contain unusual combinations of equipment that generate high one minute peak loads which may show up in a sliding window interval and not in a fixed window. After processing the data for the demand analysis, the 15 minute data can then be converted to hourly data for archiving and further analysis against hourly weather data.

5.3 INSTRUMENTATION AND MEASUREMENT TECHNIQUES

Special meters may be used to measure physical quantities or to submeter an energy flow. Example quantities which may have to be measured without the use of energy supplier meters are temperature, humidity, flow, pressure, equipment runtime, electricity and thermal energy. To determine energy savings with reasonable accuracy and repeatability, good measurement practices should be followed for these quantities. Such practices are continually evolving as metering equipment improves. It is recommended that the latest measurement practices be followed to support any savings determination. Appendix C provides a review of some common measurement techniques. The IPMVP web site contains relevant current references on measurement techniques."

5.4 CALIBRATION OF INSTRUMENTATION

It is highly recommended that instrumentation be calibrated with procedures developed by the National Institute of Standards and Technology (NIST). Primary standards and no less than third order NIST traceable calibration equipment should be utilized wherever possible. Sensors and metering equipment should be selected based in part on the ease of calibration and the ability to hold calibration. An attractive solution is the selection of equipment that is se If- calibrating.

Selected references on calibration have been provided in Chapter 6.2, including: ASTM (1992), Baker and Hurley (1984), Benedict (1984), Bryant and O'Neal (1992), Cortina (1988), Doebelin (1990), EEI (1981), Haberl et al. (1992), Harding (1982), Huang (1991), Hurley and Schooley (1984), Hurley (1985), Hyland and Hurley (1983), Kulwicki (1991), Leider (1990), Liptak (1995), Miller (1989), Morrissey (1990), Ramboz and McAuliff (1983), Robinson et al. (1992), Ross and White (1990), Sparks (1992), Wiesman (1989), Wise (1976), Wise and Soulen (1986).

5.5 DATA COLLECTION ERRORS AND LOST DATA

Methodologies for data collection differ in degree of difficulty, and consequently in the amount of erroneous or missing data. No data collection is without error. The M&V Plan should consider two aspects of data collection problems:

- establish a maximum acceptable rate of data loss and how it will be measured. This level should be part of the overall accuracy consideration. The level of data loss may dramatically affect cost.

- establish a methodology by which missing or erroneous data will be interpolated for final analysis. In such cases, baseyear and post-retrofit models may be used to calculate savings.

5.6 USE OF ENERGY MANAGEMENT SYSTEMS FOR DATA COLLECTION

The facility *energy management system*[1] (EMS) can provide much of the monitoring necessary for data collection. However, the system and software must be fully specified to provide this extra service as well as its primary realtime control function. For example, significant use of trending functions may impair the basic functions of the EMS. Some parameters to be monitored may not be required for control. These extra points must be specified in the design documents. Electric power metering is an example. Trending of small power, lighting and main feed power consumption may be very useful for high quality savings determination and operational feedback, but useless for real time control.

Other functions that can easily be incorporated into the software are automatic recording of changes in set-points.

It is not unusual for many of the trending capabilities required for verification to be incorporated in an EMS. However adequate hardware and software capability must be provided since data trending can tie up computer processing, communication bandwidth and storage.

Facility staff should be properly trained in this use of the EMS so they too can develop their own trending information for diagnosing system problems, providing the system has the capacity for extra trending. However where a contractor is responsible for some operations controlled by the system, EMS security arrangements should ensure that persons can only access functions for which they are competent and authorized.

The EMS design and monitoring team may have a direct read-only connection into the EMS via a modem link so they can easily inspect trend data in their office. However possible concerns for virus attacks and computer security should be addressed in this situation.

---

1. The terms in italics are defined in Chapter 6.1

The EMS can record energy use with its trending capability. However, most EMSs record "change of value" (COV) event recordings that are not directly used for calculating energy savings without tracking time intervals between individual COV events (Claridge et al. 1993, Heinerneier and Akbari 1993). It is possible to tighten COV limits in order to force the trending towards more regular intervals, but this can overload systems which are not designed for such data densities. Great care should be exercised to:

- control access and/or changes to the EMS trend log from which the energy data are extracted.
- develop post-processing routines for changing the EMS COV data into time series data for performing an analysis.
- get from the EMS supplier:
  - — NIST traceable calibrations of all sensors,
  - — evidence that proprietary algorithms for counting and/or totaling pulses, Bras, and kWh data are accurate. (Currently, there are no industry standards for performing this analysis (Sparks et al. 1992), and
  - — commitment that there is adequate processing and storage capacity to handle trending data while supporting the system's control functions.

Chapter 6
Definitions and References

6.1 DEFINITIONS

Baseline Adjustments—The non-routine adjustments (Chapter 3.4) arising during the post-retrofit period that cannot be anticipated and which require custom engineering analysis (see Chapter 4.8).

Baseyear Conditions—The set of conditions which gave rise to the energy use/demand of the baseyear.

Baseyear Energy Data—The energy consumption or demand during the baseyear.

Baseyear—A defined period of any length before implementation of the ECM(s).

Commissioning—A process for achieving, verifying and documenting the performance of equipment to meet the operational needs of the facility within the capabilities of the design, and to meet the design documentation and the owner's functional criteria, including preparation of operator personnel.

CV (RMSE)—Coefficient of Variation of the RMSE (see Appendix B)

Degree Day—A degree day is measure of the heating or cooling load on a facility created by outdoor temperature. When the mean daily outdoor temperature is one degree below a stated reference temperature such as 18T, for one day, it is defined that there is one heating degree day. If this temperature difference prevailed for ten days there would be ten heating degree days counted for the total period. If the temperature difference were to be 12 degrees for 10 days, 120 heating degree days would be counted. When the ambient temperature is below the reference temperature it is defined that heating degree days are counted. When ambient temperatures are above the reference, cooling degree days are counted. Any reference temperature may be used for recording degree days, usually chosen to reflect the temperature at which heating or cooling is no longer needed.

Energy Conservation/Efficiency Measure (ECM or EEM)—A set of activities designed to increase the energy efficiency of a facility. Several ECM's may be carried out in a facility at one time, each with a different thrust. An ECM may involve one or more of physical changes to facility equipment, revisions to operating and maintenance procedures, software changes, or new means of training or managing users of the space or operations and maintenance staff.

EMS or Energy Management System—A computer that can be programmed to control and/or monitor the operations of energy consuming equipment in a facility.

Energy Performance Contract—A contract between two or more parties where payment is based on achieving specified results; typically, guaranteed reductions in energy consumption and/or operating costs.

Energy Savings—Actual reduction in electricity use (kWh), electric demand (M), or thermal units (Btu).

ESPC or Energy Savings Performance Contract—A term used in the United States equivalent to Energy Performance Contract.

ESCO or Energy Services Company—A firm which provides a range of energy efficiency and financing services and guarantees that the specified results will be achieved under an energy performance contract.

M&V or Measurement & Verification—The process of determining savings using one of the four IPMVP Options.

Metering—Collection of energy and water consumption data over time at a facility through the use of measurement devices.

Monitoring—The collection of data at a facility over time for the purpose of savings analysis (i.e., energy and water consumption, temperature, humidity, hours of operation, etc.)

M&V Option—One of four generic M&V approaches defined herein for energy savings determination.

Post-Retrofit Period—Any period of time following commissioning of the ECM.

R2-R Squared (see Appendix B)

Regression Model—Inverse mathematical model that requires data to extract parameters describing the correlation of independent and dependent variables

RMSE—Root mean square error (see Appendix B)

Simulation Model—An assembly of algorithms that calculates energy use based on engineering equations and user-defined parameters.

Verification—The process of examining the report of others to comment on its suitability for the intended purpose.

6.2 REFERENCES

NOTE: The following references are meant to provide the reader with sources of additional information. These sources consist of publications, textbooks and reports from government agencies, national laboratories, universities, professional organizations and other recognized authorities on building energy analysis. For the most part care has been taken to cite the publication, publisher or source where the document can be obtained.

1 Akbari, H., Heinemeier, K.E., LeConiac, P. and Flora, D.L. 1988. "An Algorithin to Disaggregate Commercial Whole-Facility Hourly Electrical Load Into End Uses," Proceedings of the ACEEE 1988 Summer Study on Energy Efficiency in Buildings, Vol. 10, pp. 10. 14-10.26.

2 ASHRAE Guideline 1-1996. The HVAC Commissioning Process. American Society of Heating, Ventilating, and Air Conditioning Engineers, Atlanta, Georgia.

3 ASHRAE Proposed Guideline 14P, Measuring Energy and Demand Savings, draft available for public review April to June 2000. American Society of Heating, Ventilating, and Air Conditioning Engineers, Atlanta, Georgia.

4 ASHRAE. 1989. An Annotated Guide to Models and Algorithms for Energy Calculations Relating to HVAC Equipment, American Society of Heating, Ventilating, and Air Conditioning Engineers, Atlanta, Georgia.

5 ASHRAE 1997. Handbook: Fundamentals, Chapter 30—"Energy Estimating and Modeling Methods," Atlanta, Georgia.

6 ASTM 1992. Standard Test Method for Determining Air Leakage Rate by Fan Pressurization, American Society for Testing Materials, Philadelphia, Pennsylvania.

7 Baker, D. and Hurley, W. 1984. "On-Site Calibration of Flow Metering Systems Installed in Buildings," NBS Building Science Series Report No. 159, January.

8 Benedict, R. 1984. Fundamentals of Temperature, Pressure and Flow Measurement. John Wiley and Sons, New York, New York.

9 Benton, C., Chace, J., Huizenga, C., Hyderman, M. and Marcial, R. 1996. "Taking A Building's Vital Signs: A Lending Library of Handheld Instruments," Proceedings of the ACEEE 1996 Summer Study on Energy Efficiency in Buildings, Vol. 4, pp. 4.11-4.21.

10 Bourdouxhe, J.P., Grodent, M., LeBrun, J. 1995. "HVAC01 Toolkit: A Toolkit for 20 Primary HVAC System Energy System Energy Calculations," Final report submitted to ASHRAE.

11 Bourdouxhe, J.P., Grodent, M., LeBrun, J. 1994a. "Toolkit for Primary HVAC System Energy Calculation—Part 1: Boiler Model," ASHRAE Transactions, Vol. 100, Pt. 2.

12 Bourdouxhe, J.P., Grodent, M., LeBrun, J. 1994b. "Toolkit for Primary HVAC System Energy Calculation—Part 2: Reciprocating Chiller Models," ASHRAE Transactions, Vol. 100, Pt. 2.

13 Bou Saada, T.E. and Haberl, J.S. 1995a. "A Weather-Daytyping Procedure for Disaggregating Hourly End-Use Loads in an Electrically Heated and Cooled Building from Whole-facility Hourly Data." 30th Intersociety Energy Conversion Energy Conference, July 30-August 4.

14 Bou Saada, T.E. and Haberl, J.S. 1995b. "An Improved Procedure for Developing Calibrated Hourly Simulated Models," Proceedings of Building Simulation, 1995: pp. 475-484.

15 Bou Saada, T.E., Haberl, J., VaJda, J. and Harris, L. 1996. "Total Utility Savings From the 37,000 Fixture Lighting Retrofit to the USDOE Forrestal Building," Proceedings of the 1996 ACEEE Summery Study, August.

16 Brandemuchl, M. 1993. HVAC02: Toolkit: Algorithms and Subroutines for Secondary HVAC Systems Energy Calculations, American Society of Heating, Ventilating, and Air Conditioning Engineers, Atlanta, Georgia.

17 Bryant, J. and O'Neal, D. 1992. "Calibration of Relative Humidity Trans-

ducers for use in the Texas LoanSTAR Program," Proceedings of the 1992 Hot and Humid Conference, Texas A&M University, Energy Systems Laboratory Report No. ESL-PA-92/02-15.

18 Claridge, D., Haberl, J., Bryant, J., Poyner, B. and McBride, J. 1993. "Use of Energy Management and Control Systems for Performance Monitoring of Retrofit Projects," Final Summary Report, USDOE Grant #DE-FGOI-90CE21003, Submitted to the USDOE Office of Conservation and Energy, Energy Systems Laboratory Report ESL-TR-91/09/02, Texas A&M University, March.

19 Claridge, D., Haberl, J., Liu, M., Houcek, J. and Aather, A. 1994. "Can You Achieve 150% of Predicted Retrofit Savings? Is it Time for Recommissioning?," Proceedings of the 1994 ACEEE Summer Study, pp. 5.73-5.88, August.

20 Claridge, D., Haberl, J., Liu, M. and Athar, A. 1996. "Implementation of Continuous Commissioning in the Texas LoanSTAR Program: Can you Achieve 150% of Estimated Retrofit Savings: Revisited," Proceedings of the 1996 ACEEE Summery Study, August.

21 Cortina, V. (ed.) 1988. "Precision Humidity Analysis," EG&G Environmental Equipment, 151 Bear Hill Road, Waltham, Massachusetts, (IR sensors).

22 Doebelin, E. 1990. Measurement Systems. McGraw-Hill, New York, New York, ISBN 0-07-017338-9.

23 EEI 1981. Handbook for Electricity Metering, Edison Electric Institute, Washington, D.C., ISBN-0-931032-11-3.

24 EPRI 1993. "Fundamental Equations for Residential and Commercial EndUses" (Rep. #EPRI TR- 100984 V2). Palo Alto, California: Electric Power Research Institute.

25 Fels, M. (ed.)l 986. "Special Issue Devoted to Measuring Energy Savings, The Princeton Scorekeeping Method (PRISM)," Energy and Buildings, Vol. 9, Nos. 1 and 2.

26 Fels, M., Kissock, K., Marean, M.A. and Reynolds, C. 1995. "Advanced PRISM User's Guide," Center for Energy and Environmental Studies Report, Princeton University, Princeton, New Jersey, January.

27 Goldberg, M.L. 1996a. "The Value of Improved Measurements: Facing the Monsters That Won't Annihilate Each Other," Energy Services Journal, 2(1):43-56.

28 Goldberg, M.L. 1996b. "Reasonable Doubts: Monitoring and Verification for Performance Contracting," Proceedings of the ACEEE 1996 Summer Study on Energy Efficiency in Buildings, 4.133-4.143 Washington, D.C.: American Council for an Energy-Efficient Economy.

29 Haberl, J., Bronson, D. and O'Neal, D. 1995. "Impact of Using Measured Weather Data vs. TMY Weather Data in a DOE-2 Simulation," ASHRAE Transactions, V. 105, Pt. 2, June.

30 Haberl, J., Reddy, A., Claridge, D., Turner, D., O'Neal, D. and Heffington, W. 1996. "Measuring Energy-Savings Retrofits: Experiences from the Texas LoanSTAR Program," Oak Ridge National Laboratory Report No. ORNL/Sub/93-SP090/1, February.

31 Haberl, J., Turner, W.D., Finstad, C., Scott, F. and Bryant, J. 1992. "Calibration of Flowmeters for use in HVAC Systems Monitoring," Proceedings of the 1992 ASME/JSES/KSES International Solar Energy Conference.

32 Hadley, D.L. and Tomich, S.D. 1986. "Multivariate Statistical Assessment or Meteorological Influences in Residence Space Heating," Proceedings of the ACEEE 1986 Summer Study on Energy Efficiency in Buildings, Vol. 9, pp. 9.132-9.145.

33 Harding, J. (ed). 1982. "Recent Advances in Chilled Mirror Hygrometry," General Eastern Corporation Technical Bulletin, 50 Hunt Street, Watertown, Massachusetts.

34 Heinerneier, K. and Akbari, H. 1993. "Energy Management and Control Systems and Their Use for Performance Monitoring in the LoanSTAR Program," Lawrence Berkeley National Laboratory Report No. LBL-33114-UC-350, June, (prepared for the Texas State Energy Conservation Office).

35 Houcek, J., Liu, M., Claridge, D., Haberl, J., Katipamula, S. and Abbas, M. 1993. "Potential Operation and Maintenance (O&M) Savings at the State Capitol Complex," Energy Systems Lab Technical Report No. ESL-TR-93/01-07, Texas A&M University, College Station, Texas.

36 Huang, P. 1991. "Humidity Measurements and Calibration Standards," ASHRAE Transactions, Vol. 97, p.3521.

37 Hurley, CW. and Schooley, J.F. 1984. "Calibration of Temperature Measurement Systems Installed in Buildings," N.B.S. Building Science Series Report No. 153, January.

38 Hurley, W. 1985. "Measurement of Temperature, Humidity, and Fluid Flow," Field Data Acquisition for Building and Equipment Energy Use Monitoring, ORNL Publication No. CONF-8510218, March.

39 Hyland, R.W. and Hurley, CW. 1983. "General Guidelines for the On-Site Calibration of Humidity and Moisture Control Systems in Buildings," N.B.S. Building Science Series 157, September.

40 IPCC 1995. Impacts, Adaptations and Mitigation of Climate Change: Scientific-Technical Analyses. Contribution of Working Group II to the Second Assessment Report of the Intergovernmental Panel on Climate Change, Geneva, Switzerland. pp 64.

41 Katipamula, S. 1996. "The Great Energy Predictor Shootout 11: Modeling Energy Use in Large Commercial Buildings," ASHRAE Transactions, Vol. 102, Pt 2.

42 Katipamula, S. and Haberl, J. 1991. "A Methodology to Identify Diurnal Load Shapes for Non-Weather-Dependent Electric End-Uses," Proceedings of the 1991 ASME-JSES International Solar Energy Conference, ASME, New York, New York, pp. 457-467, March.

43 Kats, G., Kumar, S., and Rosenfeld, A. 1999. "The Role for an International Measurement & Verification Standard in Reducing Pollution," Proceedings of the ECEEE 1999 Summer Study, Vol. 1, Panel 1.

44 Kats, G., Rosenfeld, A., and McGaraghan, S. 1997. "Energy Efficiency as A Commodity: The Emergence of an Efficiency Secondary Market for Savings in Commercial Buildings," Proceedings of the ECEEE 1997 Summer Study, Vol. 1, Panel 2.

45 Kissock, K., Claridge, D., Haberl, J. and Reddy, A. 1992. "Measuring Retrofit Savings For the Texas LoanSTAR Program: Preliminary Methodology and Results," Solar Engineering, 1992: Proceedings of the ASME-JSES-SSME International Solar Energy Conference, Maui, Hawaii, April.

46 Kissock, K., Wu, X., Sparks, R., Claridge, D., Mahoney, J. and Haberl, J. 1994. "EModel Version, 1.4d," Energy Systems Laboratory ESL-SW-94/12-01, Texas Engineering Experiment Station, Texas A&M University System, December.

47 Knebel, D.E. 1983. "Simplified Energy Analysis Using the Modified Bin Method," ASHRAE, Atlanta, Georgia, ISBN 0-910110-39-5.

48 Kulwicki, B. 199 1. "Humidity Sensors," Journal of the American Ceramic Society, Vol. 74, pp. 697-707.

49 Landman, D. and Haberl, J. 1996a. "Monthly Variable-Based Degree Day Template: A Spreadsheet Procedure for Calculating 3-parameter Change-point Model for Residential or Commercial Buildings," Energy Systems Laboratory Report No. ESL-TR-96/09-02.

50 Landman, D. and Haberl, J. 1996b. "A Study of Diagnostic Pre-Screening Methods for Analyzing Energy Use of K- 12 Public Schools," Energy Systems Laboratory Report No. ESL-TR-96/1 1-01, November.

51 Leider, M. 1990. A Solid State Amperometric Humidity Sensor, Journal of Applied Electrochemistry, Chapman and Hill: Vol. 20, pp. 964-8.

52 Liptak, B. 1995. Instrument Engineers' Handbook, 3rd Edition: Process Measurement and Analysis. Chilton Book Company, Radnor, Pennsylvania, ISBN 0-8019-8197-2.

53 Miller, R. 1989. Flow Measurement Handbook, McGraw Hill Publishing Company, New York, New York, ISBN 0-07-042046-7.

54 Morrissey, C.J. 1990. "Acoustic Humidity Sensor," NASA Tech Brief Vol. 14, No. 19, April, (acoustic).

55 Rabl, A. 1988. "Parameter Estimation in Buildings: Methods for Dynamic Analysis of Measured Energy Use," Journal of Solar Energy Engineering, Vol. 110, pp. 52-66.

56 Rabl, A. and Riable, A. 1992. "Energy Signature Model for Commercial Buildings: Test With Measured Data and Interpretation," Energy and Buildings, Vol. 19, pp. 143-154.

57 Ramboz, J.D. and McAuliff, R. C. 19 83. "A Calibration Service for Watt-meters and Watt-Hour Meters," N.B.S. Technical Note 1179.

58 Reddy, T. and Claridge, D. 2000. "Uncertainty of "Measured" Energy Savings From Statistical Baseline Models," ASHRAE HVAC&R Research, Vol 6, No 1, January 2000.

59 Reynolds, C. and Fels, M. 1988. "Reliability Criteria for Weather Adjustment of Energy Billing Data," Proceedings of ACEEE 1988 Summer Study on Energy Efficiency in Buildings, Vol. 10, pp. 10.237-10.24 1.

60 Robinson, J., Bryant, J., Haberl, J. and Turner, D. 1992. "Calibration of Tangential Paddlewheel Insertion Flowmeters," Proceedings of the 1992 Hot and Humid Conference, Texas A&M University, Energy Systems Laboratory Report No. ESL-PA-92/02-09,

61 Ross, I.J. and White, G.M. 1990. "Humidity," Instrumentation and Measurement for Environmental Sciences: Transactions of the ASAE, 2nd ed., p. 8-01.

62 Ruch, D. and Claridge, D. 1991. "A Four Parameter Change-Point Model for Predicting Energy Consumption in Commercial Buildings," Proceedings of the ASME-JSES-JSME.

63 SEL 1996. TRNSYS Version 14.2, and Engineering Equation Solver (EES). Solar Energy Laboratory, Mechanical Engineering Department, University of Wisconsin, Madison, Wisconsin.

64 Soebarto, V. 1996. "Development of a Calibration Methodology for Hourly Building Energy Simulation Models Using Disaggregated Energy Use Data From Existing Buildings," Ph.D. Dissertation, Department of Architecture, Texas A&M University, August.

65 Sparks, R., Haberl, J., Bhattacharyya, S., Rayaprolu, M., Wang, J. and Vadlamam, S. 1992. "Testing of Data Acquisition Systems for Use in Monitoring Building Energy Conservation Systems," Proceedings of the Eighth Symposium on Improving Building Systems in Hot and Humid Climates, Dallas, Texas, pp. 19 7-204, May.

66 Vine, E. and Sathaye, J. 1999. "Guidelines for the Monitoring, Evaluation, Reporting, Verification, and Certification of Energy-Efficiency Projects for Climate-Change Mitigation," LBNL Report # 41543.

67 Violette, D., Brakken, R., Schon, A. and Greef, J. 1993. "Statistically-Adjusted Engineering Estimate: What Can The Evaluation Analyst Do About The Engineering Side Of The Analysis? " Proceedings of the 1993 Energy Program Evaluation Conference, Chicago, Illinois.

68 Wiesman, S. (ed.) 1989. Measuring Humidity in Test Chambers, General Eastern Corporation, 50 Hunt Street, Watertown, Massachusetts.

69 Wise, JA. 1976. "Liquid-In-Glass Thermometry," N.B.S. Monograph 150, January.

70 Wise, JA. and Soulen, R.J. 1986. "Thermometer Calibration: A Model for State Calibration Laboratories," N.B.S. Monograph 174, January.

6.3 SOURCES

The following organizations provide useful and relevant information:

1 Air Conditioning and Refrigeration Center, Mechanical Engineering, University of Illinois. TEL: 217-333-3115, http://acrc.me.uiuc.edu

2 American Council for an Energy Efficient Economy (ACEEE), Washington, D.C. TEL: 202-429-8873, http://www.aceee.org

3 American Society of Heating, Refrigerating, and Air Conditioning Engineers (ASHRAE), Atlanta, Georgia. TEL: 404-636-8400, http://www.ashrae.org

4 American Society of Mechanical Engineers (ASME), New Jersey. TEL: 800-843-2763. http://www.asme.org

5 Association of Energy Engineers (AEE), Lilburn, GA. TEL: 404-925-9558, http://www.aeecenter.org

6 Boiler Efficiency Institute, Department of Mechanical Engineering, Auburn University, Alabama. TEL: 334/821-3095, http://www.boilerinstitute.com

7 Center for Energy and Environmental Studies (CEES), Princeton University, New Jersey. TEL: 609-452-5445, http:Hwww.princeton.edu/-cees

8 Edison Electric Institute (EEI). Washington, DC. TEL: 202-508-5000, http://www.eei.org/resources/pubcat

9 Energy Systems Laboratory, College Station, Texas. TEL: 979-845-9213, http:Hwww-es1.tamu.edu

10 Florida Solar Energy Center, Cape Canaveral, Florida. TEL: (407) 6 3 81000, http:/Awvw.fsec.ucf.edu

11 IESNA Publications, New York, New York. TEL: 212-248-5000, http://www.iesna.org.

12 Lawrence Berkeley National Laboratory (LBNL), Berkeley CA. TEL: 5 10486-6156, Email: EETDinfo@lbl.gov, http://eetd.lbl.gov

13 National Association of Energy Service Companies (NAESCO), Washington, D.C. TEL: 202-822-0950, http://www.naesco.org

14 Energy Information Administration (EIA), Department of Energy, Washington, D.C., TEL: 202-586-8800, http://www. eia. doe. gov

15 National Renewable Energy Laboratory (NREL), Boulder, Colorado, TEL: (303) 275-3000, http://www.nrel.gov

16 National Technical Information Service (NIST), U.S. Department of Commerce (This is repository for all publications by the Federal labs and contractors), Springfield Virginia. TEL: 703-605-6000, http://WWW.ntis.gov

17 Oak ridge National Laboratory (ORNL), Oak Ridge, Tennessee, Tel: (865) 574-5206, http://www.ornl.gov/ORNL/BTC

18 Pacific Northwest National Laboratory (PNNL), Richland, Washington, Tel: (509) 372-4217, http://www.pnl.gov/buildings/

Appendix A Examples

Option A Example: Lighting Efficiency Retrofit

Situation

More efficient fixtures are installed in place of existing fixtures in a school to reduce energy requirements, while maintaining lighting levels.

M&V Plan

An M&V Plan was developed showing that Option A was to be used for savings determination because partial measurement was deemed to provide adequate accuracy. An outline of the Plan is shown below:

- The boundary of this ECM was drawn to include the ceiling mounted lighting circuits fed by the 277 volt supply, and the radiation heating system. The associated decrease in air conditioning load was considered trivial since little of the school is air conditioned, and most of it is closed for the summer months.

- The baseyear conditions are those of the 12 months immediately preceding the decision to proceed with the project. They included a lighting level survey, description location and number of lamps ballasts and fixtures.

- Engineering calculations determined that the ECM would increase boiler load by the energy equivalent of 6% of the lighting savings from November through March. This number was estimated to range between 4% and 8%.

- The boiler efficiency in winter was estimated to be 79% under typical winter conditions.

- The baseyear fuel use from the gas utility bills from November through March is $2{,}940 \times 10^3$ ft^3 (83.25×10^3 m^3).

- The lighting operating periods of the post-retrofit period are selected as the common set of conditions for the energy use terms in Equation 1 of Chapter 3.

- Baseyear lighting periods were established through one month logging of lighting in representative areas. Lighting period logs established the following annual load/duration data for the baseyear:

Baseyear Load/Duration

Fraction of Lighting Load	On Hours Per Year
9%	240
61%	1,450
15%	2,500
6%	6,100
9%	8,760

- Due to a change in occupancy patterns planned to take effect about the same time as the ECM installation, it is assumed that the load/duration profile in the post-retrofit period will be as shown below:

Stipulated Post-retrofit Load/Duration

Fraction of Lighting Load	On Hours Per Year
9%	240
61%	2,000
15%	2,500
6%	6,100
9%	8,760

- Measurements were made with a recently calibrated RMS power meter of the three phase power draw on the 277 volt lighting circuits. The manufacturer's rating on this power meter is ±2% of full scale and readings were roughly 50% of full scale. From a thirty second measurement on the input side of two lighting transformers, it was found that with all fixtures switched on, the total power draw was 28.8 kW, though 7 lamps (= 0.3 kW or 1%) were burned out at the time of the test. It was determined that the fraction burned out at the time of this measurement was normal.

- The electrical demand for lighting was assumed to be equal to the measured circuit load for ten months of the year when school is in session. This stipulation may be in error by no more than 3% since lighting is the dominant electrical load of the building. Based on the utility bills showing a demand reduction during July and August, the minimal use of the facility during these months, and the other equipment used during the summer, it was assumed that the July and August lighting circuit demand is only 5 0% of the measured circuit load.

- The possible errors in the above stipulated post-retrofit lighting load/duration profile are:

— only half of the anticipated growth from 1,450 hrs to 2,000 hrs may happen, and

— the 9% load fraction may be switched on for 400 hours.

- These possible errors could affect the post-retrofit energy use by as much as about 2,500 kWh, which represents 8.2% of the expected 30,000 kWh annual savings. The impact of assumptions of the lighting impact on the electrical demand meter for all twelve months of the baseyear and post-retrofit years might affect total reported demand savings by as much as 3%. Neither of these stipulation impacts is considered significant for the project.

- Estimated accuracy of the power measurements is ±4%.

- The savings calculation process shown below was summarized in the M&V Plan.

- Savings are to be computed annually for the subsequent year using a remeasurement of the lighting electrical load immediately after ECM completion and on each anniversary thereafter.

- The electrical power readings on the baseyear and all future years will be made by a contract electrician. All data and analyses are available for inspection. As a check on the readings, building maintenance staff will also measure the electrical load at the same times as the contractor. If there is a difference of more than 4% between staff and contractor readings, a second contractor reading will be made and the proper value selected between the two contractor readings.

- This savings determination process is expected to require an electrician 5 hours each year to make the readings and calibrate the measurement equipment. Total cost each year is expected to be $200 including reporting,

Baseyear Electricity Use/Demand

The baseyear energy use for Equation 1 is computed by multiplying the 28.8 baseyear load by baseyear load/duration data, above. The computation is shown below.

Baseyear Energy Use

Fraction of Lighting Load	kW	On Hours Per Year	kWh
9%	2.6	240	622
61%	17.6	1,450	25,474
15%	4.3	2,500	10,800
6%	1.7	6,100	10,541
9%	2.6	8,760	22,703
Total (100%)	**28.8**		**70,140**

The baseyear demand is 28.8 kW for each of 10 months and 14.4 kW for each of July and August, bringing the total demand to 317 M-mo.

Post-Retrofit Electricity Use/Demand

After installation of the ECM, the lighting circuit power was re-measured as in the baseyear. The power draw was 16.2 kW with all lights on and none burned out. With the same 1% burnout rate as in the baseyear, the post-retrofit period maximum power would be 16.0 kW (= 16.2 × 0.99). Therefore the post-retrofit annual energy use for Equation 1 is computed by multiplying the 16.0 kW post-retrofit load by-the stipulated post-retrofit load/duration data. The computation is shown below.

Post-Retrofit Energy Use

Fraction of Lighting Load	kW	On Hours Per Year	kWh
9%	1.4	240	346
61%	9.8	2,000	19,520
15%	2.4	2,500	6,000
6%	1.0	6,100	5,860
9%	1.4	8,760	12,614
Total 16.0			**44,340**

The post-retrofit demand is 16.0 kW for each of 10 months and 8.0 kW for each of July and August, bringing the total demand to 176 kW-mo.

Post-Retrofit Fuel Use

Fuel increases resulting from the lighting ECM are derived from the electrical energy savings. The unadjusted electrical savings are 70,140 – 44,340 = 25,800 kWh per year. Assuming these savings are achieved uniformly over a 10 month period, the typical winter month electrical savings are 25,800/10 = 2,580 kWh/month. The associated boiler load increase is 6% of these electrical savings for November through March, namely:

= 6% × 2,580 kWh/mo × 5 months =774 kWh equivalent

Extra boiler input energy is:

= 774 kWh * 79% =980 kWh equivalent units of fuel
= 3,344,000 Btu or 3,000 ft^3 (84.95 m^3) of natural gas

Therefore total post-retrofit fuel use is estimated to be 2,940 + 3 = 2,943 × 10^3 ft^3 (83.34 10^3 m^3).

Routine Adjustments

Routine adjustments are needed to bring baseyear energy use to the conditions of the stipulated post-retrofit period.

By applying the 28.8 kW baseyear electrical load to the stipulated post-retrofit load/duration data, the routine adjustment for the longer operating hours is derived, as shown at the top of the following page.

No adjustments are needed to electric demand since the increase in operating hours occurs during the school sessions, therefore not increasing demand.

Though adjustments are appropriate for associated fuel use, they would be trivial so are ignored.

Savings

From Equation 1, the energy savings for the first year after ECM installation are determined at shown in the center of the following page.

Baseyear Energy Use at Stipulated Post-Retrofit Conditions				At Baseyear Conditions	Adjust-ment
Fraction of Lighting Load	**kW**	**On Hours Per Year**	**kWh**	**kWh**	**kWh**
9%	2.6	240	622	622	
61%	17.6	2,000	35,144	25,474	
15%	4.3	2,500	10,800	10,800	
6%	1.7	6,100	10,541	10,541	
9%	2.6	8,760	22,703	22,703	
Total	**128.8**		**179,8101**	**70,140**	**9,670**

	Baseyear	–	Post-Retrofit	+	Adjustment	=	Savings
Electricity	70,140	–	44,340	+	9,670	=	35,470 kWh
Electric Demand	317	–	176	+	0	=	141 kW-mo
Gas	2,940,000 (83,250)	–	2,943,000 (83,340)	+	0	=	–3000 ft^3 (-84.95 m^3)

Subsequent years' savings will be computed identically, from each year's measured load on the same electrical panel.

Note that in this example the savings reported are for operations under post-retrofit period conditions. Therefore the savings can be called "avoided energy use."

Option B Example: Boiler Replacement

Situation

An office building boiler is replaced with a more efficient boiler. 95% of the load on the boiler is for building heating while 5% is for domestic water heating. There are no changes other than an improvement in boiler efficiency No other equipment in the building uses gas.

M&VPlan

An M&V Plan was developed showing that Option B was to be used for savings determination because the boiler retrofit for energy reduction was just part of many non-energy related changes planned for the building. An outline of the Plan is shown below:

- The boundary of this ECM was drawn to include only the boiler fuel systems. This boundary excludes the electricity associated with the boiler auxiliaries of burner and blower. Though less gas may be used by the boiler, the power uses of old and new blower are expected to be very similar and their operating periods will be the same. Therefore the auxiliaries are not expected to change their electricity use significantly and can be excluded from the boundary of measurement.

- The baseyear conditions were chosen to be the load pattern of typical winter periods before ECM installation.

- The conditions of the baseyear were chosen as the common set of conditions for the energy use terms in Equation 1, since it was expected that there would be significant changes in the building's heating loads in the post-retrofit period. It is recognized that the reported savings will then be for baseyear conditions, not post-retrofit conditions.

- The baseyear energy use was 35,200 × 10^3 ft^3 (1,000 × 10^3 m^3) of gas.

- Before retrofit, boiler efficiency was tested over three separate one week periods when average ambient temperature ranged from 20°F (– 6.7°C) to 24°F (– 4.4°C) and building occupancy was normal. A recently calibrated energy flow meter was installed on the boiler, measuring supply and return line temperature and supply water flow rate. This meter system with its data capture and processing has a manufacturer's rated accuracy of ±7% for the Btu ranges involved in this project. The utility's gas meter was used to measure gas use and is taken as the reference source, i.e. it has no error. The average efficiency readings for the three weekly intervals were 66%, 64% and 65%. An overall average efficiency of 65% was established. Outdoor temperature was measured by a sensor that was calibrated, twice a year and recorded by the building control system.

- It is assumed that the percentage change in efficiency measured under typical winter conditions will prevail in all other conditions. The error in this assumption is not likely to exceed 5%.

- The savings calculation process shown below was summarized in the M&V Plan.

- Savings are to be computed annually for the subsequent year using boiler efficiency data measured each year. Data from the energy flow meter and gas meter will be stored for examination by a third party if needed.

- The cost of installing and commissioning the energy flow meter, was $7,900. The cost of each year's reading of efficiency, meter calibration and reporting is $4,000.

- Gas and energy flow meter readings will be made daily by building maintenance staff through winter months until three valid weeks have been obtained. This data will be logged in the boiler room and open for inspection at any time. Ambient temperature data will be recorded by the building automation system and logs printed for the selected valid weeks.

- Energy flow meter calibration will be done annually by xyz contractor immediately before the efficiency testing period begins. Gas meter direct readings will be corrected for pressure and temperature by the utility company's factors for the corresponding period. These factors will be provided in writing by the utility.

Baseyear Energy Use

The baseyear annual energy use for Equation 1 is 35,200 × 10^3 ft^3 (1,000 × 10^3 m^3).

Post-Retrofit Energy Use

After installation and commissioning of the ECM, three separate weekly test periods were found with an average ambient temperature between 20°F (– 6.7°C) to 24°F (– 4.4°C) and normal occupancy. The efficiency results over the three one week periods were 81%, 79% and 80%, averaging 80%.

The post-retrofit annual energy use for Equation 1 is determined from the baseyear use to be:

Baseyear Condition + Correction to Post-retrofit condition

$$= \frac{35{,}200 \times 0.65}{0.80} + C$$

$$= (26{,}410 + C)\ 10^3\text{ft}^3\ ([750 + C]\ 10^3\text{m}^3)$$

C is an unknown quantity needed to convert baseyear projected use of the new boiler to post-retrofit conditions.

Routine Adjustments

Routine adjustments are needed to bring post-retrofit energy use to the conditions of the baseyear. This is exactly the correction amount C million ft^3 (m^3).

Savings

From Equation 1, energy savings are determined to be:

	Baseyear	—	Post-Retrofit	+	Adjustment	=	Savings
Gas	35,200	—	(26,410 + C)	+	C	=	$8{,}790 \times 10^3$ ft^3 ($248.9\ 10^3\ \text{m}^3$)

Note that in this example the savings reported are for operations under baseyear conditions.

Option C Example: Whole Building Multiple ECM Project

Situation

An energy efficiency project was implemented in a high school, involving six ECMs spanning lighting, HVAC, pool heating and operator training and occupant awareness campaigns. The objectives of the project were to reduce energy costs.

M& V Plan

An M&V Plan was developed showing that Option C was to be used for savings determination because total facility energy cost was the focus. An outline of the Plan is shown below:

- The boundary of this savings determination was defined as:
  - The main electricity account #766A234-593 including demand
  - The auxiliary electrical account #766B 122-601 serving the field house
  - The natural gas account #KHJR3333-597
- The baseyear conditions are those of the 12 months immediately preceding the decision to proceed with the project. Included in the documentation of these conditions is:
  - a lighting level survey, with a count of the number of burned out lamps in January and June;
  - a summary of typical space temperatures and humidities during occupied and unoccupied periods in each of four seasons;
  - a count of the number and size of all computers, monitors and printers, along with an estimate of the operating hours of each;
  - a record of the number of day pupils and evening courses each month of the year;
  - a record of the number of public rental hours of the gym, cafeteria and pool each month;
  - a count of the number of window air conditioning units installed;
  - the temperature setting of pool water, and domestic hot water serving the pool showers, the gym showers and the rest of the school;

— the volume of make-up water supplied to the pool each month, as recorded by a separate uncalibrated sub-meter;

— the cafeteria kitchen hot water temperature and the number and rating of all kitchen equipment; and

— the open hours of the cafeteria kitchen and the value of food sales each month.

- The baseyear energy use is shown on the above utility accounts spanning the period January 1998 to December 1998.

- The baseyear energy data were analyzed as follows. Multiple linear regression was performed on monthly energy use and demand, metering period length, and degree days (DD). Degree days data were derived from mean daily dry bulb temperature published monthly by the government weather service for the city where the school is located. No significant correlation with weather was found for electric demand, summer electricity use in the field house or summer gas use. Analysis found reasonable correlation between weather and winter gas use and the main electricity meter's winter consumption. Therefore no other independent variables were sought. The energy per DD and energy per day data shown below describe the characteristics of the straight line relationship found by the regression analyses:

		Gas	**Electricity**		
			Demand	**Consumption**	**Consumption**
Account Number		KHJR3333-597	766A234-593		766B122-601
Units		10^3 ft^3 (10^3 m^3)	kW-mo	kWh	kWh
Annual Total		10,238 (290)	5,782	1,243,000	62,000
	DD Base	15°C		16°C	20°C
Winter	Energy/DD	2.55		39.61	18.12
Regression	Energy/Day	9.16		2,640	20.1
Analysis	CV (RMSE)	9%		18%	5%

- Savings will be determined under post-retrofit conditions.

- The savings calculation process shown below was summarized in the M&V Plan.

- The school has provided XYZ contractor authorization to receive energy use data from the electric and gas utility companies until 2008.

- Savings are to be computed and reported monthly by XYZ contractor in a format for physical plant staff to understand and quarterly in a format for teaching staff and students to understand. This reporting is to begin immediately after ECM completion. It will continue at this rate for eight years.

- Annually the school will report any changes in the baseyear conditions listed above, within a month after the end of each school year. XYZ contractor will compute the energy impact of these changes and any others that it believes are relevant and present Non-Routine Baseline Adjustments two months before the end of the school board's fiscal year.

- This savings determination process is expected to require a data entry and utility bill analyst 10 hours each year and an engineer 5 hours to review reports for accuracy and establish suitable computations for Non-Routine Baseline Adjustments. Total cost each year is expected to be about $ 1,000 including reporting.

The CV (RMSE) of the baseyear models range from 5% to 18% and are far less than the expected savings of 35% for both fuel and electricity. No sampling or instrumentation error exists. Therefore the reported savings will be statistically significant, subject to any error introduced through non-routine baseline adjustments which may arise.

Baseyear Energy Use

The baseyear energy use for Equation 1 is taken directly from the utility bills without adjustment. The data were tabulated in the M&V Plan.

Post-Retrofit Energy Use

The post-retrofit energy use for Equation 1 is taken directly from the utility bills without adjustment.

Routine Adjustments

Routine adjustments are needed to bring baseyear energy use to the conditions of the post-retrofit period. For the first year after retrofit the routine adjustments are computed as follows.

Gas:

See figure opposite.

a. facts from the baseyear energy data
b. facts from the baseyear energy data
c. facts from the post-retrofit metering periods
d. facts from the post-retrofit metering periods
e. (c) × 9.16 for month where DD > 25
f. (d) × 2.55 for months where DD > 25
g. (a/b) × (c) for months where DD = 25 or less
h. (e) + (f) + (g)
i. (h) - (a)

Electricity Consumption:

Calculations for each of the two electricity consumption meters are performed separately in the same fashion as the gas meter above, using the relevant baseyear data, regression factors, metering periods and degree days. The net routine adjustments for each month are shown in the Savings section below.

Electric Demand:

No routine adjustments are made since no correlation was found with weather.

Non-Routine Adjustments

During the first post retrofit period extra computer equipment was added, partially replacing older computers. The following monthly energy and demand estimates were made from nameplate ratings, typical loading and operating hours for the ten months when school is in session:

	Baseyear Energy Use		Post-Retrofit Conditions		Baseyear Energy Use Projected to Post-Retrofit Conditions				
	Consumption[a]	Days[b]	Days[c]	DD[d]	Winter Base[e]	Winter Heating[f]	Summer[g]	Total[h]	Adjustment[i]
Jan	2,239.1	29	31	742	284.0	1,892.1		2,176.1	-63.0
Feb	1,676.3	31	30	551	274.8	1,405.1		1,679.9	3.5
Mar	1,223.1	31	32	401	293.1	1,022.6		1,315.7	92.6
Apr	723.3	30	28	208	256.5	530.4		786.9	63.6
May	399.6	30	30	41	274.8	104.6		379.4	-20.3
Jun	240.1	28	30	12			257.3	257.3	17.2
Jul	201.2	31	32	0			207.7	207.7	6.5
Aug	193.6	30	30	2			193.6	193.6	0.0
Sep	288.7	30	30	20			288.7	288.7	0.0
Oct	439.1	30	31	99	284.0	252.5		536.4	97.3
Nov	1,023.6	31	30	302	274.8	770.1		1,044.9	21.3
Dec	1,591.1	33	33	521	302.3	1,328.6		1,630.8	39.7
Total	**10,238.8**	**364**	**367**					**10,497.2**	**258.4**

	Computer	Monitor	Printer	Total
Net Number Added	23	23	5	
Nameplate Watts	150	120	175	
Average Watts	70	110	50	
Hours Use/month	150	150	120	
kWh/month	242	380	30	652 kWh
Demand diversity	90%	90%	70%	
kW demand	1.45	2.28	0.23	3.96 kW

Though there may be a 50% error in these estimates, their impact is small relative to the savings report.

Savings

From Equation 1, the energy savings for the first year after ECM installation are determined for each account to be:

1 Gas Account #KHJR3333-597 Thousand ft^3 or Thousand m^3

	Baseyear Energy Use	–	Post-Retrofit	+	Adjustment	=	Savings
Jan	2,239.1	–	1,839.1	+	–63.0	=	337.0 (9.54)
Feb	1,676.3	–	1,233.6	+	3.5	=	446.3 (12.64)
Mar	1,221.1	–	932.1	+	92.6	=	383.6 (10.86)
Apr	723.3	–	621.1	+	63.6	=	165.8 (4.69)
May	399.6	–	301.0	+	–20.3	=	78.4(2.22)
Jun	240.1	–	160.2	+	17.2	=	97.1 (2.75)
Jul	201.2	–	120.1	+	6.5	=	87.6(2.48)
Aug	193.6	–	150.9	+	0.0	=	42.7(1.21)
Sep	288.7	–	202.3	+	0.0	=	86.4(2.45)
Oct	439.1	–	339.1	+	97.3	=	197.3 (5.59)
Nov	1,023.6	–	678.4	+	21.3	=	366.5 (10.38)
Dec	1,591.1	–	1,123.2	+	39.7	=	507.6 (14.27)
Total	**10,238.8**	–	**7,701.1**	+	**258.4**	=	**2,796.1 (79.16)**

2 Electricity Account #766A234-593 Consumption (kWh)

	Baseyear Energy Use	–	Post-Retrofit	+	Routine Adjustment	+	Non-Routine Adjustment	=	Savings
Jan	122,400	–	81,200	+	3,740	+	652	=	45,592
Feb	118,600	–	76,200	+	2,780	+	652	=	45,832
Mar	132,200	–	83,200	+	-1,220	+	652	=	48,432
Apr	110,800	–	77,600	+	1,890	+	652	=	35,742
May	106,000	–	65,400	+	2,120	+	652	=	43,372
Jun	101,200	–	61,200	+	120	+	652	=	40,772
Jul	30,200	–	20,800	+	-3,600	+	0	=	5,800
Aug	36,200	–	23,800	+	2,480	+	0	=	14,880
Sep	105,200	–	66,800	+	2,260	+	652	=	41,312
Oct	110,200	–	70,600	+	200	+	652	=	40,452
Nov	126,600	–	83,200	+	5,320	+	652	=	49,372
Dec	128,400	–	81,000	+	-2,240	+	652	=	45,812
Total	**1,228,000**	–	**791,000**	+	**13,850**	+	**6,520**	=	**457,370**

3 Electricity Account #766A234-593 Demand (kW)

	Baseyear Energy Use	–	Post-Retrofit	+	Routine Adjustment	+	Non-Routine Adjustment	=	Savings
Jan	561	–	402	+	0	+	4	=	163
Feb	521	–	381	+	0	+	4	=	144
Mar	502	–	352	+	0	+	4	=	154
Apr	490	–	328	+	0	+	4	=	166
May	472	–	310	+	0	+	4	=	146
Jun	470	–	336	+	0	+	4	=	138
Jul	300	–	222	+	0	+	0	=	78
Aug	470	–	324	+	0	+	0	=	146
Sep	476	–	336	+	0	+	4	=	144
Oct	480	–	350	+	0	+	4	=	134
Nov	500	–	362	+	0	+	4	=	142
Dec	540	–	390	+	0	+	4	=	154
Total	**5,782**	–	**4,113**	+	**0**	+	**40**	=	**1,709**

4 Electricity Account #766BI22-601 Consumption (kWh)

	Baseyear Energy Use	–	Post-Retrofit	+	Adjustment	=	Savings
Jan	12,200	–	10,200	+	-1,200	=	800
Feb	9,600	–	11,200	+	2,320	=	720
Mar	8,800	–	7,800	+	-200	=	800
Apr	4,400	–	4,800	+	1,280	=	880
May	3,800	–	5,100	+	2,120	=	820
Jun	1,200	–	500	+	120	=	820
Jul	800	–	400	+	23	=	423
Aug	600	–	300	+	48	=	252
Sep	1,200	–	400	+	41	=	841
Oct	4,400	–	3,800	+	140	=	740
Nov	6,600	–	5,400	+	-290	=	910
Dec	8,400	–	9,000	+	1,400	=	800
Total	62,000	–	58,900	+	5,706	=	8,806

Note that in this example the savings reported are for operations under post-retrofit period conditions. Therefore the savings can be called "avoided energy use."

Option D Example: Calibrated Simulation Multiple ECM Project

Situation

An energy efficiency project was implemented in a university library building, involving four ECMs spanning lighting, HVAC, operator training and occupant awareness campaigns. The building is part of a multiple building campus without individual building meters. As part of the energy management program steam, electricity and electric demand meters were installed on the main supply lines to the library. The objectives of the project were to reduce energy costs in the library.

M&V Plan

An M&V Plan was developed showing that Option D was to be used for savings determination because baseyear data did not exist for the library on its own. An outline of the Plan is shown below:

- The boundary of this project was defined as the total energy use of the library as it affects the main campus energy and demand purchases, assuming:

  — a pound of steam at the library requires 1.5 ft^3 (0.04 m^3) of natural gas at the campus heating plant's gas meter,

  — a kWh of electricity at the library requires 1.03 kWh of electricity at the campus electricity meter, and

  — a kW of demand at the library is coincident with 1.03 kW of electric demand at the campus electric demand meter.

- The baseyear conditions are those of the 12 months immediately preceding the decision to proceed with the project, 1999. Light levels were surveyed during this period and recorded. However the library use and occupancy is assumed to be the same in the baseyear and post-retrofit periods.

- No baseyear energy data exist so it will be simulated using DOE-2 software, version 2.1 calibrated against actual meter data from the first year of post-retrofit operations.

- The common set of conditions selected for use in the energy use terms in Equation 1 consists of the library use and occupancy in the first year of the post-retrofit period, and the weather conditions of a "normal" year for the city, as published by the National Renewable Energy Lab in 1989.

- Recordings were made of the following load and operating conditions during the post-retrofit period:

  — turnstile data, producing hourly occupancy data for each hour of the year, averaging a peak daily occupancy of 300 persons;

  — a library open hours: 8: 00 AM to midnight, seven days a week, except for statutory holidays when it is closed;

— operating staff measurements of space temperature and humidity at twenty five locations, mid morning and mid afternoon on the first day of each of the 12 months; and

— continuous power draw on the 120 volt circuits supplying library equipment, for five typical days and a statutory holiday. A total of 801 kWh/occupied day was recorded and an hourly profile was developed for typical occupied and unoccupied days.

- the input file of data including assumptions and the above measured data were printed and saved electronically for use by any other person.

- ABC consulting engineering firm designated J. Smith as the professional engineer to conduct the simulation and calibration because of his experience in this field.

- The intended savings calculation process shown below was summarized in the M&V Plan.

- Savings are to be computed after the end of the first post-retrofit year. To ensure that savings remain in place the building operating staff will regularly report the status of the key operating parameters which were used in the calibrated simulation model. If operating conditions change, the savings will not be adjusted since they are computed at a fixed set of conditions.

  — Savings are to be determined using the following marginal prices derived from the respective energy supply contracts: electricity consumption = $.079 1 /kWh electric demand = $9.93/kW-month steam = $14.23/10^3 lbs ($31.34/10^3 kg)

- This savings determination process is expected to require a consulting professional engineer one month to set up and calibrate an appropriate simulation model, costing about $20,000. A review of the work by DEF consultant is planned and may cost a further $8,000.

Baseyear Energy Use

The following steps were followed to compute baseyear energy use after the first post-retrofit year:

1 The newly installed meters were calibrated before installation. Operating staff read the meters monthly and recorded monthly total steam and electricity use, as well as monthly demand, for each of 12 months throughout the post-retrofit year.

2 A model was developed of the building with the ECM's installed. This model used actual weather of the post-retrofit period and the operating profiles recorded in the same period. The modeled space temperatures and humidities were examined to ensure they reasonably matched the typical range of indoor conditions during occupied and unoccupied days. Initially the model did not model energy use well, so farther site investigations were undertaken. During these investigations it was found that during unoccupied night periods, there was no effective indoor temperature change, so the thermal mass characteristics of the model were adjusted. With this correction the model was determined to adequately match the calibration data. The modeled results compared to the monthly data as follows

	MBE	CV (RMSE)
Electricity Consumption	8%	10%
Electric Demand	12%	15%
Steam	5%	8%

3 This accuracy of calibration is good enough to allow reasonable confidence in the relative results of two runs if the model. However the model should not be used to compare simulated results to actual data.

4 The calibrated model was archived, with both printed and electronic copy of input data, diagnostic reports and output data.

5 The calibrated model was then adjusted to remove the ECMs, and the weather data file was changed to correspond to the actual

weather of the baseyear, 1999. The modeled space temperatures and humidities were again examined to ensure they reasonably matched the typical range of indoor conditions during occupied and unoccupied days. This baseyear model was archived, with both printed and electronic copy of input data, diagnostic reports and output data. The energy consumption of this model was:

Baseyear Energy Data

Electricity use	=	2,97 1,000 kWh
Electric Demand	=	6,132 kW-months
Steam	=	10.67×10^6 lbs (4.84×10^6 kg)

Post-Retrofit Energy Use

The calibrated model showed the following energy use with the ECMs in place:

Post Retrofit Energy Data

Electricity use	=	1,711,000 kWh
Electric Demand	=	5,050 kW-months
Steam	=	6.26×10^6 lbs (2.84×10^6 kg)

Routine Adjustments

Routine adjustments are needed to bring baseyear and post-retrofit energy use to the agreed standard set of conditions: post-retrofit operations and weather of a "normal"' year. The following steps were followed:

1 The calibrated model was re-run with the 'normal' weather data. The modeled space temperatures and humidities were again examined to ensure they reasonably matched the typical range of indoor conditions during occupied and unoccupied days.

2 This calibrated model with "normal" weather was archived, with both printed and electronic copy of input data, diagnostic reports and output data.

3 The difference between the two versions of the calibrates model were computed as the Adjustment term, and is shown below.

4 The baseyear model was re-run with the "normal" weather data. The modeled space temperatures and humidities were again ex-

amined to ensure they reasonably matched the typical range of indoor conditions during occupied and unoccupied days.

5 This baseyear model with "normal" weather was archived, with both printed and electronic copy of input data, diagnostic reports and output data.

6 The difference between the two versions of the baseyear model were computed as the Adjustment term, and is shown below.

	Baseyear Model Adjustment	**Calibrated Post-Retrofit Model Adjustment**	**Total Adjustment**
Electricity consumption (kWh)	122,000	50,000	172,000
Electric Demand (kW-months)	200	100	300
Steam (10^3 lbs) or (10^3 kg)	521 (-236.3)	1,096 (497.1)	575(260.8)

Savings

From Equation 1, the energy savings at the standard set of conditions are:

	Baseyear	**–**	**Post**	**+**	**Adjustment**	**=**	**Savings**
Electricity	2,971,000	–	1,711,000	+	172,000	=	1,432,000 kWh
Electric Demand	6,132	–	5,050	+	300	=	1,382 kW–mo
Steam (10^3 lbs or 10^3 kg)	10,673 (4,841)	–	6,261 (2,840)	+	575 (261)	=	4,987 (2,262)

The value of these energy/demand savings are computed from the marginal prices as:

electricity consumption	=	\$113,300
electric demand	=	\$13,700
steam	=	\$70,970
Total	=	\$197,970

Appendix B
Uncertainty

Note: Use of statistical techniques such as sampling in determining energy savings is relatively unsophisticated compared to the exact science of statistics. Nonetheless, relatively simple statistical methods are helpful in explaining the results of an energy saving program and securing confidence and financing. The MVP uses the language of statistics, such as confidence levels and sampling, in a way that reflects best industry practices, and not as prescribed in statistics textbooks. These methods may not be statistically rigorous, but do provide sufficient confidence to complete and finance projects.

INTRODUCTION

Instrumentation Error

The magnitude of instrumentation errors is given by manufacturer's specifications. Typically instrumentation errors are small, and are not the major source of error in estimating savings.

Modeling Error

Modeling error refers to errors in the models used to estimate parameters of interest from the data collection. Biases in these models arise from model miss-specification. Miss-specification errors include:

- omitting important terms from the model.

- assigning incorrect values for "known" factors.
- extrapolation of the model results outside their range of validity.

Non systematic errors are the random effects of factors not accounted for by the model variables.

The most common models are linear regressions of the form

$$y = b_0 + b_1 x_1 + b_2 x_2 + \ldots + b_p x_p + e$$

Eq. 2

where:

y and x_k, $k = 1, 2, 3, \ldots, p$	observed variables
b_k, $k = 0, 1, 2, \ldots, p$	coefficients estimated by the regression
e	residual error not accounted for by the regression equation

Models of this type can be used in two ways:

1. To estimate the value of y for a given set of x values. An important example of this application is the use of a model estimated from data for a particular year or portion of a year to estimate consumption for a normal year.
2. To estimate one or more of the individual coefficients b_k.

In the first case, where the model is used to predict the value of y given the values of the x_ks, the accuracy of the estimate is measured by the *root mean squared error* (RMSE) of the predicted mean. This accuracy measure is provided by most standard regression packages. The MSE of prediction is the expected value of:

$$(y|_x - \hat{y}|_x)^2 \qquad \textbf{Eq. 3}$$

where $y|_x$ is the true mean value of y at the given value of x, and $\hat{y}|_x$ is the value estimated by the fitted regression line. The RMSE of prediction is the square root of the MSE.

In the second case, where the model is used to estimate a particular coefficient bk, the accuracy of the estimate is measured by the standard error of the estimated coefficient. This standard error is also provided by standard regression packages. The variance of the estimate ~ is the expected value of

$$(b - \hat{b})^2 \qquad \textbf{Eq. 4}$$

where b is the true value of the coefficient, and $\hat{b}$ is the value estimated by the regression. The standard error is the square root of the variance.

Whether the quantity of interest is the predicted value of y or a particular coefficient b_k, the accuracy measures provided by the standard statistical formulas are valid characterizations of the uncertainty of the estimate only if there are no important biases in the regression model.

Three statistical indices that can be used to evaluate regression models are defined below (SAS 1990):

1. The Coefficient of Determination, R^2(%):

$$R^2 = \left(1 - \frac{\sum_{i=1}^{n} \left(y_{pred,i} - y_{data,i} \right)^2}{\sum_{i=1}^{n} \left(\bar{y}_{data} - y_{data,i} \right)^2} \right) \times 1 \qquad \textbf{Eq. 5}$$

2. The Coefficient of Variation CV (%):

$$CV = \frac{\sqrt{\frac{\sum_{i=1}^{n} \left(y_{pred,i} - y_{data,i} \right)^2}{n-p}}}{\bar{y}_{data}} \times 100 \qquad \textbf{Eq. 6}$$

3. The Mean Bias Error, MBE

$$MBE = \frac{\frac{\sum_{i=1}^{n} \left(y_{pred,i} - y_{data,i} \right)}{n-p}}{\bar{y}_{data}} \times 100 \qquad \textbf{Eq. 7}$$

where:

$y_{data,\ i}$ data value of the dependent variable corresponding to a particular set of the independent variables,

$y_{pred,\ i}$ predicted dependent variable value for the same set of independent variables above,

$\bar{y}_{data}$ mean value of the dependent variable of the data set,

n number of data points in the data set.

p total number of regression parameters in the model.

Sampling Error—Sampling error refers to errors resulting from the fact that a sample of units were observed rather than observing the entire set of units under study. The simplest sampling situation is that of a simple random sample. With this type of sample, a fixed number n of units is selected at random from a total population of N units. Each unit has the same probability n/N of being included in the sample. In this case, the standard error of the estimated mean is given by:

$$SE(y) = \sqrt{\left(-\frac{n}{N}\right)\left(\left[\sum_{i=1}^{n} \frac{(y_1 - \bar{y}^2)}{n-1}\right] / n\right)} \qquad \textbf{Eq. 8}$$

For more complicated random samples, more complex formulas apply for the standard error. In general, however, the standard error is proportional to $1/\sqrt{n}$). That is, increasing the sample size by a factor "f" will reduce the standard error (improve the precision of the estimate) by a factor of $\sqrt{f}$.

Combining Components of Uncertainty

If the savings (S) estimate is a sum of several independently estimated components (C), then

$$S = C_1 + C_2 + C_3 + \ldots + C_p \qquad \textbf{Eq. 9}$$

the standard error of the estimate is given by

$$SE(S) \sqrt{\left[SE(C_1)^2 + (C_2)^2 + (C_3)^2 + \ldots + \left(C_p\right)^2\right]}$$ **Eq. 10**

If the savings (S) estimate is a product of several independently estimated components (C), then

$$S = C_1 \times C_2 \times C_3 \times \ldots \times C_p$$ **Eq. 11**

the relative standard error of the estimate is given approximately by

$$\frac{SE(S)}{S} \approx \sqrt{\left[\frac{SE(C_1)^2}{(C_1)} + \frac{SE(C_2)^2}{(C_2)} + \frac{SE\ (C_3)^2}{(C_3)} + \ldots + \frac{SE\left(C_p\right)^2}{\left(C_p\right)}\right]}$$ **Eq. 12**

The requirement that the components be independently estimated is critical to the validity of these formulas. Independence means that whatever random errors affect one of the components are unrelated to errors affecting the other components. In particular, different components would not be estimated by the same regression fit, or from the same sample of observations.

The above formulae for combining error estimates from different components can serve as the basis for a Propagation of Error analysis. This type of analysis is used to estimate how errors in one component will affect the accuracy of the overall estimate. Monitoring resources can then be designed cost-effectively to reduce error in the final savings estimate. This assessment takes into account:

- the effect on savings estimate accuracy of an improvement in the accuracy of each component.

- the cost of improving the accuracy of each component.

This procedure is described in general terms in ASHRAE 1991 and EPRI 1993. Applications of this method have indicated that, in many cases, the greatest contribution to savings estimate uncertainty is the uncertainty in baseyear conditions. The second greatest source of error tends to be the level of use, typically measured by hours (Violette et al. 1993). Goldberg (1996a) describes how to balance sampling errors against errors in estimates for individual units in this type of analysis.

Establishing a Level of Quantifiable Uncertainty

Determining savings means estimating a difference in level rather than measuring the level of consumption itself In general, calculating a difference with a given relative precision requires greater absolute precision, therefore a larger sample size than measuring a level with the same relative precision. For example, suppose the average load is around 500 kW, and the anticipated savings is around 100 kW. A 10% error with 90% confidence (90/10) criterion applied to the load would require absolute precision of 50 kW at 90 percent confidence. The 90/10 criterion applied to the savings would require absolute precision of 10 kW at the same confidence level.

In M&V, the precision criterion may be applied not only to demand or energy savings, but also to parameters that determine savings. For example, suppose the savings amount is the product of number (N) of units, hours (H) of operation and change (C) in watts:

$$S = N \times H \times C \qquad \textbf{Eq. 13}$$

The 90/10 criterion could be applied separately to each of these parameters. However, achieving 90/10 precision for each of these parameters separately does not imply that 90/10 is achieved for the savings, which is the parameter of ultimate interest. On the other hand, if number of units and change in watts are assumed to be known without error, 90/10 precision for hours implies 90/10 precision for savings.

In the M&V context, the precision standard could be imposed at various levels. The choice of level of disaggregation dramatically affects the sample size requirements and associated monitoring costs. Possible choices include the following:

- For individual sites, where sampling is conducted within each site
- For all savings associated with a particular type of technology, across several sites for a given project, where both sites and units within sites may be sampled
- For all savings associated with a particular type of technology in a particular type of usage, across several sites for a project
- For all savings associated with all technologies and sites for a given ESCO

In general, the finer the level at which the precision criterion is imposed, the greater the data collection requirement. If the primary goal is to ensure savings accuracy for a project or group of projects as a whole, it is not necessary to impose the same precision requirement on each subset. In fact, a uniform relative precision target for each subset is in conflict with the goal of obtaining the best precision possible for the project as a whole.

Appendix C Measurement Techniques

ELECTRICITY

The most common way of sensing alternating electrical current (AC) for energy efficiency and savings applications is with a current transformer or current transducer (CT). CTs are placed on wires connected to specific loads such as motors, pumps or lights and then connected to an ammeter or power meter. CTs are available in split core and solid torroid configuration. Torroids are usually more economical than split-core CTs, but require a load to be disconnected for a short period while they are installed. Split-core CTs allow installation without disconnecting the load. Both types of CTs are typically offered with accuracies better than one percent.

Voltage is sensed by a direct connection to the power source. Some voltmeters and power measuring equipment directly connect voltage leads, while others utilize an intermediate device, a potential transducer (PT), to lower the voltage to safer levels at the meter.

Though electrical load is the product of voltage and current, separate voltage and current measurements should not be used for inductive loads such as motors or magnetic ballasts. True RMS power digital sampling meters should be used. Such meters are particularly important if variable frequency drives or other harmonic-producing devices are on the same circuit, resulting in the likelihood of ham-ionic voltages at the motor terminals. True RMS power and energy metering technology, based on digital sampling principles, is recommended due to its ability to accurately measure distorted waveforms and properly record load shapes.

It is recommended that power measurement equipment meeting the IEEE Standard 519-1992 sampling rate of 3 kHz be selected where harmonic issues are present. Most metering equipment has adequate sampling strategies to address this issue. Users should, however, request documentation from meter manufacturers to ascertain that the equipment is accurately measuring electricity use under waveform distortion.

Power can be measured directly using watt transducers. Watt-hour energy transducers that integrate power over time eliminate the error inherent in assuming or ignoring variations in load over time. Watt-hour transducer pulses are typically recorded by a pulse-counting data logger for storage and subsequent retrieval and analysis. An alternate technology involves combining metering and data logging functions into a single piece of hardware.

Hand-held wattmeters, rather than ammeters, should be used for spot measurements of watts, volts, amps, power factor or waveforms.

Regardless of the type of solid-state electrical metering device used, it is recommended that the device meet the minimum performance requirements for accuracy of the American National Standards Institute standard for solid state electricity meters, ANSI C12.16-1991, published by the Institute of Electrical and Electronics Engineers. This standard applies to solid-state electricity meters that are primarily used as watt-hour meters, typically requiring accuracies of one to two percent based on variations of load, power factor and voltage.

RUNTIME

Determination of energy savings may involve measuring the time that a piece of equipment is on, then multiplying it by a short-term power measurement. Constant load motors and lights are examples of equipment that may not be continuously metered with recording watt-hour meters to establish energy consumption. Self-contained battery-powered monitoring devices are available to record equipment runtime and, in some cases, time-of-use information. This equipment provides a reasonably priced, simple to install approach for energy savings calculations.

TEMPERATURE

The most commonly used computerized temperature measurements devices are resistance temperature detectors (RTDs), thermocouples, thermistors, and integrated circuit (IC) temperature sensors.

Resistance Temperature Detectors or RTDs

These are common equipment for measuring air and water temperature in the energy management field. They are among the most accurate, reproducible, stable and sensitive thermal elements available. An RTD measures the change in electrical resistance in special materials.

RTDs are economical and readily available in configuration packages to measure indoor and outdoor air temperatures as well as fluid temperatures in chilled water or heating systems. Considering overall performance, the most popular RTI)s are 100 and 1,000 Ohm platinum devices in various packaging including ceramic chips, flexible strips and thermowell installations.

Depending on application, two, three and four-wire RTI)s are available. Required accuracy, distance, and routing between the RTD and the data logging device can determine the specific type of RTD for a project. Four-wire RTI)s offer a level of precision seldom required in the energy savings determination, and are most commonly found in high precision services or in the laboratory. Three-wire RTDs compensate for applications where an RTD requires a long wire lead, exposed to varying ambient conditions. The wires of identical length and material exhibit similar resistance-temperature characteristics and can be used to cancel the effect of the long leads in an appropriately designed bridge circuit. Two-wire RTI)s must be field-calibrated to compensate for lead length and should not have lead wires exposed to conditions that vary significantly from those being measured.

Installation of RTI)s is relatively simple with the advantage that conventional copper lead wire can be used as opposed to the more expensive thermocouple wire. Most metering equipment allows for direct connection of RTDs by providing internal signal conditioning and the ability to establish offsets and calibration coefficients.

Thermocouples

They measure temperature using two dissimilar metals, joined together at one end, which produce a small unique voltage at a given temperature that is measured and interpreted by a thermocouple thermometer. Thermocouples are available in different combinations of metals, each with a different temperature range. Apart from temperature range, consider chemical abrasion and vibration resistance and installation requirements while selecting a thermocouple.

In general, thermocouples are used when reasonably accurate temperature data are required. The main disadvantage of thermocouples is their weak output signal, making them sensitive to electrical noise and always requiring amplifiers. Few energy savings determination situations, except for thermal energy metering, warrant the accuracy and complexity of thermocouple technology.

Thermistors

These are semiconductor temperature sensors usually consisting of an oxide of manganese, nickel, cobalt or one of several other types of materials. One of the primary differences between thermistors and RTDs is that thermistors have a large resistance change with temperature. Thermistors are not interchangeable, and their temperature-resistance relationship is non-linear. They are fragile devices and require the use of shielded power lines, filters or DC voltage. Like thermocouples, these devices are infrequently used in savings determination.

Integrated Circuit Temperature Sensors

Certain semiconductor diodes and transistors also exhibit reproducible temperature sensitivities. Such devices are usually ready-made Integrated Circuit (IC) sensors and can come in various shapes and sizes. These devices are occasionally found in HVAC applications where low cost and a strong linear output are required. IC sensors have a fairly good absolute error, but they require an external power source, are fragile and are subject to errors due to self-heating.

HUMIDITY

Accurate, affordable and reliable humidity measurement has always been a difficult and time-consuming task. Equipment to measure relative humidity is available from several vendors, and installation is relatively straightforward. However, calibration of humidity sensors continues to be a major concern (see Chapter 5.4) and should be carefully described in the M&V Plan and documented in savings reports.

FLOW

Different types of flow measurement may be used for quantities such as natural gas, oil, steam, condensate, water, or compressed air.

This section discusses the most common liquid flow measurement devices. In general, flow sensors can be grouped into two different types of meters:

1 Intrusive Flow Meters (Differential Pressure and Obstruction)

2 Non-Intrusive Flow Meters (Ultrasonic and Magnetic)

Choosing a flow meter for a particular application requires knowing the type of fluid being measured, how dirty or clean it is, the highest and lowest expected flow velocities, and the budget.

Differential Pressure Flow Meters

The calculation of fluid flow rate by measuring the pressure loss across a pipe restriction is perhaps the most commonly used flow measurement technique in building and industrial applications. The pressure drops generated by a wide variety of geometrical restrictions have been well characterized over the years, and, these primary or "head" flow elements come in a wide variety of configurations, each with specific application strengths and weaknesses. Examples of flow meters utilizing the concept of differential pressure flow measurement include Orifice Plate meter, Venturimeter, and Pitot Tube meter. Accuracy of differential pressure flow meters is typically in the vicinity of 1-5% of the maximum flow for which each meter is calibrated.

Obstruction Flow Meters

Several types of obstruction flow meters have been developed that are capable of providing a linear output signal over a wide range of flow rates, often without the severe pressure loss penalty incurred with an orifice plate or venturi meters. In general, these meters place a small target, weight or spinning wheel in the flow stream that allows fluid velocity to be determined by the rotational speed of the meter (turbine) or by the force on the meter body (vortex).

Turbine meters

They measure fluid flow by counting the rotations of a rotor that is placed in a flow stream. Turbine meters can be an axial-type or insertion-type. Axial turbine meters usually have an axial rotor and a housing that is sized for an appropriate installation. An insertion turbine meter allows the axial turbine to be inserted into the fluid stream and uses existing pipe as the meter body. Because the insertion turbine meter only measures fluid velocity at a single point on the cross-sectional area of the pipe, total volumetric flow rate for the pipe can only be accurately inferred if the meter is installed according to manufacturer's specifications. Most important with insertion turbine meters is installation in straight sections of pipe removed from internal flow turbulence. This type of meter can inserted without having to shut down the system. Insertion meters can be used on pipelines in excess of four inches with very low pressure loss. Turbine meters provide an output that is linear with flow rate. Care must be taken when using turbines as they can be damaged by debris and are subject to corrosion. Insertion meters can be damaged during insertion and withdrawal.

Vortex meters

They utilize the same basic principle that makes telephone wires oscillate in the wind between telephone poles. This effect is due to oscillating instabilities in a low pressure field after it splits into two flow streams around a blunt object. Vortex meters require minimal maintenance and have high accuracy and long-term repeatability. Vortex meters provide a linear output signal that can be captured by meter/ monitoring equipment.

Non-Interfering Flow Meters

They are well suited to applications where the pressure drop of an intrusive flow meter is of critical concern, or the fluid is dirty, such as sewage, slurries, crude oils, chemicals, some acids, process water, etc.

Ultrasonic flow meters

They measure clean fluid velocities by detecting small differences in the transit time of sound waves that are shot at an angle across a fluid stream. Accurate clamp-on ultrasonic flow meters facilitate rapid measurement of fluid velocities in pipes of varying sizes. An accuracy rate from 1 % of actual flow to 2% of full scale are now possible, although this technology is still quite expensive. Recently, an ultrasound meter that uses the Doppler principle in place of transit time has been developed. In such a meter a certain amount of particles and air are necessary in order for the signal to bounce-off and be detected by the receiver. Doppler-effect meters are available with an accuracy between

2% and 5% of full scale and command prices somewhat less than the standard transit time-effect ultrasonic devices. Meter cost is independent of pipe size. Ultrasonic meters can have low installation costs since they do not require shutting down systems to cut pipes for installation.

Magnetic flow meters

They measure the disturbance that a moving liquid causes in a strong magnetic field. Magnetic flow meters are usually more expensive than other types of meters. They have advantages of high accuracy and no moving parts. Accuracy of magnetic flow meters are in the 1-2% range of actual flow.

PRESSURE

Mechanical methods of measuring pressure have been known for a very long time. U-tube manometers were among the first pressure indicators. But manometers are large, cumbersome, and not well suited for integration into automatic control loops. Therefore, manometers are usually found in the laboratory or used as local indicators. Depending on the reference pressure used, they can indicate absolute, gauge, or differential pressure. Things to keep in mind while selecting pressure measurement devices are: accuracy, pressure range, temperature effects, outputs (millivolt, voltage or current signal) and application environment

Modem pressure transmitters have come from the differential pressure transducers used in flow meters. They are used in building energy management systems and are capable of measuring pressures with the necessary accuracy for proper building pressurization and air flow control.

THERMAL ENERGY

The measurement of thermal energy flow requires the measurement of flow and some temperature difference. For example, the cooling provided by a chiller is recorded in Btu and is a calculated value determined by measuring chilled water flow and the temperature difference between the chilled water supply and return lines. An energy flow meter performs an internal Btu calculation in real time based on input from a flow meter and temperature sensors. It also uses software constants for the specific heat of the fluid to be measured. These electronic energy flow meters offer an accuracy better than 1%. They also provide other useful data on flow rate and temperature (both supply and return).

When a heating or cooling plant is under light load relative to its capacity there may be as little as a 5°F difference between the two flowing streams. To avoid significant error in the thermal energy measurement the two temperature sensors should be matched or calibrated to the tightest tolerance possible. It is more important that the sensors be matched, or calibrated with respect to each another, than for their calibration to be traceable to a standard. Suppliers of RTDs can provide sets of matched devices when ordered for this purpose. Typical purchasing specifications are for a matched set of RTD assemblies (each consisting of an RTD probe, holder, connection head with terminal strip and a stainless steel thermowell), calibrated to indicate the same temperature, for example within a tolerance of 0. 1°F over the range of 25°F to 75°F. A calibration data sheet is normally provided with each set.

The design and installation of the temperature sensors used for thermal energy measurements should consider the error caused by: sensor placement in the pipe, conduction of the thermowell, and any transmitter, power supply or analog to digital converter. Complete error analysis through the measurement system is suggested, in recognition of the difficulty of making accurate thermal measurements.

Thermal energy measurements for steam can require steam flow measurements (e.g., steam flow or condensate flow), steam pressure, temperature and feedwater temperature where the energy content of the steam is then calculated using steam tables. In instances where steam production is constant, this can be reduced to measurement of steam flow or condensate flow (i.e., assumes a constant steam temperature-pressure and feedwater temperature-pressure) along with either temperature or pressure of steam or condensate flow.

REPORT DOCUMENTATION PAGE

Form Approved
OMB NO. 0704-0188

Public reporting burden for this collection of information is estimated to average 1 hour per response, including the time for reviewing instructions, searching existing data sources, gathering and maintaining the data needed, and completing and reviewing the collection of information. Send comments regarding this burden estimate or any other aspect of this collection of information, including suggestions for reducing this burden, to Washington Headquarters Services, Directorate for Information Operations and Reports, 1215 Jefferson Davis Highway, Suite 1204, Arlington, VA 22202-4302, and to the Office of Management and Budget, Paperwork Reduction Project (0704-0188), Washington, DC 20503.

1. AGENCY USE ONLY (Leave blank)

2. REPORT DATE
Revised March 2002

3. REPORT TYPE AND DATES COVERED
Technical Report

4. TITLE AND SUBTITLE
International Performance Measurement & Verification Protocol Volume I: Concepts and Options for Determining Energy Savings

5. FUNDING NUMBERS
IS414140

6. AUTHOR(S)
International Performance Measurement & Verification Protocol Technical Committee

7. PERFORMING ORGANIZATION NAME(S) AND ADDRESS(ES)
National Renewable Energy Laboratory
1617 Cole Blvd.
Golden, CO 80401-3393

8. PERFORMING ORGANIZATION REPORT NUMBER
NREL/TP-710-31505

9. SPONSORING/MONITORING AGENCY NAME(S) AND ADDRESS(ES)
U.S. Department of Energy

10. SPONSORING/MONITORING AGENCY REPORT NUMBER
DOE/GO-102002-1554

11. SUPPLEMENTARY NOTES
Technical coordinator: Satish Kumar, Lawrence Berkeley National Laboratory, 202-646-7953

12a. DISTRIBUTION/AVAILABILITY STATEMENT
National Technical Information Service
U.S. Department of Commerce
5285 Port Royal Road
Springfield, VA 22161

12b. DISTRIBUTION CODE

13. ABSTRACT *(Maximum 200 words)*

This protocol serves as a framework to determine energy and water savings resulting from the implementation of an energy efficiency program. It is also intended to help monitor the performance of renewable energy systems and to enhance indoor environmental quality in buildings.

14. SUBJECT TERMS
buildings, energy savings, water savings, energy efficiency, indoor environmental quality

15. NUMBER OF PAGES

16. PRICE CODE

17. SECURITY CLASSIFICATION OF REPORT
Unclassified

18. SECURITY CLASSIFICATION OF THIS PAGE
Unclassified

19. SECURITY CLASSIFICATION OF ABSTRACT
Unclassified

20. LIMITATION OF ABSTRACT
UL

NSN 7540-01-280-5500

Standard Form 298 (Rev. 2-89)
Prescribed by ANSI Std. Z39-18
298-102

Appendix C

International Performance Measurement & Verification Protocol

Concepts and Practices for Improved Indoor Environmental Quality Volume II

International Performance Measurement & Verification Protocol Committee

Revised March 2002
DOE/GO-102002-1517

NOTICE

Table of Contents

Acknowledgments

PARTICIPATING ORGANIZATIONS

Brazil
- Institute Nacional De Eficiencia Energetica (INEE)
- Ministry of Mines and Energy

Bulgaria Bulgarian Foundation for Energy Efficiency (EnEffect)

Canada
- Canadian Association of Energy Service Companies (CAESCO)
- Natural Resources Canada (NRC)

China
- State Economic and Trade Commission
- Beijing Energy Efficiency Center (BECON)

Czech Republic Stredisko pro Efektivni Vyuzivani Energie (SEVEn7)

India Tata Energy Research Institute (TERI)

Japan Ministry of International Trade and Industry (MITI)

Korea Korea Energy Management Corporation (KEMCO)

Mexico
- Comision Nacional Para El Ahorro De Energia (CONAE)
- Fideicomiso De Apoyo Al Programa De Ahorro De Energia

Del Sector Electrico (FIDE)

Poland
- The Polish Foundation for Energy Efficiency (FEWE)

Russia
- Center for Energy Efficiency (CENEf)

Sweden
- Swedish National Board for Technical and Urban Development

Ukraine
- Agency for Rational Energy Use and Ecology (ARENA—ECO)

United Kingdom
- Association for the Conservation of Energy

United States
- American Society of Heating, Refrigerating and Air-Conditioning Engineers (ASHRAE)
- Association of Energy Engineers (AEE)
- Association of Energy Services Professionals (AESP)
- Building Owners and Managers Association (BOMA)
- National Association of Energy Service Companies (NAESCO)
- National Association of State Energy Officials (NASEO)
- National Realty Association
- U.S. Department of Energy (DOE)
- U.S. Environmental Protection Agency (EPA)

IPMVP COMMITTEES

EXECUTIVE COMMITTEE

1 Gregory Kats (Chair), Department of Energy, USA
2 Jim Halpern (Vice Chair), Measuring and Monitoring Services Inc., USA
3 John Armstrong, Hagler Bailly Services, USA
4 Flavio Conti, European Commission, Italy
5 Drury Crawley, US Department of Energy, USA
6 Dave Dayton, HEC Energy, USA
7 Adam Gula, Polish Foundation for Energy Efficiency, Poland
8 Shirley Hansen, Kiona International, USA
9 Leja Hattiangadi, TCE Consulting Engineers Limited, India
10 Maury Hepner, Enron Energy Services, USA
11 Chaan-min Lin, Hong Kong Productivity Council, Hong Kong
12 Arthur Rosenfeld, California Energy Commission, USA

TECHNICAL COMMITTEE

1 John Cowan (Co-chair), Cowan Quality Buildings, Canada
2 Steve Kromer (Co-chair), Enron Energy Services, USA
3 David E. Claridge, Texas A&M University, USA
4 Ellen Franconi, Schiller Associates, USA
5 Jeff S. Haberl, Texas A&M University, USA
6 Maury Hepner, Enron Energy Services, USA
7 Satish Kumar, Lawrence Berkeley National Laboratory, USA
8 Eng Lock Lee, Supersymmetry Services Pvt. Ltd., Singapore
9 Mark Martinez, Southern California Edison, USA
10 David McGeown, NewEnergy, Inc., USA
11 Steve Schiller, Schiller Associates, USA

INDOOR ENVIRONMENTAL QUALITY COMMITTEE

1 Bill Fisk (Chair), Lawrence Berkeley National Laboratory, USA
2 Jim Bailey, Building Dynamics, USA
3 Dave Birr, Synchronous Energy Solutions, USA
4 Steve Brown, CSIRO, Australia
5 Andrew Cripps, Building Research Establishment, UK
6 John Doggart, ECD Energy and Environment, UK
7 Dale Gilbert, Built Environment Research, Australia
8 Mike Jawer, Building Owners & Managers Association, USA
9 Gregory Kats, Department of Energy, USA
10 Ken-ichi Kimura, Waseda University, Japan
11 Satish Kumar, Lawrence Berkeley National Laboratory, USA
12 Hal Levin, Hal Levin & Associates, USA
13 Richard Little, National Academy of Sciences, USA
14 David Mudarri, Environmental Protection Agency, USA
15 Tedd Nathanson, Public Works and Government Services, Canada
16 Andy Persily, National Institute of Standards and Technology, USA
17 Arthur Rosenfeld, California Energy Commission, USA
18 Sumeet Saksena, Tata Energy Research Institute, India
19 Olli Seppanen, Helsinki University of Technology, Finland
20 Jan Sundell, Int 1 Center for Indoor Environment and Energy, Denmark
21 Steve Turner, Chelsea Group, USA
22 Ole Valbjoern, Danish Building Research Institute, Denmark

INDOOR ENVIRONMENTAL QUALITY REVIEWERS

1 Ed Arens, University of California, Berkeley, USA
2 Jan W Bakke, Norway
3 Carl-Gustaf Bornehag, Hibbitt, Karlsson & Sorensen, Inc., Sweden
4 Terry Brennan, Camroden Associates, USA
5 Graham Charlton, Australian Institute of Refrigeration, Air-Conditioning & Heating, Australia
6 Susan Clampitt, General Services Administration, USA
7 Geo Clausen, Technical University, Denmark
8 Jim Coggins, Energy Applications, Inc., USA
9 Catherine Coombs, Steven Winter Associates, Inc., USA
10 Joan Daisey, Lawrence Berkeley National Laboratory, USA
11 Richard deDear, Macquarian University, Australia
12 Angelo Delsante, CSIRO, Australia
13 David Grimsrud, University of Minnesota, USA
14 Shirley Hansen, Kiona International, USA
15 Sten Olaf Hansen, Norwegian University of Science and Technology, Norway
16 Jim Harven, Custom Energy, USA
17 Matti J. Jantunen, National Public Health Institute, Finland
18 Erkki Kahkonen, Finnish Institute of Occupational Health, Finland
19 Pentti Kalliokoski, University of Kuopio, Finland
20 Greg Kurpiel, Johnson Controls, USA
21 Marianna Luoma, VTT Building Technology, Finland
22 Mark Mendell, NIOSH, USA
23 Jack McCarthy, Environmental Health and Engineering Inc., USA
24 Aino Nevalainen, National Public Health Institute, Finland
25 Rob Obenreder, General Services Administration, USA
26 Bud Offermann, Indoor Environment Engineering, San Francisco, USA
27 Deo Prasad, University of New South Wales, Australia
28 Dan Prezernbel, The RREEF Funds, USA
29 John Reese, Department of Energy, USA
30 Ed Reid, American Gas Cooling Center, USA
31 James Rogers, Energy Consultant, USA
32 Jorma Sateri, Finnish Society of Indoor Air Quality and Climate, Finland
33 Christine Schweizer, Environment Australia, Australia

34 Jaikrishna Shankavaram, Carnegie Mellon University, USA
35 Staffan Skerfving, Lund University, Sweden
36 John Talbott, Department of Energy, USA
37 Steve Taylor, Taylor Engineering, USA
38 John Tiffany, Tiffany Bader Environmental, USA
39 Gene Tucker, Environmental Protection Agency, USA
40 Dave Warden, DWT Engineering, Canada
41 Davidge Warfield, Tri-Dim Filter Corporation, USA
42 Jim Woods, HP Woods Research Institute, USA

TECHNICAL COORDINATOR

Satish Kumar, Lawrence Berkeley National Laboratory, USA
Email: SKumar@lbl.gov, Phone: 202-646-7953

We would like to gratefully acknowledge the many organizations that made the IPMVP possible. In particular we would like to thank the Office of Building Technology, State and Community Programs in the U.S. Department of Energy's Office of Energy Efficiency and Renewable Energy, which has provided essential funding support to the IPMVP, including publication of this document.

The reprinting of Volume I and II of the IPMVP has been made possible through a generous grant from the US. Department of Energy's Federal Energy Management Program (FEMP). FEMP's gesture is greatly appreciated by the IPMVP.

DISCLAIMER

This Protocol serves as a framework to determine energy and water savings resulting from the implementation of an energy efficiency program. It is also intended to help monitor the performance of renewable energy systems and to enhance indoor environmental quality in buildings. The IPMVP does not create any legal rights or impose any legal obligations on any person or other legal entity. IPMVP has no legal authority or legal obligation to oversee, monitor or ensure compliance with provisions negotiated and included in contractual arrangements between third persons or third parties. It is the responsibility of the parties to a particular contract to reach agreement as to what, if any, of this Protocol is included in the contract and to ensure compliance.

1
Introduction

This year 2000 edition of International Performance Measurement and Verification Protocol (IPMVP) has been expanded to address indoor environmental quality (IEQ) issues. It provides information that will help energy efficiency professionals and building owners and managers maintain or improve IEQ and occupant health and comfort during the implementation of building energy conservation measures in retrofits or new construction of commercial and public buildings.

Volume II focuses exclusively on indoor environmental quality issues (See Preface of Volume I for overview of IPMVP). This volume starts with a general introduction to IEQ. Best practices for maintaining a high level of IEQ are then reviewed. The potential positive and negative influences of specific energy conservation measures on IEQ are summarized in a tabular format in *Section 5.* The remainder of the document addresses IEQ measurement and verification procedures that may be used to address the following goals: 1) ensure that the energy conservation measures have no adverse influence on IEQ, 2) quantify the improvements in IEQ resulting from implementation of energy conservation measures, and 3) verify that selected IEQ parameters satisfy the applicable IEQ guidelines or standards. A multi-step procedure for IEQ measurement and verification is presented, followed by a discussion of general approaches for measurement and verification and then by a table of measurement and verification alternatives linked to specific IEQ parameters.

This document has been prepared by an international team of IEQ and building energy efficiency experts and reflects the current state of knowledge. The IPMVP, including the IEQ volume will be updated every two years.

2
Purpose

The International Performance Measurement and Verification Protocol (IPMVP) provides a framework and guidance for the measurement and verification of building energy performance, emphasizing the changes in energy performance that result from implementation of energy conservation measures in buildings. Many building energy conservation measures have the potential to positively or negatively affect indoor pollutant concentrations, thermal comfort conditions, and lighting quality. These and other indoor environmental characteristics are collectively referred to as indoor environmental quality[1].

IEQ can influence the health and productivity of building occupants. Small changes in occupant health and productivity may be very significant financially, potentially exceeding the financial benefits of energy conservation (Fisk and Rosenfeld, 1998). It is important that these IEQ considerations be explicitly recognized prior to selection and implementation of energy conservation measures[2]. Consequently, the primary purpose of this document is to provide information that will help energy conservation professionals and building owners and managers maintain or improve IEQ when they implement building energy conservation measures in non-industrial commercial and public buildings[3]. This document also describes some practical IEQ and ventilation measurements that, in certain circumstances, can also help energy conservation professionals maintain or improve IEQ.

1. This document is not intended as substitute for local codes or as a guide to diagnosis and solving specific IAQ and health problems. Hospitals and other health care facilities have special indoor air quality requirements that are not addressed in this document.

---

1. The term indoor air quality (IAQ) is sometimes used for the same purpose, although IAQ is also used more narrowly in reference to the levels of pollutants in indoor air.
2. Conversely, when selecting measures to improve IEA, the potential impacts of these measures on building energy use should be considered. This document does not directly address this situation but it includes some relevant information.
3. This document is not intended as substitute for local codes or as a guide to diagnosis and solving specific IAQ and health problems. Hospitals and other health care facilities have special indoor air quality requirements that are not addressed in this document.

3
How to Use

Section 4 and 5 of this document introduce IEQ through discussions of IEQ parameters, indoor pollutant sources, IEQ standards, and methods of maintaining a high level of IEQ. These sections are designed to educate building energy professionals about IEQ and to provide a foundation for subsequent sections of the document. Professionals with a strong working knowledge of IEQ issues can skip *Section 4* and 5. *Section 6* describes the linkage between specific energy conservation measures and IEQ, highlighting measures that are particularly attractive because they often improve IEQ while saving energy. This information is provided in a table so readers can quickly identify the information relevant to their situation. Some users of this document may not need to read beyond *Section 6* because their energy conservation efforts are not likely to significantly influence IEQ. *Section 7* provides a discussion of the application of *Section 6* to specific buildings. *Section 8* addresses IEQ measurement and verification (M&V). An overall IEQ M&V procedure is provided along with a description of basic M&V approaches and a table of M&V alternatives for specific IEQ parameters. The most practical and useful M&V alternatives are highlighted. *Section 9* discusses how to implement the concepts laid out in this document. Concluding remarks and references are in *Sections 10* and II, respectively.

4
General Introduction to Indoor Environmental Quality

4.1 IMPORTANT ENERGY-RELATED IEQ PARAMETERS

Many characteristics of the indoor environment may influence comfort, health, satisfaction, and productivity of building occupants. The following indoor environmental characteristics are most likely to be influenced by building energy conservation measures:

- indoor thermal conditions such as air temperature and its vertical gradient, mean radiant temperature, air velocity, and humidity;

- concentrations of pollutants and odors in indoor air and amount of pollutants on surfaces;

- lighting intensity and quality;

This document provides more detailed information on indoor air pollution than on thermal comfort and lighting because the users of the IPMVP are less likely to be knowledgeable about indoor air pollution.

4.2 INDOOR THERMAL CONDITIONS

The influence of the indoor thermal environment on thermal comfort is widely recognized. Thermal comfort has been studied for decades resulting in thermal comfort standards and models for predicting the level of satisfaction with the thermal environment as a function of the occupants clothing and activity level

(ASHRAE 1997). Despite the significant attention placed on thermal comfort by building professionals, dissatisfaction with indoor thermal conditions is the most common source of occupant complaints in office buildings (Federspiel 1998). In a large field study (Schiller et al. 1988), less than 25% of the subjects were moderately satisfied or very satisfied with air temperature. Also, 22% of the measured thermal conditions in the winter, and almost 50% of measured thermal conditions in the summer, were outside of the boundaries of the 1988 version of the ASHRAE thermal comfort zone. These findings indicate that greater effort should be placed on maintaining thermal conditions within the prescribed comfort zones. Even in laboratory settings with uniform clothing and activity levels, it is not possible to satisfy more than 95% of occupants by providing a single uniform thermal environment (Fanger 1970) because thermal preferences vary among people. Task conditioning systems that provide occupants limited control of the air temperature and velocity in their workstation are being explored as a means to maximize thermal comfort (e.g., Arens et al. 1991, Bauman et al. 1993).

Extremes in humidity will adversely influence thermal comfort (ASHRAE 1997, Chapter 8). ASHRAE s thermal comfort zones for summer and winter have a lower absolute humidity boundary of 0.045 g H_2O per kg dry air corresponding approximately to a 30% RH at 20.5°C and a 20% RH at 27°C. Relative humidities below approximately 25% have been associated with complaints of dry skin, nose, throat, and eyes. At high humidities, discomfort will increase due substantially to an increase of skin moisture. The upper humidity limits of ASHRAE s thermal comfort zone vary with temperature from approximately 60% RH at 26°C to 80% RH at 20°C.

Air temperature and humidity also influence perceptions of the quality of indoor air and the level of complaints about non-specific building-related health symptoms (often called sick building syndrome symptoms). Higher air temperature has been associated with increased health symptom prevalences in several studies (Skov et al. 1989, Jaakkola et al. 1991, Wyon 1992, Menzies et al. 1993). Occupants perceived acceptability of air quality has been shown to decrease as temperature and humidity increase in the range between 18°C, 30% RH and 28°C, 70% RH (Fang et al. 1997, Molhave et al. 1993).

4.3 INDOOR LIGHTING

The quality of the indoor environment depends significantly on several aspects of lighting (JES 1993, Veitch and Newsham 1998) including the illuminance (intensity of light that impinges upon a surface), the amount of glare, and the spectrum of the light[1]. There is evidence that a decrease in the amount of flicker in light, i.e., the magnitude of the rapid cyclic change in illuminance over time, may be associated with a decrease in headache and eyestrain (Wilkens et al. 1988) and with an increase in worker performance (Veitch and Newsham 1997). In many indoor spaces, the indoor environment is influenced by both daylight and by artificial lighting. Characteristics of windows and skylights and their shading affect the daylighting of indoors. The quality of electric indoor lighting is a function of the types, locations, and number of luminaires, and the optical characteristics of indoor surfaces such as their spectral reflectivity and color.

The method of lighting control, such as no control, automatic dimming of artificial light, and manual control of overhead or task lighting may also influence lighting quality.

Lighting characteristics influence the quality of vision and can have psychological influences on mood and on perceptions about the pleasantness of a space. Because extremes in lighting have a clear impact on performance, indoor lighting in commercial buildings is usually maintained within the limits specified in guidelines or standards. The

---

1. The evidence for effects of spectrum of light on satisfaction and performance is mixed (e.g., Veitch 1994, Berman 1992)

recommended range of illuminance is a function of the type of visual activity and the age of the occupants. Guidelines also provide recommendations for the maximum luminance ratio, i.e., range of luminance in the visual field. Occupants satisfaction with lighting may vary with illuminance and with the characteristics of the lighting system (Katzev 1992).

4.4 INDOOR POLLUTANTS, THEIR SOURCES, AND HEALTH EFFECTS

There are a large number of indoor air pollutants that can influence occupant health and the perceived acceptability of indoor air. The following paragraphs introduce these pollutants.

Gaseous human bioeffluents—Humans release a variety of odorous gaseous bioeffluents, e.g., body odors, which influence the perceived acceptability of indoor air. Historically, most standards and guidelines for minimum ventilation rates in buildings have been based primarily on the ventilation needed to maintain indoor air acceptable to a large proportion (e.g., 80%) of visitors when they initially enter a space with occupants as the only indoor pollutant source. In the last decade, concerns about other sources of odors and adverse health effects from air pollutants have increasingly influenced building ventilation standards.

Carbon Dioxide (CO_2) is one of the gaseous human bioeffluents in exhaled air. Humans are normally the main indoor source of carbon dioxide. Unvented or imperfectly vented combustion appliances can also increase indoor CO_2 concentrations. The outdoor CO_2 concentration is often approximately 350 ppm[1], whereas indoor concentrations are usually in the range of 500 ppm to a few thousand ppm. At these concentrations, CO_2 is not thought to be a direct cause of adverse health effects; however, CO_2 is an easily-measured surrogate for other occupant-generated pollutants, such as body odors.

The indoor CO_2 concentration is often used, sometimes inappropriately, as an indicator of the rate of outside air supply per occupant. If the number of occupants and the rate of outside air supply are constant and the CO_2 generation rate of occupants is known, the rate of outside air supply per occupant is related to the equilibrium indoor CO_2 concentration in a straightforward manner as predicted by a steady-state

1. In urban areas, outdoor CO_2 concentrations may substantially exceed 350 ppm and vary considerably with time.

mass balance calculation (Persily and Dols 1990). However, in many buildings, CO_2 concentrations never stabilize during a workday because occupancy and ventilation rates are not stable for a sufficient time period. If the CO_2 concentration has not stabilized at its equilibrium value and the steady-state relationship between CO_2 and ventilation rate is used to estimate the rate of outside air supply, the estimated outside air ventilation rate may be substantially in error.

Carbon monoxide (CO) and nitrogen oxides (NO_x): Indoor concentrations of CO and NO_x may be higher than outdoor concentrations due to indoor unvented combustion (e.g., unvented space heaters), failures in the combustion exhaust vent systems of vented appliances, and leakage of air from attached parking garages into the building. Tobacco smoking can cause a small increase in indoor CO concentrations. Short-term exposures to highly elevated concentrations of CO can cause brain damage or death (NRC 1981). Lower concentrations can cause chest pain among people with heart disease (NRC 1981). NO_2 is usually considered to be the most important of the indoor nitrogen oxides. High concentrations (e.g., 0.5 ppm) of NO_2 can cause respiratory distress in individuals with asthma and concentrations of approximately 1 ppm cause increased airway resistance in health individuals (NRC 1981). Long term exposure to much lower concentrations of NO_2 may be associated with increased respiratory illness among children (Vedal 1985).

Volatile organic compounds (VOCs): VOCs are a class of gaseous pollutants containing carbon. The indoor air typically contains dozens of VOCs at concentrations that are measurable. VOCs are emitted indoors by building materials (e.g., paints, pressed wood products, adhesives, etc.), furniture, equipment (photocopying machines, printers, etc.), cleaning products, pest control products, and combustion activities (cooking, unvented space heating, tobacco smoking, indoor vehicle use). Humans also release VOCs as a consequence of their metabolism and use of personal products such as perfumes. The outdoor air entering buildings also contains VOCs. VOCs in contaminated soil adjacent to the building can also be drawn into buildings.

New building materials and furnishings generally emit VOCs at a much higher rate than older materials. Emission rates for many VOCs may decline by an order of magnitude during the first few weeks after the materials are installed in the building. However, the emission rates of some VOCs, such as formaldehyde emissions from pressed wood products, decline much more slowly. Because of concerns about the health effects of VOCs, many manufacturers have worked to reduce the VOC emissions of their products, and some will provide emission information to their customers.

Some VOCs are suspected or known carcinogens or causes of adverse reproductive effects. Some VOCs also have unpleasant odors or are irritants. VOCs are thought to be a cause of non-specific health symptoms that are discussed subsequently.

The total volatile organic compound (TVOC) concentration, often used as a simple, integrated measure of VOCs, is defined as the total mass of measured VOCs per unit volume of air, exclusive of very volatile (e.g., formaldehyde) organic compounds. Laboratory studies in which humans have been exposed to mixtures of VOCs under controlled conditions (Molhave et al. 1986 and 1993) have documented increased health symptoms at TVOC concentrations of the order of milligrams per cubic meter of air. A panel of 12 Nordic researchers reviewed the literature on VOCs/TVOCs and health and concluded that indoor pollution including VOC is most likely a cause of health effects and comfort problems and that the scientific literature is inconclusive with respect to TVOC as a risk index for health and comfort (Andersson et al. 1997). As an indicator of health effects, the TVOC concentration is inherently flawed because the potency of individual VOCs to elicit irritancy symptoms varies by orders of magnitude (Tenbrinke 1995). The potency for other potential health effects such as cancer or reproductive effects is also highly variable among compounds.

Despite these limitations, unusually high TVOC concentrations in commercial buildings, above one or two mg m-3 (Daisey et al. 1994), do indicate the presence of strong VOC sources. Further investigations to determine the composition of the VOCs and/or to identify the sources may be warranted. The probability of adverse heath or comfort effects caused by the high TVOC exposures will depend on the composition of the VOC mixture and on the concentrations of odorous or harmful compounds.

Radon is a naturally occurring radioactive gas. The primary source of radon in most buildings is the surrounding soil and rock. Radon enters buildings from soil as soil gas is drawn into buildings and also enters by diffusion through the portions of buildings that contact soil. Earth-based building materials and water from wells can also be a source of radon. Radon exposure increases the risk of lung cancer (BEIR VI 1998).

Ozone is brought into buildings with outdoor air. Certain types of office equipment, such as photocopy machines and laser printers, can also be a source of indoor ozone. Ozone causes pulmonary inflammation and other pulmonary health effects. Ozone is removed from indoor air by reaction with indoor surfaces; thus indoor ozone concentrations are usually lower than outdoor concentrations. If indoor sources of ozone are limited, increasing the ventilation rate, while decreasing concentrations of indoor-generated pollutants, will usually increase the indoor ozone concentration.

In addition to the direct effects of ozone on health, ozone can react chemically with VOCs in the indoor air or with surface materials. These reactions may produce VOCs that may be a source of chemical irritation (Weschler and Schields 1997).

Moisture is not a pollutant but it has a strong influence on indoor environmental quality. Water vapor is generated indoors due to human metabolism and human activities involving water use, as well as due to unvented combustion activities and by humidifiers. Moist soil may be a source of moisture in indoor air and in the flooring materials that contact the soil. The implications of high humidity for human health are complex and still a subject of debate (Baughman and Arens 1996, Arens and Baughman 1996). In some situations, high relative humidities may contribute to growth of fungi and bacteria that can adversely affect health.

Condensation of water on cool indoor surfaces (e.g., windows) may damage materials and promote the growth of microorganisms. Water leaks, such as roof and plumbing leaks, and exposure of building materials to rain or snow during building construction are a frequent source of material damage and growth of microorganisms. There is quite strong evidence that moisture problems in buildings lead to adverse respiratory health effects such as a higher prevalence of asthma or lower-respiratory-tract symptoms (e.g., Brunekreef 1992, Dales et al. 1991, Spengler et al. 1993; Smedje et al. 1996, Division of Respiratory Disease Studies 1984). There are many case studies of moisture-related microbiological problems in commercial buildings. The presence of humidifiers in commercial building HVAC systems has been associated with an increase in various respiratory health symptoms.

Particles are present in outdoor air and are also generated indoors from a large number of sources (Owen et al. 1992) including tobacco smoking and other combustion processes. Some particles and fibers may be generated by indoor equipment (e.g. copy machines and printers).

Mechanical abrasion and air motion may cause particle release from indoor materials. Particles are also produced by people, e.g., skin flakes are shed and droplet nuclei are generated from sneezing and coughing. Some particles may contain toxic chemicals. Some particles, biologic in origin, may cause allergic or inflammatory reactions or be a source of infectious disease. Increased morbidity and mortality is associated with increases in outdoor particle concentrations (EPA 1996), even when concentrations are in the vicinity of the U.S. national ambient air quality standard (50 μgÆm^3 for particles smaller than 10 micrometer). Of particular concern are the particles smaller than 2.5 micrometers in diameter, which are more likely to deposit deep inside the lungs (EPA 1996). A national ambient air quality standard for particles (http://www.epa.gov/airs/criteria. html) smaller than 2.5 micrometers was established by the US Environmental Protection Agency (EPA) in 1997 (15 μgÆm^3 for the three-year average of the annual arithmetic mean concentration; 65 μgÆm^3 24-hour average).

Particle size is important because it influences the location where particles deposit in the respiratory system (EPA 1996), the efficiency of particle removal by air filters, and the rate of particle removal from indoor air by deposition on surfaces. The large majority of indoor particles are smaller than 1 μm. Particles smaller than approximately 2.5 μm are more likely to deposit deep inside the lungs. Many of the **bioaerosols** are approximately 1 μm and larger, with pollens often larger than 10 μm. These larger particles deposit preferentially in the nose.

Non-infectious bioaerosols include pollens, molds, bacteria, dust mite allergens, insect fragments, and animal dander. The sources are outdoor air, indoor mold and bacteria growth, insects, and pets. These bioaerosols may be brought into buildings as air enters or may enter buildings attached to shoes and clothing and subsequently be resuspended in the indoor air. The health effects of non-infectious bioaerosols include allergy symptoms, asthma symptoms, and hypersensitivity pneumonitis which is characterized by inflamation of the airway and lung (Gammage and Berven 1997).

Infectious non-communicable bioaerosols are airborne bacteria or fungi that can infect humans but that have a non-human source (Gammage and Berven 1997). The best known example is Legionella, a bacterium that causes Legionnaires Disease and Pontiac Fever. Cooling towers and other sources of standing water (e.g., humidifiers) are thought to be a source of aerosolized Legionella in buildings. Legionella may also be present in potable water systems and aspiration of potable water is also thought to be a potential source of infection with Legionella. Some fungi, from sources within a building, can also infect individuals who are immune compromised.

Infectious communicable bioaerosols generated by one person may cause disease in others. These bioaerosols contain bacteria or virus within small droplet nuclei produced from the drying of larger liquid droplets, often expelled during coughing or sneezing. Examples of respiratory diseases transmitted, at least in part, by bioaerosols include tuberculosis, influenza, measles, and some types of common colds. Several studies, as reviewed in Fisk and Rosenfeld (1997), have indicated that building characteristics may significantly influence the incidence of respiratory disease among building occupants.

Fibers in indoor air include those of asbestos, and man-made mineral fibers such as fiberglass, and glass wool. The primary indoor sources are building materials, especially insulation products. Exposures to asbestos in industrial settings have been shown to cause lung cancer and other lung disease. Fiberglass and glass wool fibers are a source of skin irritation. The link between fiberglass and glass wool fibers and lung cancer remains uncertain.

Environmental tobacco smoke (ETS) is the diluted mixture of pollutants caused by smoking of tobacco and emitted into the indoor air by a smoker (as opposed to the mainstream smoke inhaled by a smoker). Constituents of ETS include submicron-size particles composed of a large number of chemicals, plus a large number of gaseous pollutants. ETS is a source of odor and irritation complaints. Panels of experts have reviewed the scientific evidence pertaining to the health effects of ETS and concluded that ETS is causally associated with lung cancer and heart disease in adults and asthma induction, asthma exacerbation, acute lower respiratory tract infections, and middle ear infection in children (EPA 1992, California EPA 1997).

4.5 CONTROLLING INDOOR POLLUTANT CONCENTRATIONS

The indoor concentration of a particular air pollutant depends on the outdoor concentration, the indoor pollutant generation rate, and on the total rate of pollutant removal by ventilation, air cleaning, and other removal processes. A simple mass balance equation can be used to illustrate the relationship among these variables at steady state in a space with well-mixed air:

$$\text{Indoor Conc - Outdoor Conc} + \frac{\text{Indoor Pollutant Generation Rate}}{\text{Ventilation Rate + Air Cleaning Rate + Rate of Other Removal Processes}} \qquad \textbf{Eq. 1}$$

4.5.1 Pollutant Sources

The indoor pollutant generation rate is a function of the type and quantity of indoor pollutant sources. For pollutants that have primarily indoor sources, excluding human bioeffluents, the indoor pollutant source strength tends to vary over a wider range than the other parameters that affect indoor pollutant concentrations. The indoor pollutant generation rate is often considered to be most important determinant of the indoor pollutant concentration. Eliminating or minimizing the emissions of pollutants from indoor sources is a highly effective and energy efficient means of reducing indoor pollutant concentrations.

4.5.2 Ventilation

In addition to minimizing the emissions of pollutants from indoor sources, to maintain acceptable IEQ, ventilation with outside air must be provided at an adequate rate. The ventilation rate, i.e., the rate of outside air supply, is usually normalized by the floor area (LÆ$^{-1}$sÆm^{2}), number of occupants (LÆ$^{-1}$s per person), or indoor air volume (h^{-1} or air changes per hour). The outside air supplied to a building must be distributed properly to the various rooms to maintain acceptable IEQ throughout the building. The required rate of outside air supply often changes with time because of changes in occupancy and indoor pollutant emission rates.

Often, local exhaust ventilation is used in rooms with high pollutant or odor sources. Exhaust ventilation is more efficient in controlling indoor pollutant concentrations than general ventilation of the entire space (general ventilation is often called dilution ventilation). Exhaust ventilation is a means of controlling pressure differentials, as described below.

4.5.3 Pressure Control

HVAC systems are often used in commercial buildings to maintain pressure differences between indoors and outdoors or between different indoor spaces. Maintaining buildings under positive pressure relative to outdoors can help to maintain IEQ by limiting the infiltration of outdoor air that may adversely affect thermal comfort and that may contain moisture and pollutants. Maintaining pressure differences between different indoor spaces can limit the rate of pollutant transport between these spaces. For example, smoking rooms, bathrooms, and laboratories are often depressurized so that pollutants generated within these rooms do not leak into the surrounding rooms.

4.5.4 Air Cleaning

Gaseous or particulate air cleaning (ASHRAE 1996, Chapters 24 and 25) may be used to remove air pollutants from recirculated indoor air or from incoming outdoor air. Most commercial buildings use particle filters in the HVAC system. Commonly, these filters have a low particle removal efficiency for particles smaller than approximately one micrometer in diameter (Hanley et al. 1994). However, particle filters with a wide range of efficiencies for submicron-size particles are readily available for use in commercial buildings and indoor particle concentrations may be lowered substantially by the use of higher efficiency air filters. To maintain the efficiency of filter systems, the filter installation method must prevent significant air leakage between or around the filters. In contrast to particle filters, gaseous air cleaners, such as beds of activated carbon, are used in only a small minority of buildings because of their higher costs and uncertain performance; however, considerable effort is being devoted to the development of new technologies for gaseous air cleaning.

Air cleaning systems require regular maintenance. For example, air filters must be periodically replaced to prevent reductions in air flow and to limit odor emissions and microbiological growth from soiled filters.

4.5.5 Natural Pollutant Removal Processes

For some pollutants there are other natural removal processes. Examples are loss of ozone due to its reaction with indoor surfaces and the deposition of particles on surfaces. These removal processes can substantially influence the indoor concentrations of these pollutants.

4.5.6 Air Recirculation and Air Flow Patterns

The indoor air flow pattern also influences IEQ. Mechanical recirculation of air spreads pollutants emitted from localized sources throughout a building so that a larger population is exposed; however, recirculation may decrease the pollutant concentration near the source. Also, recirculation of air through air cleaning systems can reduce indoor pollutant concentrations.

The pattern of airflow within rooms also influences indoor pollutant exposures. A floor-to-ceiling indoor airflow pattern, called displacement ventilation, can reduce pollutant concentrations in the breathing zone. A short-circuiting air flow pattern between supply and return air grills at ceiling level can increase pollutant concentrations in the breathing zone.

4.5.7 Maintenance, Cleaning, and Commissioning

Maintenance, cleaning, and commissioning of the HVAC system and building are generally thought to be significant determinants of indoor pollutant concentrations. Poor HVAC system maintenance may lead to poor control of indoor thermal comfort or outside air supply and to growth of microorganisms inside the HVAC system. Air filters may become sources of odors or a substrate for microbiological growth (e.g., Elixmann et al. 1990, Martikainen et al. 1990). HVAC system commissioning and air balancing help to assure that the system performs as intended. In a large study of office buildings with health complaints (Sieber et al. 1996), significantly increased prevalences of respiratory health symptoms were associated with poorer HVAC maintenance and less frequent building cleaning.

The methods and quality of building cleaning affects odor emissions from surfaces and the quantity of particles on surfaces that may become suspended in the air.

While increased maintenance and cleaning is generally considered beneficial, maintenance activities and products can be sources of indoor pollutants (e.g., cleaning compounds, waxes) and cleaning activities can be a source of resuspended particles.

4.6 ACUTE NON-SPECIFIC HEALTH SYMPTOMS AND BUILDING RELATED ILLNESS

The general public is familiar with many of the health effects that may be influenced by IEQ (and by energy conservation measures), such as acute respiratory diseases, allergies, asthma, and cancer. This document will not discuss the nature of these health effects. Rather, this section will briefly describe two less familiar classes of building-related health effects: acute non-specific health symptoms associated with buildings and building related illness (BRI). More comprehensive discussions of the health effects associated with IEQ factors are provided in the published literature (e.g., Gammage and Berven 1996)

4.6.1 Non-Specific Health Symptoms

The most common health symptoms attributed by building occupants to their indoor environments are non-specific health symptoms that do not indicate a specific disease, such as irritation of eyes, nose, and skin, headache, fatigue, chest tightness, and difficulty breathing. These symptoms are commonly called sick building syndrome symptoms; however, we use the term non-specific health symptoms because the term sick-building syndrome can be misleading (i.e., the building is not sick and the building is not always the cause of symptoms). People commonly experience these non-specific health symptoms; however, their prevalence or severity varies considerably among buildings and, in some buildings, the symptoms coincide with periods of occupancy in the building. Buildings within which occupants experience unusually high levels of these symptoms are sometimes called sick buildings. Some non-specific health symptoms are experienced frequently by a substantial fraction of all office workers (e.g., Brightman et al. 1997, Fisk et al. 1993; Nelson et al. 1995). The causes of non-specific health symptoms appear to be multifactorial and are not thoroughly understood. Although psychosocial factors such as the level of job stress are known to influence non-specific health symptoms, several characteristics of buildings and indoor environments are also known or suspected to influence these symptoms including: the type of ventilation system, type or existence of humidifier, rate of outside air ventilation, the chemical and microbiological pollution in the indoor air and on indoor surfaces, and indoor temperature and humidity (Mendell 1993; Sundell 1994). On average, occupants of sealed air-conditioned buildings report more symptoms than occupants of naturally ventilated buildings. Humidifiers increase the likelihood that occupants report these symptoms possibly because they can be a source of bioaerosols. Most studies have found that lower indoor air temperatures are associated with fewer non-specific health symptoms. Symptoms have been reduced through practical measures such as increased ventilation, decreased temperature, and improved cleaning of floors and chairs (Mendell 1993).

Non-specific health symptoms are a distraction from work and can lead to absence from work (Preller et al. 1990) and visits to doctors. When problems are severe and investigations of the building are required, there are financial costs to support the investigations and considerable effort is typically expended by building management staff, by health and safety personnel and by building engineers. Responses to non-specific health symptoms have included costly changes in the building.

4.6.2 Building-Related Illness

In contrast to non-specific health symptoms, the term building related illness (BRI) is sometimes used to describe a specific building-related health effect with known causes and objective clinical findings. Examples of BRIs include Legionnaires Disease, hypersensitivity pneumonitis, lung cancer from radon exposure, and health effects known to be a consequence of exposures to specific toxic compounds in buildings. Allergies and asthma are considered by some to be building related illnesses.

4.7 SENSITIVE POPULATIONS

Significant subsets of the total population have an increased sensitivity to indoor pollutants. Approximately 20% of people have environmental allergies and approximately 10% experience asthma (Committee on Health Effects of Indoor Allergens 1993). Peoples sensitivity to chemical irritants and their ability to detect odors also vary. To maintain low levels of building-related health complaints and health effects, the indoor environment must be maintained satisfactory for a substantial majority of occupants, many of whom are more sensitive than the average person to indoor pollutants.

A very small portion of the population report severe health effects when exposed to extremely low concentrations of a large variety of chemicals in the air. Their very high sensitivity to these chemicals may follow a period of sensitization caused by exposure to a higher concentration of one or more chemicals. The term multiple chemical sensitivity (MCS) is commonly used to describe this phenomenon. There is considerable uncertainty and controversy within the medical community about the concept of MCS. The current state of knowledge about MCS and its causes, physiological and psychological, is very limited. Owners and operators of buildings and building design and energy professionals usually do not have sufficient information to eliminate health symptoms in individuals with MCS.

4.8 STANDARDS, CODES AND GUIDELINES FOR VENTILATION AND INDOOR ENVIRONMENTAL QUALITY

A variety of standards, codes, or guidelines for minimum acceptable ventilation rates, lighting, thermal conditions, pollutant concentrations, tobacco smoking restrictions, pollutant sources, and building characteristics that influence IEQ have been adopted by national, state, regional, or municipal governments. Professional organizations also write voluntary or model standards or guidelines that are sometimes adopted, in whole or part, by code-making organizations. Some of these standards or codes constrain building and HVAC designs but do not strictly apply for building operation. A limitation of building design standards, applied at the time of new construction, is that these standards often fail to stimulate adjustments in building operation as building characteristics and building occupancy vary over time. However, failure to operate buildings in a manner that meets minimum design code requirements is considered poor practice and is likely to be a consideration in the event of IEQ-related litigation.

Thermal and lighting standards generally specify acceptable ranges for thermal and lighting conditions. Ventilation standards may specify minimum acceptable rates of outside air supply. Maximum pollutant concentrations are specified in some standards for a small number of air pollutants; however, at present there are no maximum concentration limits specified in standards for many of the pollutants present in indoor air.

IEQ-related codes and standards vary qualitatively and in their quantitative specifications. This document does not recommend specific standards to the exclusion of others because standards must reflect the regional climate, economic situation, and culture. This document does recommend that building professionals be familiar with all applicable standards and that the requirements in applicable standards be considered minimum requirements. In many situations, exceeding the minimum requirements is desirable. Examples of prominent codes or standards are listed and briefly described in *Table 1*.

4.9 RELATIONSHIP OF IEQ TO PRODUCTIVITY

Improvements in IEQ have the potential to improve worker pro-

Table 1: Examples of Standards, Codes, or Guidelines Pertinent to IEQ in Non-industrial, Commercial Buildings

Title	Organization	Primary Content
ASHRAE / ANSI Standard 55-1992, Thermal Environmental Conditions for Human Occupancy (ASHRAE 1992b)	ASHRAE[a] ANSI[b]	Acceptable range for temperature, humidity, and air velocity
ASHRAE Standard 62-1989, Ventilation for acceptable indoor air quality (ASHRAE 1989)	ASHRAE ANSI	Minimum acceptable rates of ventilation per occupant; Alternate IAQ performance path maintains selected pollutant concentrations below limits and subjective satisfaction to air above a limit; Includes a few pollutant concentration limits
ISO 7730: 1994 Moderate thermal environments determination of the PMV and PPD indices and specification of the conditions for thermal comfort (ISO 1994)	International Organization for Standardization (ISO)	Acceptable range for temperature, humidity, and air velocity
ANSI/IESNA —RP-1-1993	ANSI IESNA[c]	For office lighting, topics covered include office tasks, design process, lighting criteria for visual performance and comfort, luminous environmental factors, the lighting system, maintenance, light areas, energy and energy management
European concerted action, indoor air quality and its impact on man, report no. 11: guidelines for ventilation requirements in buildings (ECA 1992)	Commission of the European Communities, Directorate General for Science, Research, and Development	Minimum ventilation rates per unit of sensory indoor pollutant emission in olfs[d] to satisfy 70%, 80%, or 90% of people based on initial judgments when they enter space; Alternate minimum ventilation rate for protecting health
Air quality guidelines for Europe (WHO 1987)	WHO, Regional Office for Europe	Guideline concentrations for 25 chemicals, apply for outdoor and indoor air
Indoor climate-air quality: NKB publication no. 61e (NKB 1991)	Nordic Committee on Building Regulations (NKB)	General guidelines regarding: minimum quality of intake air; limiting spread of pollutants indoors (air recirculation is discouraged); use of low emission building materials, furnishings, processes and activities; assuring cleanability of buildings and HVAC; air balancing; minimum air change efficiency; necessary documentation and operation and maintenance. Also: minimum rates of outside air supply that are a sum of a minimum rate per unit floor area and a minimum rate per person
Law for Maintenance of Sanitation of Buildings, 1970 (Ministers 1970)	Ministers of Justice, Health and Welfare, Labor and Construction, Japan	Specifies limits or acceptable ranges for indoor temperature, relative humidity, air velocity, carbon dioxide concentration, carbon monoxide concentration, and suspended particle concentration. Specifies training and testing requirements for building sanitation engineers and oversight of building maintenance and management by building sanitation engineers. Requires maintenance of documents including those on state of regulation of air, management of water supply and wastewater, cleaning, and control of rodents, insects and other pests. Specifies fines for violations of law.

a. American Society of Heating, Refrigerating, and Air Conditioning Engineers
b. American National Standards Institute
c. Illuminating Engineering Society of North America
d. One olf is the rate of sensory pollutant emission from a standard person (non-smoker). The total sensory indoor pollutant emission rate, from people and other sources, is expressed in olfs

ductivity, in part by reducing: a) costs for health care, b) sick leave, c) performance decrements at work caused by illness or adverse health symptoms; and d) costs of responding to occupant complaints and costs of IEQ investigations. Some characteristics of the indoor environment, such as temperatures and lighting quality, may also influence worker performance without impacting health. In many businesses, such as office work[1], worker salaries plus benefits dominate total costs; therefore, very small percentage increases in productivity, even a fraction of one percent, are often sufficient to justify expenditures for building improvements that increase productivity.

At the present time, the linkages between specific building and IEQ characteristics and productivity have not been well quantified. However, in a critical review and analysis of existing scientific information, Fisk and Rosenfeld (1997, 1998) have developed estimates of the potential to improve productivity in the U.S. through changes in indoor environments. The review indicates that building and HVAC characteristics are associated with prevalences of acute respiratory infections and with allergy and asthma symptoms and non-specific health symptoms. For the normal range of indoor lighting conditions, the effects of improved lighting on the performance of typical office work is poorly understood. Several studies have found performance to be affected only by unusually low lighting levels or have found performance to change only with small low contrast type (Fisk and Rosenfeld 1997). However, a recent laboratory study with computerized performance tests suggests that high frequency ballasts may increase performance (Veitch and Newsham 1998). There is evidence that improvements in lighting quality can improve the performance of work that is very visually demanding, such as mail sorting or detailed product inspections, by several percent. Finally, there is evidence that quite small changes in temperature, a couple of degrees centigrade, may increase or decrease the performance of office work; however, the optimal temperature varies with the type of work. Also, the optimal thermal conditions for work performance may differ from the optimal conditions for comfort. From analyses of existing scientific literature and calculations using statistical data, the estimated *potential annual nationwide benefits* of improvements in IEQ include the following:

1. In office buildings, salaries are typically about 100 times larger than building energy or maintenance costs.

- a 10% to 30% reduction in acute respiratory infections and allergy and asthma symptoms;
- a 20% to 50% reduction in acute non-specific health symptoms;
- a 0.5% to 5% increase in the performance of office work; and
- associated annual cost savings and productivity gains of $30 billion to $170 billion.

5
Best Practices for Maintaining IEQ

The recommended best practices for maintaining a high quality indoor environment vary throughout the world; however, there are several widely accepted elements of best practice. This section describes best practices that are widely accepted as beneficial.

5.1 MAINTAINING GOOD INDOOR AIR QUALITY

5.1.1 Limiting Pollutant Emissions from Indoor Sources

1 Limiting the rates of pollutant emission from indoor sources is essential to cost-effectively maintain good IAQ. While the rates of pollutant emission from occupants are not generally controllable, emissions from many other pollutant sources can be minimized by adhering to the following practices. Tobacco smoking should be restricted to enclosed and depressurized[1] smoking rooms from which air is exhausted to outdoors. Alternately, indoor smoking should be entirely eliminated.

2 Building materials, furnishings, office equipment, and cleaning and pest control products and practices with low VOC and odor emission rates should be selected. In general, however, selection of low-emitting products is a difficult task because of limited information on VOC emissions, the lack of standard methods

1. Smoking rooms should be maintained depressurized relative to adjoining rooms.

for determining emission rates, and because of the health effects resulting from exposures to many VOCs and VOC mixtures are poorly understood. Many cleaning, painting, and pest control activities should be performed when the building is unoccupied. Cleaning and pest control products must be properly diluted and applied.

3 To limit indoor growth of fungi and bacteria, water leaks from plumbing and the building envelope should be eliminated. Building materials that become wet should be dried rapidly (e.g., within 1-2 days) or removed and replaced Construction materials, including concrete, should be dry before they are covered (e.g., with carpet) or enclosed (e.g., in a wall cavity).

4 To limit microbiological sources in HVAC systems, sources of moisture in HVAC systems must be controlled. The entry of water droplets and snow into outside air inlets must be restricted by proper design of the air intake system, e.g., use of rain-proof louvers and limiting of intake air velocities. Condensate drain pans must drain fully, be cleaned periodically, and have proper traps in the condensate drainage systems. Air velocities through cooling coils and humidifiers must be restricted to prevent water droplets from being entrained in the airstream and wetting downstream surfaces.

5 To limit microbiological sources in buildings, condensation of water vapor inside the building envelope and on interior building surfaces, including slab floors in contact with the soil, must be limited by controlling indoor humidity, proper use of moisture barriers and thermal insulation in the building envelope, and through control of indoor-to-outdoor pressure differences. Specific requirements will vary with climate.

6 A building should not be constructed on soil with unusual levels of hazardous contaminants (VOCs or radon) or special measures should be used to prevent soil gas from being drawn into the building. When building slabs are constructed on soils that are unusually wet, water proofing should be used to limit moisture transport from the soil into the slab.

7 Pollutants emitted from strong sources may need to be exhausted directly to outdoors. Combustion products from appliances should be vented to outdoors. The rates of air exhaust from janitor storage rooms and restrooms should be sufficient to maintain these rooms depressurized relative to surrounding rooms and the air should be exhausted directly to outdoors. Equipment with high emission rates of pollutants or odors should be isolated in rooms (e.g., copy machine rooms or kitchens) with high air exchange rates with air exhausted directly to outdoors. These rooms should also be maintained depressurized. Parking garages should be physically separated from occupied spaces and maintained under negative pressure relative to the adjoining occupied spaces.

8 Mechanical equipment rooms should not be used for storage of building materials, solvents, cleaning supplies, pesticides, adhesives, or other pollutant- or odor-emitting materials.

9 During building construction and major renovation projects, unusually strong sources of particles and VOCs may be present in the building. (Products or processes with low pollutant emissions should be employed when feasible.) Occupants must be isolated from these sources of pollutants using temporary internal walls as necessary and by maintaining the air pressure in the construction area less than the pressure in the adjoining occupied spaces. Measures should be taken to ensure that ventilation systems serving the construction zone do not become contaminated by construction dusts, or the systems should be cleaned thoroughly prior to occupancy (SMACNA 1995). A short delay (e.g., few days to weeks) in the occupancy of newly constructed or renovated spaces can help to prevent odor and irritancy complaints associated with the VOC emissions of new materials.

10 Good housekeeping practices should be employed to limit the accumulation of pollutants on surfaces and to reduce the potential for microbiological growth on these surfaces.

5.1.2 Assuring Adequate Quality of Intake Air

Assuring adequate quality of intake air is essential. When possible, buildings should be located in areas with acceptable outdoor air quality. Outside air intakes should not be located near strong sources of pollutants such as combustion stacks, exhausts from fume hoods, sanitary vents, busy streets, loading docks, parking garages, standing water, cooling towers, and vegetation. The outside air intake must be separated sufficiently from locations where ventilation air is exhausted to prevent significant re-entrainment of the exhaust air.

Incoming air should be filtered to remove particles. The recommended minimum particle filtration efficiencies vary among IAQ and ventilation standards and guidelines; however, use of filters that exceed minimum requirements is an option to improve IAQ, often with a small or negligible incremental cost. If unacceptable concentrations of gaseous pollutants are present in outside air, gaseous air cleaning may be required.

5.1.3 Maintain Minimum Ventilation Rates

The minimum ventilation rates specified in the applicable code requirements should be maintained or exceeded. The HVAC system should be designed so that rates of outside air intake can be measured using practical measurement techniques. In buildings with variable air volume (VAV) ventilation systems, special controls may be needed to ensure the minimum outside air intake into the air handling unit is maintained during all operating conditions (Cohen 1994, Drees and Wenger 1992, Solberg et al. 1990).

In addition to maintaining the minimum rate of outside air intake into the building, the ventilation system must be designed and balanced to assure the proper air delivery to each major room or section of the building. In VAV systems, VAV control units must have a minimum open position[1] to ensure the required outside air supply to specific regions of the building.

When the air supply and return air registers are placed in or near the ceiling and the supply air is warmer than room air, the supply air may short circuit to the return registers resulting in poor ventilation at breathing level; i.e., a poor air change effectiveness (e.g., Fisk et al. 1997b, ASHRAE 1998). The ventilation system should be designed to assure a high air change effectiveness or the rate of outside air supply should be increased to correct for poor air change effectiveness.

---

1. When internal cooling loads are low, the supply air temperatures may need to be increased to prevent overcooling of the conditioned space.

5.1.4 Recirculation of Indoor Air

Recirculation of indoor air is standard practice in some countries, such as the U.S., and discouraged in other countries such as those of Scandinavia. When air is recirculated, it should be filtered to remove particles. However, filters are often used only to prevent soiling and fouling of the HVAC equipment. These filters have a very low efficiency for respirable-size particles (smaller than 2.5 micrometers). Use of filters that exceed minimum requirements is an option to improve IAQ, often with a small or negligible incremental cost.

5.1.5 Maintenance of the HVAC System

Regular preventive maintenance of the HVAC system is necessary to assure proper delivery of outside air throughout the building and to limit growth of microorganisms in the system. A written plan for periodic maintenance should be developed and followed and maintenance activities should be documented. Elements of periodic maintenance that are important for maintaining good IEQ include changing of filters[1], cleaning of drain pans and cooling coils, checking fan operation, and checking dampers that influence air flow rates. Testing and balancing of an HVAC system may be necessary in the following circumstances: (a) after significant changes in the building, HVAC system, occupancy; or activity within the building (b) when control settings have been readjusted by maintenance personnel; and c) when accurate records are not available.

5.1.6 Integrated Approach to IEQ

It is widely recognized that an integrated or whole-building approach is the most effective means of saving energy in buildings because the energy performance of the building depends on the interactions of building systems and their control and operation. The IEQ performance of a building also depends on the interactions among building design, building materials, and building operation, control, and maintenance. Therefore, an integrated or whole building approach is recommended to maximize IEQ. Such an integrated approach may include the integrated consideration of the following: a) IEQ targets or objectives, b) occupancy and indoor pollutant sources and pollutant sinks and their

1. Filters become sources of odors. Also, microorganisms may colonize fil-

variation over time, c) building and HVAC design, d) commissioning, e) training and education, and f) building operation and maintenance.

5.2 IEQ MANAGEMENT PLANS AND RELATED PROGRAMS

The establishment and implementation of IEQ management plans is recommended to help maintain high quality IEQ and a high level of occupant satisfaction with IEQ. Common elements of IEQ management plans include the following[2]:

1 Selecting an IEQ manager, responsible for management and coordination of all aspects of IEQ

2 Developing an IEQ profile of the building

3 Assigning responsibilities and training of staff

4 Development of an IEQ checklist

5 Periodic building and HVAC system inspections

6 Facility operation and maintenance practices to maintain IEQ

7 Specific procedures to record and respond to occupant complaints

8 Special practices to maintain IEQ during building renovation, painting, pesticide use, or other periods of high indoor pollutant generation

9 Maintaining IEQ documentation

In the US, an independent non-profit private sector program called the Building Air Quality Alliance (BAQA) offers recognition to building owners and managers who commit to maintaining IAQ primarily through implementation of a proactive management plan. BAQA (http://www.baqa.org) has a practical step-by-step protocol in checklist form that is reviewed annually with the assistance of an IEQ professional.

An IEQ insurance policy is becoming available for buildings that follow BAQA guidelines. Other IEQ insurance policies under development are not specifically linked to the BAQA process but reportedly

2. For a more detailed description of elements in an IEQ management plan, see Building Air Quality Action Plan, U.S. EPA and NIOSH, http://www.epa.gov/IAQ/base/actionpl.html

will require an integrated building assessment.

The principles that underlie the BAQA process are derived substantially from a guidance document entitled Building Air Quality prepared by the US. EPA and NIOSH (EPA/NIOSH 1991; http://www. epa. gov/reg5oair/radon/healbld1.htm#bldgaqguinfo). The US EPA/NIOSH have also prepared and an accompanying Building Air Quality Action Plan (http://www.epa.gov/iaq/base/actionpl.html).

The US EPA/DOE (Department of Energy) Energy Star Program have recently announced a new Energy Star Building labeling program (http://www.epa.gov/buildinglabel). Based on current plans, commercial buildings whose energy performance is among the top 25% in the nation and which meet specific requirements for comfort and IEQ will be awarded and Energy Star Building Plaque.

Another private sector organization that promotes improved IEQ, as well as energy efficiency, is the US Green Buildings Council (http://www.usgbc.org). The Green Building Council has a building rating system called LEED (http://www.usgbc.org/programs/index.htm).

6

Linkages Between Energy Conservation Measures and IEQ

This section lists common energy conservation measures for commercial buildings, describes their potential influence on IEQ, and identifies precautionary actions or mitigations that can help to assure acceptable IEQ. The primary information is provided in *Table* 2. For many energy conservation measures, the cited references provide additional information on the IEQ impacts or on the related precautions and mitigations. The measures marked with "♦" in *Table* 2 deserve special consideration because they will often simultaneously improve IEQ and save energy. Because of the growing interest in IEQ, energy efficiency proposals that are expected to protect or improve IEQ will have a competitive advantage relative to proposals that ignore IEQ.

The last column of *Table* 2 links each energy conservation measure to the most directly relevant IEQ measurement and verification (M&V) alternatives provided subsequently in *Table 4*.

Table 2: List of Specific ECMs, Their Potential Influence on IEQ, and Related Precautions or Mitigation Measures

Energy Conservation Measure	Potential Influence on IEQ	IEQ Precautions or Mitigations	*Table 4* Rows
Lighting			
◆ Energy efficient lamps, ballasts, fixtures (IES 1993)	Improved lighting quality is common if lighting system is properly designed and installed.	Emphasize lighting quality in design. Check lighting levels and reflected images in VDT screens. Provide task lighting. Ensure that lighting retrofits do not result in disturbance and release of asbestos, fiberglass, or irritating dusts.	18-20
Automated lighting controls: occupancy sensors, dimming (IES 1993)	Improved lighting quality possible. Improperly operating control systems can degrade lighting quality	Emphasize lighting quality in design. Commission control systems. Provide task lighting where practical.	18-20
Removing lamps and fixtures	Risk of insufficient local or overall lighting level.	Ensure appropriate light levels and distribution. Provide occupant controllable task lighting where appropriate.	18-20
Use of window, skylights, light shelves, and light tubes to provide natural lighting. (IES 1993)	Improved or degraded lighting quality are possible depending on placement and optical characteristics of building elements that provide daylight. There is some evidence that proximity to windows, even when they can not be opened, is associated with a decreased frequency of acute non-specific building-related health symptoms (Fisk et al. 1993).	Ensure proper design of natural lighting system to prevent lighting problems such as glare, or incorrect or uneven lighting levels. Check lighting levels. Provide occupant controllable task lighting.	18-20
Automatic and manually-adjustable shading controls, fixed shading, window films (IES 1993)	Improved lighting quality possible. Improperly operating shading controls can degrade lighting quality. Shading can reduce potential pollutant emissions from indoor materials caused (or increased) by exposure to direct sunlight.	Commission shading control systems to ensure proper operation. Provide occupant controllable task lighting.	18-20
HVAC energy conservation measures			
Improve efficiency of HVAC components (motors, pumps, fans, chillers)	Adverse influence on IEQ unlikely if components have sufficient capacity.	Commission HVAC system to ensure proper performance under full- and part-load in heating and cooling modes. (ASHRAE 1996c)	NA
◆ Heat recovery from exhausted ventilation air or other source of waste heat.	If heat recovery system allows increase in rate of outside air supply, IEQ will usually be improved. See outside air economizer discussion related to outdoor air pollutants. Some heat recovery systems transfer moisture or pollutants from exhaust to supply airstream. (ASHRAE 1992)	See outside air economizer discussion related to outdoor air pollutants. Ensure that heat recovery system does not transfer unwanted moisture or pollutants to supply airstream.	2, 4, 5, 8, 12, 13, 15, 21, 22

Table 2: List of Specific ECMs, Their Potential Influence on IEQ, and Related Precautions or Mitigation Measures

Energy Conservation Measure	Potential Influence on IEQ	IEQ Precautions or Mitigations	*Table 4* Rows
Reducing operating time of HVAC components (e.g., fans, chillers) to save energy or limit peak energy demand.	Risk of degraded indoor thermal environment and / or increased indoor air pollutant concentrations if components are not operated during periods of occupancy. Also, when HVAC systems are not operating, indoor pressure differences and the associated transport of pollutants between zones or between outdoors and indoors are not controlled.	Operating periods must be sufficient to ensure acceptable thermal comfort and ventilation during occupancy. Ventilation with outside air should precede occupancy to reduce concentrations of air pollutants emitted from building materials and furnishings during unoccupied / low ventilation periods. Minimize indoor pollutant sources to reduce the pollution burden on the ventilation system. Equipment shutdown to limit peak energy demands should be infrequent and of limited duration. Use energy efficient HVAC systems or thermal energy storage to limit peak energy demands without sacrificing thermal comfort. Sequencing the startup of HVAC equipment may also reduce peak demands, often without adverse influence on IEQ.	1-8, 21, 22
◆ Use of outside air economizer[a] for free cooling.	Generally, IEQ will improve due to the increase in average ventilation rate. (Seppanen et al. 1989; Mudarri et al. 1996) During periods of elevated outdoor pollutant concentrations, economizer use may increase indoor concentrations of outdoor pollutants. In humid climates, economizer use may increase indoor humidities and the potential moisture-related IEQ problems	Locate outside air intakes as far as practical from strong sources of pollutants such as vehicle exhausts, HVAC exhausts, trash storage, and restaurant exhausts. (ASHRAE 1996b) If outdoor air is highly polluted with particles, use high efficiency air filters. If outdoor air is highly polluted with ozone, consider checking indoor ozone levels and/or use of activated charcoal filters. Design and control HVAC economizer to prevent moisture problems. Economizer controls and associated minimum outside air flow rates should be regularly calibrated and maintained.	2, 4, 5, 8, 12, 13, 15, 21, 22
◆ Nighttime pre-cooling using outdoor air. (ASHRAE 1995)	Nighttime ventilation may result in decreased indoor concentrations of indoor-generated pollutants when occupants arrive at work. Nighttime ventilation with humid air may result in condensation on HVAC equipment or building components increasing the risk of growth of microorganisms.	Design and operate nighttime ventilation systems to prevent moisture problems. Often, controls prevent nighttime cooling when outdoor dew-point temperature is excessive.	2, 15, 21, 22

Table 2: List of Specific ECMs, Their Potential Influence on IEQ, and Related Precautions or Mitigation Measures

Energy Conservation Measure	Potential Influence on IEQ	IEQ Precautions or Mitigations	*Table 4* Rows
Use of variable air volume (VAV) ventilation system in place of constant volume system.	Risk of insufficient outside air supply when indoor cooling or heating loads are low (Mudarri et al. 1996). See discussion of minimum outside air supply with VAV systems in *Section 4.1.* Particularly in HVAC systems with fixed outside air fraction, risk of excessive cooling and thermal discomfort when cooling loads are low if minimum outside air supply is maintained and supply air temperature is not increased. Increased risk of thermal comfort problems from supply air dumping[b].	Maintain outside air intake into air handler at or above minimum requirement for all supply air flow rates. (Solberg et al. 1990, Cohen 1994; Janu et al. 1995, Utterson and Sauer 1998) Avoid VAV control units that close fully when space temperatures are satisfied. Supply air temperatures may need to be increased when cooling loads are low. Check total and local outside air supply and indoor temperatures for a range of cooling loads. Use supply registers, minimum supply flow rates and temperatures that do not cause supply air dumping and thermal discomfort.	2, 3 — 5, 8, 21, 22
Use of variable speed drives in place of dampers for flow control.	No influence on IEQ expected.	Use flow controllers that depend on measured air flows rather than theoretical or design flows.	4, 5
Use of computerized digital HVAC control systems, Energy monitoring and control systems	Flexibility and ease of HVAC control are enhanced with properly operating control systems, hence, IEQ may be improved. Digital controls facilitate use of demand-controlled ventilation based on pollutant sensors.	Assure proper function of control system via commissioning (ASHRAE 1996c). Assure adequate training of building operators.	2-5, 8
Reduce air pressure drops and air leakage in duct systems.	May allow improved air supply and thermal control. May reduce noise generated in duct systems.	Air system balancing may be necessary after retrofits. Ensure quality of duct assembly to reduce noise.	2, 3
Use of hydronic radiant heating and cooling[c] with consequent fan energy savings	Mean radiant temperature is affected. Thermal comfort improvement or degradation are possible (e.g., the risk of draft from high air motion decreases but the risk of thermal discomfort from low air motion increases). Hydronic systems increase risk of water leaks or water condensation leading to microbiological growth. Average rate of outside air supply is reduced if radiant cooling precludes use of outside air economizer because of a reduction in capacity of fans and ducts.	Design, operating and maintenance practices of hydronic radiant heating and cooling systems should assure acceptable thermal comfort and outside air supply and low risks of water leakage and condensation. Periodic cleaning of radiant panels or radiators may be necessary. Water leaks should be immediately repaired. Water damaged materials must be quickly dried or replaced.	1 - 5, 8, 21, 22

Table 2: List of Specific ECMs, Their Potential Influence on IEQ, and Related Precautions or Mitigation Measures

Energy Conservation Measure	Potential Influence on IEQ	IEQ Precautions or Mitigations	*Table 4* Rows
Reduction in average or minimum rate of outside air supply, (especially closure of outside air dampers).	Primary effect is that concentrations of indoor-generated air pollutants will increase potentially leading to complaints and adverse health effects even though indoor concentrations of pollutants from outdoor air may be reduced (especially pollutants like ozone and particles that react with or deposit on indoor surfaces). In air conditioned buildings, indoor humidity may also be reduced.	Maintain rates of outside air supply specified in applicable codes and standards. Do not fully close outside air dampers during occupancy. Minimize indoor pollutant sources to decrease pollution burden on ventilation system. Use improved particle and gaseous air cleaning.	4, 5, 8, 21, 22
◆ Increase supply air temperature when cooling space (may decrease chiller energy but increase fan energy).	Higher supply air temperatures in VAV ventilation systems will increase supply air flow rates. In many VAV systems, outside air flow rates will also increase leading to reduced concentrations of indoor-generated air pollutants. Increasing chilled water temperatures often reduces the moisture removal by the HVAC system resulting in higher indoor humidities.	Maintain chilled water temperatures sufficiently low for control of indoor humidity.	1 - 3, 4, 5, 8
Increasing thermostat setpoints during periods of space cooling or decreasing thermostat setpoints during periods of space heating to save energy or limit peak energy demand. (ASHRAE 1992b, ISO 1994).	Air temperatures near or outside of the boundaries of locally applicable thermal comfort zones are likely to increase complaints of thermal discomfort, especially in air-conditioned buildings without provisions for occupant control. Occupants' perceived acceptability of air quality decreases as temperature increases between 18 °C and 28 °C (Fang et al. 1997). Increased air temperature is associated with increased prevalences of acute building related health symptoms in some studies (Mendell 1993)	Maintain temperatures within the bounds of applicable thermal comfort standards. Provide occupant-controllable fans and space heaters . Thermally efficient windows and walls may help to maintain thermal comfort (see Energy Efficient Building Envelope). Resetting of space temperature to limit peak energy demands should be infrequent and of limited duration. Use energy efficient HVAC systems or thermal energy storage to limit peak energy demands without sacrificing thermal comfort.	1 -3
Increase interior or exterior thermal insulation of piping and duct systems	Increased insulation will usually have negligible influence on IEQ. Potential for improved thermal comfort if insulation enables HVAC system to satisfy peak thermal loads. Increased insulation with vapor barriers can reduce moisture condensation and potential for microbiological growth. Potential increase in irritation symptoms if fibers or particles from insulation enter occupied space or if insulation releases VOCs at a high rate. Interior insulation in ducts can reduce fan noise (ASHRAE 1995). Interior insulation in ducts can be colonized by microorganisms (Morey 1991) potentially resulting in increased indoor concentrations of bioaerosols and microbiological VOCs.	Ensure that fibrous insulation is isolated from indoor air. Minimize fiber or particle release during installation of insulation and perform space cleanup prior to occupancy. Use insulation products with low emission rates of VOCs, especially of odors. The surface of insulation installed on the interior of ducts should prevent release of fibers or particles and not degrade. Interior duct insulation should not be located where it is likely to become wet or damaged. Promptly remove or repair damaged or wet insulation.	1 - 3

Table 2: List of Specific ECMs, Their Potential Influence on IEQ, and Related Precautions or Mitigation Measures

Energy Conservation Measure	Potential Influence on IEQ	IEQ Precautions or Mitigations	*Table 4* Rows
CO_2 - based demand controlled ventilation (DCV). (Carpenter 1996; De Almeida, and Fisk 1997, International Energy Agency 1990, 1992; Emmerich and Persily 1997, Persily 1993, ASTM 1998)	IEQ may improve or degrade depending on the reference condition and on the outside air control strategy used for DCV. Improved IEQ is most likely in spaces with high occupancy where occupant-generated pollutants dominate. DCV systems that provide outside air only after CO_2 concentrations exceed a setpoint may lead to substantially increased indoor concentrations of pollutants from building components and furnishings during the first few hours of occupancy	Avoid CO_2 - based DCV when building has strong pollutant emissions from sources other than occupants. Ventilate prior to occupancy to reduce concentrations of pollutants from non-occupant sources. CO_2 measurement locations must provide data representative of concentrations in occupied spaces. Consider advanced DCV control strategies that supply outside air in proportion to the indoor CO_2 generation rate, which is a better surrogate for occupancy than the CO_2 concentration (Federspiel 1996). Periodically check calibration of CO_2 sensors.	4, 5, 8, 15, 21, 22
◆ Displacement ventilation (Displacement ventilation systems usually supply 100% outside air with improved IEQ as the main goal. The addition of heat recovery system may be necessary to achieve energy savings relative to some other methods of HVAC.)	Generally, concentrations of indoor-generated air pollutants at breathing zone are reduced. (Seppanen et al. 1989; Yuan et al. 1998) Reduced transport of pollutants from sources to other rooms. Reduced risk of thermal drafts. Increased risk of thermal discomfort due to large vertical gradients in air temperature. Potential increased indoor concentrations of pollutants from outdoors, especially pollutants like ozone and particles that react with or deposit on indoor surfaces.	See outside air economizer discussion related to outdoor air pollutants. Design and operate displacement ventilation systems to avoid drafts in vicinity of supply diffusers and to avoid excessive gradients in air temperature. Displacement ventilation, without radiant cooling panels, may be effective only with internal heat generation less than 40 W m^{-2}.	1 — 5, 8, 21, 22
◆ Use of natural ventilation with operable windows as a substitute for air conditioning. (Olgyay 1963, Koenigsberger et al. 1973; Watson and Labs 1983, Awbi 1991, Givoni 1997, Busch 1992)	In some climates, thermal acceptance of environment may increase because occupants in naturally ventilated buildings are tolerant of a wider range of thermal conditions (De Dear and Brager 1998). Thermal comfort may decrease because of elevated indoor temperatures and humidities. On average, occupants of buildings with natural ventilation and operable windows report fewer acute non-specific health symptoms. Open windows admit sounds from outdoors, potentially degrading the indoor acoustic environment.	Building design, e.g., size, layout, openings for outside air, position of openings, and shading must assure adequate natural ventilation and thermal conditions throughout building. Generally, cross-ventilation is desirable. Occupant - controllable fans can enhance thermal comfort. Openable windows should not be located near concentrated outdoor sources of pollutants or annoying sounds.	1 — 3, 6, 8, 21, 22

Table 2: List of Specific ECMs, Their Potential Influence on IEQ, and Related Precautions or Mitigation Measures

Energy Conservation Measure	Potential Influence on IEQ	IEQ Precautions or Mitigations	*Table 4* Rows
◆ Preventive maintenance of HVAC system	Preventive maintenance will help to assure proper HVAC operation, sometimes also saving energy and improving IEQ. Preventive maintenance measures that may save energy and improve IEQ include calibration of temperature and humidity sensors, periodic replacement of air filters, maintenance of airflow and pressures control systems, balancing of airflows to provide proper air distribution, and cleaning of coils and other components to reduce airflow resistance and pollutant sources in the HVAC system.	Ensure that preventive maintenance does not disturb or release asbestos fibers (asbestos is present in mechanical rooms of many older buildings).	2, 3, 21, 22
Energy efficient building envelope:			
Increased thermal insulation in building envelope.	Often negligible influence on IEQ. Potential increase in thermal comfort because insulation helps HVAC system satisfy thermal loads and because of reduced radiant heat transfer between occupants and building envelope. Potential increase in irritation symptoms if fibers or binders from insulation enter occupied space or if insulation releases VOCs at a high rate.	Assure that fibrous insulation is isolated from indoor air. Minimize fiber or particle release during installation of insulation and perform space cleanup prior to occupancy. Use insulation products with low emission rates of VOCs, especially of odors.	1 - 3
Light color roof and walls to reduce solar loads	Often negligible influence on IEQ. Potential increase in thermal comfort because reduced loads help HVAC system satisfy thermal loads and because of reduced radiant heat transfer between occupants and envelope.	NA	1 - 3
◆ Thermally efficient windows	Improvements in thermal comfort possible from reductions of drafts and reduced radiant heat exchange between occupants and windows (ASHRAE 1995b, Heiselberg 1994, Heiselberg et al 1995). Reduces risk of condensation on windows and associated risks from growth of microorganisms. Thermally efficient windows help to isolate the indoor space from outdoor sounds.	NA	1 - 3

Table 2: List of Specific ECMs, Their Potential Influence on IEQ, and Related Precautions or Mitigation Measures

Energy Conservation Measure	Potential Influence on IEQ	IEQ Precautions or Mitigations	*Table 4* Rows
Reduce air leakage through building envelope (e.g., install infiltration barriers)	Thermal comfort may increase due to reduced entry of unconditioned air. Reducing air leakage may help to isolate the indoor space from outdoor sounds. Reducing envelope leakage facilitates room or building pressure control via the HVAC system. Indoor concentrations of outdoor pollutants or pollutants from adjoining spaces (e.g., parking garages) may decrease because leakage of outdoor pollutants to indoors decreases. Improperly placed infiltration and vapor barriers can lead to condensation and associated microbiological problems in building envelope. Reduced infiltration of outside air will generally increase concentrations of indoor-generated air pollutants; however, magnitude of increase may be insignificant, particularly if adequate outside air ventilation is provided mechanically.	Ensure adequate mechanical or intentional natural ventilation. Infiltration and vapor barriers should be located near warm side of building envelope.	1 — 3, 8, 21, 22
Reducing indoor heat generation or heat gain through building envelope			
Reduced indoor heat generation through use of energy efficient lighting and equipment or through reduced heat gain through building envelope.	Increased thermal comfort possible if measures enable HVAC system to provide adequate cooling. Decreased thermal comfort possible if building has inadequate heating system. In buildings with VAV ventilation systems, supply air flow rates will decrease when building is being cooled; in turn, rate of outside air supply may decrease (see prior information and references on VAV systems). Reduced heat loads without compensating changes in supply flows may lead to increased indoor humidity because control systems may increase cooling coil discharge temperatures. Excessive cycling and control problems with oversized refrigeration systems may cause discomfort.	Use a control system that maintains outside air intake into air handler at or above minimum requirement for all supply air flow rates. Avoid VAV control units that close fully. Check total and local outside air supply and indoor temperatures for a range of cooling loads. Check for and eliminate control system problems associated with oversized refrigeration systems.	1 — 5, 8

a. To save energy, economizer systems automatically increase the rate of outside air supply above the minimum setpoint during mild weather.

b. The term supply air dumping refers to the tendency for the jet of cool supply air exiting a supply register located at ceiling level to drop toward the floor without sufficient mixing between the jet and the warm air within the room. Supply air dumping, which is more common with low supply flow rates, lower supply temperatures, and certain supply diffuser designs, is a source of thermal discomfort.

c. A hydronic radiant heating or cooling system uses a heated or chilled liquid to create a heated or cooled radiant panel or surface. The occupants thermal comfort is determined, in part, by radiant heat exchange between occupant and the radiant panel.

7
Influence of Energy Conservation Measures on IEQ in Specific Buildings

7.1 BACKGROUND

Many of the energy conservation measures in *Table* 2 have multiple potential impacts on IEQ. When considering the application of these measures in specific buildings, energy conservation professionals may be faced with three important questions:

1 Which of the potential IEQ outcomes will occur in this building and what is the expected magnitude of the changes in IEQ?

2 Will the change in IEQ significantly affect occupants health, comfort, or productivity?

3 Can the impacts on IEQ be assessed with measurements?

A working knowledge of IEQ and its effects on occupants is an essential first step in addressing these questions. *Section 4* summarizes much of the critical background information. This section addresses the first two questions. IEQ measurements are reviewed in *Section 8.*

7.2 IDENTIFYING THE PROBABLE IEQ OUTCOMES AND PREDICTING THEIR MAGNITUDE

The impact of an energy conservation measure on IEQ may depend on several factors including the climate, outdoor air quality, characteristics of the building, and the details of implementation of the energy conservation measure.

Warm humid climates increase certain IEQ risks associated with high indoor humidity. In humid climates, energy conservation measures that increase the rate of outside air supply, such as openable windows or a poorly-controlled economizer system, are more likely to result in unacceptable indoor humidities. Also, reducing the capacity of HVAC equipment and increasing temperatures at cooling coils can result in insufficient moisture removal and excess indoor humidity. Condensation of water vapor on radiant cooling panels is much more likely in high humidity climates. Maintaining thermal comfort with natural ventilation is much more difficult in warm humid climates.

In cold climates, thermally efficient windows, high levels of thermal insulation in the building envelope, and measures that reduce air infiltration are more likely to significantly improve thermal comfort by reducing drafts and radiant heat losses from occupants to walls and windows. The potential to have very low indoor humidities is also increased in cold climates.

Elevated levels of pollutants in the outdoor air increase certain IEQ risks. Energy conservation measures that increase outside air supply, will generally improve IEQ by reducing concentrations of indoor-generated air pollutants. However, when outdoor air quality is poor, e.g., when outdoor pollutant concentrations exceed applicable standards, these same energy conservation measures may increase the indoor concentration of outdoor pollutants.

Characteristics of the building and building HVAC system also influence the IEQ outcomes from energy conservation. Only a few of the many possible interactions are described here. The magnitude of indoor pollutant emission rates is one consideration. If the building contains strong indoor sources of air pollutants, energy conservation measures that reduce outside air supply are much more likely to result in elevated concentrations of these indoor-generated pollutants. (The converse is also true.) Similarly, if a building has a low rate of outside air supply, pollutants emitted from energy conservation products such as sealants are more likely to significantly degrade indoor air quality. Energy conservation measures that reduce outside air ventilation are more likely to lead to IEQ problems if the initial rate of outside air supply is low.

Proper implementation of the energy conservation measures can prevent many of the potential adverse impacts on IEQ. Most of the possible implementation errors are obvious: faulty design, installation, calibrations, control methods, commissioning, operation, or maintenance of the energy conservation systems or practices can lead to IEQ problems. Proper design, training of users, etc. can prevent problems.

Engineering calculations and computer modeling are the primary tools for predicting levels of IEQ or the magnitude of changes in IEQ. Indoor temperatures are determined from energy balances, and indoor humidity and pollutant concentrations from mass balances. Algorithms for equipment performance are often required. The ASHRAE handbooks (e.g., ASHRAE 1992, 1995, 1997) are one source for many of the engineering calculations. Simple steady state and transient mass balance calculations for estimating pollutant concentrations in a single well-mixed zone can be implemented by the user. Several computer programs, as reviewed by National Laboratories (1997), are available for predicting air infiltration rates, airflows between zones, and indoor pollutant concentrations. Proper use of these models generally requires considerable expertise in IEQ and experience with the model. The difficulties in obtaining model inputs are a major obstacle to IEQ modeling in multizone buildings.

7.3 SIGNIFICANCE OF PREDICTED CHANGES IN IEA

Once an expected change in IEQ is identified and, to the degree possible, quantified, the resulting influence on occupants health, comfort, or productivity should be considered. The main approach for evaluating the significance of predicted (or measured) changes in IEQ parameters is to compare the initial and final values of these parameters to the values listed as acceptable in the applicable standards or guidelines (see *Table 1*). When assessing significance, the following points should be kept in mind: 1) Small changes in indoor temperature, on the order of IT, may significantly influence thermal comfort, the prevalence of acute non-specific building-related health symptoms experienced by workers, and perceptions of air quality. 2) Occupants satisfaction with thermal conditions may be estimated using thermal comfort models (e.g., ASHRAE 1992b, Fountain and Huizenga 1996, ISO 1994); however, recent research suggests that these models are imperfect because they do not account for peoples behavioral, physiological, or psychosocial adaptations to their thermal environments. 3) Changes in lighting are much more likely to influence work performance if the work is unusually visually demanding.

For many pollutants, there are no published maximum concentration limits for non-industrial workplaces. The pollutant concentration limits published for industrial workplaces, such as the Threshold Limit Values (TLVs) of the American Conference of Governmental Industrial Hygienists (ACGIH 1998), should not be directly applied to non-industrial settings and workers.

In assessments of the significance of changes in IEQ, the sensitivity of occupants to IEQ is a consideration. Older workers tend to have more stringent thermal comfort requirements and the vision of older workers is more likely to be adversely affected by low lighting levels. Women report non-specific health symptoms more frequently than men do (Mendell 1993, Menzies and Bourbeau 1997). Workers in buildings with prior or ongoing IAQ or thermal comfort problems may be more likely to respond negatively to small decrements in IEQ.

8
Measurement And Verification Alternatives for IEQ[1]

8.1 BACKGROUND

As indicated in *Table 2,* most of the impacts of energy projects on IEQ relate to thermal comfort or ventilation. Therefore, the most common parameters subject to IEQ M&V will be thermal comfort parameters (e.g., temperatures and humidities) and ventilation rates. For lighting retrofits, lighting-related M&V may sometimes be warranted. Occasionally, M&V of other IEQ parameters may be warranted. Such situations include, for example, building retrofits which move the outside air intake close to an outdoor pollutant source or measures, such as installation of an economizer, that cause a large increase of ventilation rate in a building located in a heavily polluted region. In such situations, M&V for specific pollutants (e.g., ozone, particles, carbon monoxide, or VOCs) may be desirable. Many IEQ measurements are expensive. Measurements should be performed only when there are clear objectives and the capabilities and intent to interpret the measurement results.

1. IEQ measurement and verification (M&V) activities may be hazardous due to the potential for pollutant exposures, falls, and contact with high voltage or rotating equipment. Staff performing IEQ M&V should receive training in safe work practices. Reference documentation is available in the US EPA Orientation to Indoor Air Quality: Instructor Kit (or Student Manual). Ordering information can be found at http://www.epa.gov/iaq/base/baqapp2.html. These documents may also be purchased from the National Technical Information Service (NTIS), US Department of Commerce, 5285 Port Royal Road, Springfield. VA 22161 (telephone 1-800-553-6847). The NTIS reference numbers are AVA 19276SS00 and AVA 19277B00 for the Instructor Kit and Student manual, respectively.

8.2 GOALS OF IEQ M&V

The appropriate M&V approach will depend on the goals of the IEQ M&V. Therefore, a definition of the M&V goals is a critical first step. Examples of IEQ M&V goals follow:

Goal 1: To ensure that the energy conservation measures have no adverse influence on IEQ.

Goal 2: To quantify the improvements in IEQ resulting from implementation of energy conservation measures.

Goal 3: To verify that selected IEQ parameters satisfy the applicable IEQ guidelines or standards.

8.3 CONTEXT FOR IEQ M&V

This document can be applied under two basic M&V contexts. The first basic context, and the primary focus of this volume, is the implementation of energy conservation retrofits in existing commercial buildings. Generally, in this situation, pre-retrofit and post-retrofit IEQ measurements are compared. A broad range of M&V approaches may be applied (e.g., measurements, modeling, and surveys).

The second basic context is the anticipated future application of energy conservation features in new commercial building construction. For this second context, when M&V goal 1 or goal 2 are selected, it is necessary to define the relevant characteristics of a reference building without the energy conservation features and to define the relevant characteristics of the building with the energy conservation features implemented. Modeling is the only approach available prior to construction to estimate changes in IEQ (M&V goals 1 and 2). After construction is completed, measurements may be performed to verify that IEQ parameters satisfy applicable codes or standards (M&V goal 3).

8.4 M&V PROCEDURE

Table 3 presents a recommended basic procedure for IEQ M&V The M&V Approaches and M&V Alternatives referenced in this table are described later.

8.5 BASIC M&V APPROACHES

This section identifies basic approaches for IEQ M&V and identifies situations for which each general approach may be applicable.

Table 3: The M&V Procedure

Steps in the M&V Procedure	**Comments**
1. Define M&V goals	See *Section 8.2*
2. Select M&V staff	a) Staff performing measurements must have necessary skills and knowledge; outside consultants with IEQ expertise are generally available. b) Owner should consider whether or not measurement staff should be independent of the organization (e.g., ESCO) that benefits financially from positive findings or if there should be independent oversight.
3. Select general M&V approach	See *Section 8.5*
4. Select specific IEQ parameters and M&V alternatives for measuring and predicting values of these parameters	See *Section 4, 6,* and *8.6*
5. Establish plans for interpreting M&V data	a) Generally, measured IEQ data or predictions are compared to data or predictions from another time period, to baseline data from a representative set of buildings, or to values listed in standards. b) See *Section 4, 8.3, 8.5*
6. Define requirements for M&V accuracy and quality control procedures for measurements	Often the required accuracy depends on the expected magnitude of the change in IEQ

1. A building with occupants that have excessive health or comfort complaints is a third potential context for IEQ M&V. However, this document is not intended as a guide for diagnosis or investigations of the causes of such complaints. Several existing documents may be consulted for such situations (e.g., EPA/NIOSH 1991; ISIAQ 1996 (in draft), ECA 1989, Weekes and Gammage 1990, Raferty 1993, Nathanson 1995). Usually, a phased investigative approach is recommended Early phases include building inspections and discussions with building occupants. Expensive measurements are recommended only if necessary in subsequent phases of the investigation. These guides point out the importance of maintaining open channels of communication about the complaints, about IAQ, and about the investigation.

Table 3: The M&V Procedure (*Continued*)

7. Select measurement	a) Often, pre-retrofit values of IEQ parameters are compared to post-retrofit values. b) Many IEQ parameters change with time because of changes in building operation, indoor pollutant emission rates, or outdoor air quality. Measurements should take place under the relevant conditions (e.g., minimum outside air supply) or they should average over the range of conditions. If weather, outdoor air quality, or occupancy differ significantly between the pre- and or assessment periods post-retrofit measurement periods, a comparison of pre-retrofit and post-retrofit measurements may not accurately indicate the effect of the energy conservation on IEQ. c) Indoor thermal conditions and pollutant concentrations in buildings do not respond instantly to changes in the controlling factors. d) Instantaneous measurements of temperatures, humidities, and air pollutant concentrations will often not be valuable.
8. Select measurement locations	a) Thermal comfort standards provide guidance for measurement locations. b) Measurement locations should yield data representative of conditions experienced by occupants. Worst-case locations may also be monitored. c) The breathing zone is most important location for air pollutant measurements. d) Pollutant concentrations in HVAC return airstreams may approximately represent average concentration in section of building from which the return air is drawn (except with displacement ventilation).
9. Define acceptable M&V costs	Expenditure should generally be small compared to expected savings from energy conservation measures.
10. Select M&V instrumentation	Instrumentation must satisfy accuracy, cost, and data logging requirements
11. Inform occupants of M&V plans	Unexplained measurements may cause occupants concern about IEQ.
12. Implement measurements, surveys, or modeling	Methods employed should meet accuracy, space, and temporal specifications and cost constraints.
13. Analyze and report results	Results should generally be available to occupants.

8.5.1 Approach 1: No IEQ M&V

Table 2 lists several energy conservation measures (e.g., chiller upgrades) that are either unlikely to affect IEQ or that are likely to have only a beneficial influence on IEQ. In general, no IEQ M&V will be necessary when the energy conservation measures are judged highly unlikely to result in a significant adverse IEQ impact. However, IEQ M&V may still be performed if the M&V goal *is* to quantify an anticipated improvement in IEQ.

8.5.2 Approach 2: IEQ M&V Based on Modeling

IEQ modeling is usually the only method available to predict the magnitude of changes in IEQ associated with implementation of energy conservation measures during new construction. Additionally, modeling may be appropriate to estimate changes in indoor pollutant concentrations associated with changes in ventilation rates or when measurement methods are too expensive or not available.

Very simple mass balance models for single zones with well-mixed air can provide a useful estimate of the change in indoor pollutant concentrations that are expected when ventilation rates or indoor pollutant emission rates are modified. Many scientific papers describe such models which can be implemented using spreadsheet software (e.g., Persily and Dols 1990, Nazaroff et al. 1993, Fisk and deAlmeida. 1998). Several much more complex multi-zone models are available (National Laboratories 1997, Chapter 3). Many of the complex models require extensive model inputs and considerable IEQ modeling expertise.

Lighting simulation tools such as the Radiance program (Ward and Rubinstien 1998) can model the resultant luminance distribution from most lighting and daylighting systems; however, only a few lighting quality parameters, such as glare, are computed by these tools. Extensive model inputs and considerable modeling expertise are required.

8.5.3 Approach 3: Short-Term Measurements of Selected IEQ Parameters

Short term measurements of IEQ parameters (e.g., measurements for a month or less) may be used for IEQ parameters that do not vary significantly with season or with the mode of building operation. Examples of such parameters are light levels in core zone of building and minimum outside air flow rates in constant volume HVAC systems. Short term measurements may also be appropriate when the outcome of interest is an IEQ parameter for a defined set of climatic and building operating conditions, such as conditions that leads to worst case IEQ. In this instance, short-term measurements can be performed only under these conditions.

8.5.4 Approach 4: Long-term Continuous Measurements of Selected IEQ Parameters

Long-term continuous measurements are often affordable and useful for tracking selected IEQ parameters including indoor temperatures and humidities, carbon dioxide concentrations, carbon monoxide concentrations, and rates of outside air intake into air handlers. For most other IEQ parameters, long-term continuous measurements are prohibitively costly or unavailable. The costs of sensors, and sensor maintenance and calibration may be lower in continuous monitoring systems that use single sensors to analyze samples drawn sequentially from multiple locations, However, drawing some pollutants, e.g., particles and many VOCs, through long sampling tubes will lead to substantial measurement errors.

8.5.5 Approach 5: Surveys to Assess Occupant Perceptions and Ratings of IEQ

In many instances, occupant perceptions determined in a survey are as relevant an outcome as measured IEQ conditions. Survey costs can be lower or higher than costs of physical measurements.

There are two basic uses of surveys in the context of IEQ M&V First, administration of a survey before and after implementation of an energy conservation measure can provide information on the perceived change in IEQ and occupant reports of health symptoms[1]. The second method of using surveys in the context of IAQ M&V is to perform a one-time survey and to compare the results to baseline data obtained previously in other buildings with the same survey instrument.

Survey data are subjective; hence, these data may be influenced by psychosocial factors such as job satisfaction. A portion of the occupants may express dissatisfaction even when IEQ is above average. Addition-

1. There is evidence that responses to surveys of non-specific health symptoms tend to vary even with no apparent change in building conditions or IEQ. Often, occupants report fewer building-related health symptoms on a second of two surveys administered a week or two apart in time. Thus, it may be necessary to correct the change in survey results obtained from the space with the energy conservation retrofit by subtracting the change in survey results from a control population.

ally, surveys can only assess the responses to IEQ that are sensed by the human sensory systems. IEQ exposures that increase the risk of some chronic health effects, such as lung cancer from radon exposures, will not be detected using surveys.

The survey design and methods of administration can affect the outcome; therefore, surveys should be based on established questionnaires and on survey administration methods developed by staff with suitable expertise. High response rates (e.g., > 80%) to surveys are necessary to reduce risk of bias (e.g., a higher response rate from unhappy occupants could bias the overall results).

Well-established survey methods are available for thermal comfort. Thermal sensation is commonly assessed using a seven-point scale (e.g., ASHRAE 1992b, ISO 1994). Baseline data are being compiled by de Dear and Brager (1998) from thermal comfort surveys performed throughout the world.

Several survey instruments (questionnaires) have included broad assessments of the level of satisfaction with, or perception of, multiple IEQ parameters such as lighting level, lighting quality, acoustical quality, air movement, acceptability of indoor air quality, ventilation, etc. These same surveys have generally collected data on the prevalence or severity of non-specific health symptoms experienced by office workers.

A U.S. EPA-supported survey is collecting data on non-specific symptoms from 100 office buildings in the US (EPA 1994; Brightman et al. 1997). A European Audit Project has collected symptom data from 56 office buildings (Bluyssen et al. 1996). Stenberg et al. (1993) and Sundell (1994) describe similar data obtained from approximately 5000 office workers in 210 buildings in Sweden.

Questionnaires to evaluate occupant satisfaction with lighting are available (Collins et al. 1990, Dillon and Vischer 1987, Eklund and Boyce 1995, Hygge and Lofberg 1998) although customization for specific applications may be required.

8.6 M&V ALTERNATIVES FOR SPECIFIC I EQ PARAMETERS

In tabular form, this section identifies M&V alternatives for specific IEQ parameters and provides comments on these alternatives. The M&V alternatives for thermal comfort and ventilation are listed first because energy conservation retrofits more often affect these IEA parameters. The tables do not include all possible M&V alternatives. The alternatives judged to be the most practical and valuable are marked with "♦." General guidance for measurement of indoor air pollutant concentrations is available in several publications (e.g., ACGIH 1995, Nagda and Harper 1989)

Table 4: Specific M&V Alternatives for IEQ Parameters.

Row	IEQ Parameter / M&V Alternative	Comments
Thermal comfort		
1	Alternative 1. Multi-parameter measurements specified in thermal comfort standards (ASHRAE 1992b, ISO 1994)	a) Generally one measures air temperature, mean radiant temperature, relative humidity, and air velocity at multiple heights in multiple (e.g., 20 - 30) workspaces. b) Measurement system cost is at least US $5K. The system may be moved between locations (e.g., see De Dear and Fountain 1994). c) This method provides only short-term data at each location. d) Thermal comfort instruments are commercially available.
2	◆ Alternative 2. Multipoint measurement and logging of air temperatures and/or humidities using portable or permanent inexpensive instrumentation	a) In many common situations air temperature measurements at a single height (without air velocities, humidity, and mean radiant temperature) are adequate. b) This alternative is not appropriate for situations with high air velocities (e.g., fans used for cooling), high temperature stratification (e.g., displacement ventilation) or large radiant heat gains or losses (e.g., near cold windows). c) Temperature sensor accuracy should be ~ 0.25 °C because temperature differences < 1 °C may significantly influence thermal comfort. d) Humidity sensor accuracy should be ~ 5% RH. e) The cost of battery powered sensor with data logger for a single point measurement is ~ US$100 to $200.
3	◆ Alternative 3. Thermal comfort surveys (Schiller et al. 1988, De Dear and Fountain 1994)	a) In many instances, surveys are the best alternative. b) Surveys may ask about current level of thermal comfort or about thermal comfort during prior extended period.
Outside air ventilation rate		
4	Alternative 1. Outside airflow into air handler measured using anemometry (ASHRAE 1988, SMACNA 1993, Utterson and Sauer 1998, Solberg et al 1990)	a) Accuracy is often questionable when measurements are made near outside air louvers or dampers. b) ◆ Better accuracy obtained if measurements can be made in sufficient length of straight outside air duct.
5	◆ Alternative 2. Outside airflow into air handler measured based on HVAC supply flow and% outside air (Drees et al. 1992)	a) Supply air flow is typically from balometers (airflow hoods), pitot tube or hot wire anemometer traverses in supply airstream, or permanently-installed supply airflow measurement stations. b) Percent outside air is determined from carbon dioxide or tracer gas mass balance. c) The temperature method of determining percent outside air is often inaccurate.
6	Alternative 3. Outside air ventilation rates determined using tracer gas methods (ASTM 1995, NORDTEST 1982, 1988, Lagus and Persily 1985, ASHRAE 1998, Faulkner et al. 1998, Charlesworth 1988)	a) Procedures include tracer gas decays, stepups, and methods based on continuous release of tracer to indoors. b) Measurements can account for both natural and mechanical ventilation. c) Reasonable (e.g., 15-25%) accuracy is possible in many buildings. d) For most tracer gas methods, ventilation rates must be reasonably stable during the measurement period. e) Measurements are often expensive and require considerable expertise (instrumentation costs usually > US $10000). f) Most procedures give ventilation rates representative of a short (few hour) time period; some provide an average effective ventilation rate for an extended period.
7	Alternative 4: Outside air ventilation rates determined using outside air flow measurement stations in air handlers	a) Usually airflow straighteners and multiple point velocity sensors are installed in vicinity of outside air inlet dampers / louvers. b) Commercially available products are relatively new; therefore, limited performance data are available.

Table 4: Specific M&V Alternatives for IEQ Parameters.

Carbon dioxide concentrations		
8	◆ Alternative 1. Real time CO_2 infrared analyzers with output logged over time (Persily and Dols 1990; Persily 1993, ASTM 1998)	a) Peak or time average carbon dioxide concentrations are useful indicators of how effectively occupant-generated bioeffluents are controlled by ventilation and may be fair indicator for other pollutants associated with occupancy. b) Several procedures have been used (and often misused) to estimate ventilation rates from measured carbon dioxide concentrations. In many buildings, it is difficult to accurately determine the rate of outside air supply from carbon dioxide data because of uncertainties and temporal variations in occupancy, uncertain rates of carbon dioxide generation by occupants, and because concentrations change slowly after changes in ventilation or occupancy. c) The cost of a carbon dioxide analyzer is typically US $700 to $3000.
Carbon monoxide concentrations		
9	Alternative 1. Real time infrared analyzer with output logged over time	a. Elevated indoor carbon monoxide concentrations, relative to outdoor concentrations, may indicate failure of the venting of a combustion appliance or leakage of auto exhaust into the building. b) Typical instrumentation cost is a few thousand dollars US.
10	◆ Alternative 2. Use of low cost carbon monoxide alarms.	a. Elevated indoor carbon monoxide, relative to outdoor concentrations, may indicate failure of the venting of a combustion appliance or leakage of auto exhaust into the building. b) Typical alarm cost is ~ US $100.
Ozone concentrations		
11	Alternative 1. Use real time electrochemical analyzer with data logger	a) May be useful in high-ozone cities to determine if energy conservation measures that change ventilation rates significantly influence indoor-outdoor ozone concentration ratios or time average indoor ozone concentrations. b) Indoor and outdoor concentrations are highly variable with time. c) Measurement equipment is relatively expensive, > US $6000, so measurement will often be impractical.
Particle concentrations		
12	Alternative 1. Real time particle counting using light scattering instruments.	a) Indoor particle concentrations may be compared to limits specified in standards or the influence of the energy conservation measure on the indoor-outdoor concentration ratio may be assessed. b) Outdoor air is a significant, sometimes dominant, source of indoor particles. c) Indoor and outdoor concentrations vary over time. d) Real—time instruments typically cost at least a few thousand dollars (US).
13	Alternative 2. Draw air samples through filters at constant and known rates and weigh filters using precision balance to determine collected particle mass. (ACGIH 1993)	a) Indoor particle concentrations may be compared to limits specified in standards or the influence of the energy conservation measure on the indoor-outdoor concentration ratio may be assessed. b) Outdoor air is a significant, sometimes dominant, source of indoor particles. c) Indoor and outdoor concentrations vary over time. d) Sampling instrumentation costs ~ US $1000 for a single measurement site. Precision balance may cost a few thousand dollars (US). Substantial sample handling costs.

Table 4: Specific M&V Alternatives for IEQ Parameters.

Bioaerosol concentrations		
14	Alternative 1. Use single or multi-stage impactor to collect samples on culture media, incubate and count and identify microbial colonies (ACGIH 1990, 1995b, 1995c)	a) Measurements are expensive; concentrations may vary a great deal with time while sample collection periods are less than 15 minutes. b) Measurement results depend on culture media and incubation conditions. c) Culture based methods do not detect non-culturable (e.g., dead) organisms that may elicit health effects. d) High level of expertise required.
Airborne volatile organic compounds		
15	◆ Alternative 1: Collect samples on solid sorbents and have samples analyzed in a laboratory for TVOC using gas chromotography with a flame ionization detector or gas chromotography—mass spectrometry (Hodgson 1995, ECA 1997)	a) See comments in *Section 3* on use of TVOC data. b) Cost of sampling equipment for a single measurement site is usually < US $1000. c) Analysis cost for TVOC is ~ US $100 per sample.
16	Alternative 2. Use sensitive (e.g., photo acoustic) infrared analyzer to measure TVOC concentration (Hodgson 1995)	Value of TVOC measurements made with infrared analyzers is uncertain because analyzer response varies depending on the mixture of compounds in the air.
17	Alternative 3: Collect samples on solid sorbents and have samples analyzed in a laboratory using gas chromotography — mass spectrometry for concentrations of several individual VOCs (e.g., quantify the most abundant VOCs or those from known sources) (Hodgson 1995, ECA 1997, ASTM 1994, 1994b, 1994c, 1995b, 1996, 1997, 1998b)	a) See comments in *Section 3* VOCs. b) Cost of sampling equipment for a single measurement site is usually < US $1000. Analysis cost for a set of 10 to 15 VOCs is ~ US $500 per sample.
Lighting parameters and satisfaction		
18	Alternative 1: Light intensity measurements made in accordance with existing standards (IES 1993)	a) One time measurements of lighting intensity (illuminance at task height) and uniformity in typical spaces. b) Instrumentation (light meter) is relatively inexpensive (~ US $200 to $400). c) Expertise required to determine which spaces are typical and to perform measurements correctly.
19	Alternative 2: Multi-parameter measurements made in accordance with existing standards (IES 1993)	a) Measurements of luminance distribution (obtainable with an image capture system) in typical spaces. b) Existing image capture systems are still in infancy. Some software is available to analyze images after capturing; however, existing analysis tools do not directly compute lighting quality parameters from luminance data. c) Measurements of direct glare from sources in field of view (both electric light and daylight). Requires luminance meter. Difficult to document exact location of luminance measurements, thus reproducibility is questionable. d) Some light sources (e.g., daylight) are time variant.
20	◆ Alternative 3. Occupant comfort or satisfaction with lighting assessed via a survey (Collins et al. 1990, Dillon and Vischer 1987, Eklund and Boyce 1995, Hygge and Lofberg 1998)	a) Often the least expensive method. b) Survey results can integrate over time. c) A well-designed survey can assist occupants in identifying particular sources of lighting problems.

Table 4: Specific M&V Alternatives for IEQ Parameters.

Occupant satisfaction with IEQ		
21	◆ Assess using survey (see *Section 8.5.5*)	a) See comments on surveys in *Section 8.5.5*. b) Survey results can integrate over time. c) High satisfaction with air quality does not assure that pollutants do not pose a health risk. d) Occupants sometimes attribute problems to an incorrect source or to unrelated environmental conditions (e.g., noise from computers attributed to lighting system ballasts).
Prevalence or severity of acute non-specific health symptoms		
22	◆ Assess using survey (see *Section 8.5.5*)	a) See comments on surveys in *Section 8.5.5*. b) Survey results can integrate over time. c) Low level of self-reported health symptoms does not assure that pollutants do not pose a health risk.

9 IMPLEMENTING THE GUIDELINE

To implement this guideline, the following actions are recommended:

1 Develop a general knowledge of IEQ through a review of *Section 4* and *5* or equivalent documentation.

2 For the energy conservation measures that will be implemented, use *Table 2, Section 7,* and supplemental information as necessary to determine: a) the potential impacts of the energy conservation measures on IEQ; and b) the associated precautions or mitigation measures.

3 Select a goal for IEQ M&V from *Section 8.2*.

4 Based on the energy conservation measures and goal, select an IEQ M&V approach from *Section 8.5*.

5 Assuming that the selected IEQ M&V approach is not Approach I (no IEQ M&V), select and implement an IEQ M&V alternative from *Table 4*. During implementation, utilize, as appropriate, steps 4-10 of the M&V procedure described in *Table 3*. If *Table 4* does not include an acceptable IEQ M&V Alternative, other alternatives may be developed and utilized.

6 Prepare and distribute written documentation of the IEQ M&V process that includes descriptions and justifications of important decisions and procedures plus a summary and interpretation of findings.

10 CONCLUDING REMARKS

Awareness of the significant influences of IEQ on the comfort, health, satisfaction and productivity of building occupants is increasing. The implementation of energy conservation projects in commercial buildings will often influence IEQ positively or negatively; therefore, IEQ should be considered during the selection and implementation of energy conservation measures. For many projects, IEQ problems are easily avoided through proper implementation of the energy conservation measures and the application of general knowledge about IEQ. In some situations, IEQ M&V is warranted to assure that IEQ remains acceptable or to quantify improvements in IEQ. This document serves to educate building energy professionals about the most relevant aspects of IEA and also provides guidance on IEQ M&V.

11 REFERENCES

1 ACGIH (1990) *Guidelines for the assessment of bioaerosols in the indoor environment, Publication 3180,* American Conference of Governmental Industrial Hygienists, Inc., Cincinnati, OH

2 ACGIH (1993)Aerosol *measurement: principles, techniques, and applications,* Klaus Willeke and Paul A. Baron, Eds. American Conference of Governmental Industrial Hygienists, Inc., Publication 9400, Cincinnati, OH

3 ACGIH (1998) 1998-1999 *Threshold limit values for chemical substances and physical agents and biological exposure indices.* American Conference of Governmental Industrial Hygienists, Cincinnati.

4 ACGIH (1995) *Air sampling instruments,* 8th Ed., Publication 0030, American Conference of Governmental Industrial Hygienists, Inc., Cincinnati, OH

5 ACGIH (1995b) *Bioaerosols,* Harriet Burge, Ed., Publication 9612, American Conference of Governmental Industrial Hygienists, Inc., Cincinnati, OH

6 ACGIH (I 995c) *Bioaerosols handbook,* Christopher S. Cox and Christopher M. Wathes, Eds., Publication 9611, American Conference of Governmental Industrial Hygienists, Inc., Cincinnati, OH

7 Andersson, K.; Bakke J.V., Bjorseth, O., Bornehag, C.-G., Clausen, G., Hongslo, J.K., Kjelhnan, M., Kjaergaard, S., Levy, F., Mothave, L., Skerfving, S., and Sundell, J. (1997) TVOC and health in non-industrial indoor environments: report from a Nordic Scientific Consensus Meeting at Larigholmen in Stockholm, 1996, *Indoor* Air 7(2): 78-91.

8 Arens, E and Baughman, A. (1996) Indoor humidity and human health, part 2: buildings and their systems, *ASHRAE Transactions 102(l):* 212-221.

9 Arens, E.A. Bauman, F.S., Johnston, L.P., and Zhang, H. (1991) Testing of localized ventilation systems in a new controlled-environment chamber, *Indoor Air](3),* pp. 263-281.

10 ASHRAE (1988) *ASHRAE/ANSI Standard 111-1988, Practices for measurement, testing, adjusting, and balancing of building heating, ventilating, and air-conditioning systems,* American Society of Heating, Refrigerating, and Air Conditioning Engineers, Inc., Atlanta.

11 ASHRAE (1989) *ASHRAE standard 62-1989, ventilation for acceptable indoor air quality,* American Society of Heating, Refrigerating, and Air Conditioning Engineers, Atlanta.

12 ASHRAE (1992) *1992 ASHRAE handbook, HVAC systems and equipment,* American Society of Heating, Refrigerating, and Air Conditioning Engineers, Atlanta.

13 ASHRAE (1992b) *ASHRAE/ANSI Standard 55-1992, Thermal environmental conditions for human occupancy,* American Society of Heating, Refrigerating, and Air Conditioning Engineers, Atlanta.

14 ASHRAE (1995) *1995 ASHRAE handbook, HVAC applications,* American Society of Heating, Refrigerating, and Air Conditioning Engineers, Atlanta.

15 ASHRAE (19 95b) *ANSI/ASHRAE Standard 55a-1995, Addendum to thermal environmental conditions for human occupancy,* American Society of Heating, Refrigerating, and Air Conditioning Engineers, Atlanta.

16 ASHRAE (1996) *1996 ASHRAE Handbook, HVAC Systems and Equipment,* American Society of Heating, Refrigerating, and Air Conditioning Engineers, Atlanta.

17 ASHRAE (I 996b) *ASHRAE Standard 62-1989 Ventilation for acceptable indoor air quality, public review draft,* American Society of Heating, Refrigerating, and Air Conditioning Engineers, Atlanta.

18 ASHRAE (I 996c) *Guideline 1-1996—The HVAC commissioning process,* American Society of Heating, Refrigerating, and Air Conditioning Engineers, Atlanta.

19 ASHRAE (1997) *1997 ASHRAE handbook, fundamentals,* American Society of Heating, Refrigerating, and Air Conditioning Engineers, Atlanta.

20 ASHRAE (1998) *ASHRAE Standard 129, measuring air change effectiveness, To* be published by the American Society of Heating, Refrigerating, and Air Conditioning Engineers, Inc., Atlanta.

21 ASTM (1994) *ASTM Standard D 5014-94, standard test methods for measurement of formaldehyde in indoor air (passive sampler methodology),* American Society for Testing and Materials, West Conshohocken, PA.

22 ASTM (I 994b) *ASTM Standard D 4947-94, standard practice for chlorodane and heptachlor residues in indoor air,* American Society for Testing and Materials, West Conshohocken, PA.

23 ASTM (1994c) *ASTM Standard D 4861-94a, standard practice for sampling and selection of analytical techniques for pesticides and polychlorinated biphenyls in indoor air,* American Society for Testing and Materials, West Conshohocken, PA.

24 ASTM (1995) *ASTM Standard E 741-95, standard test methods for determining air change in a single zone by means of a tracer gas dilution,* American Society for Testing and Materials, West Conshohocken, PA.

25 ASTM (1995b) *ASTM Standard D 5466-95, standard test method for determination of volatile organic compounds in atmospheres (canister sampling methodology),* American Society for Testing and Materials, West Conshohocken, PA.

26 ASTM (1996) *ASTM Standard D 5075-96E01 standard test method for nicotine and 3-ethenylpyridine in indoor air,* American Society for Testing and Materials, West Conshohocken, PA.

27 ASTM (1997) *ASTM Standard D 5197-97 standard test method for formaldehyde and other carbonyl compounds in air (active sampler methodology),* American Society for Testing and Materials, West Conshohocken, PA.

28 ASTM (1998) *ASTM Standard D 6245-98, standard guide for using indoor carbon dioxide concentrations to evaluate indoor air quality and ventilation,* American Society for Testing and Materials, West Conshohocken, PA.

29 ASTM (I 998b) *ASTM Standard D 6306-98, standard guide for placement and use of diffusion controlled passive monitors for gaseous pollutants in indoor air,* American Society for Testing and Materials, West Conshohocken, PA.

30 Awbi, H.B. (1991) *Ventilation of buildings,* Chapman & Hall, Northway, Andover, Hants, United Kingdom.

31 Baughman, A. and Arens, E (1996) Indoor humidity and human health, part 1: literature review of health effects of humidity-influenced indoor pollutants, *ASHRAE Transactions* 102(l): 193-211.
32 Bauman, F., Baughman, A., Carter, G., and Arens, E. (1997) A field study of PEM (Personal Environment Module) performance in Bank of America s San Francisco office buildings, CEDR-01-97, University of California, Berkeley (cedr@ced.berkeleyedu).
33 Bauman, F.S., Zhang, H., Arens, E.A., and Benton, C. (1993) Localized comfort control with a desktop task conditioning system *ASHRAE Transactions* 99(2), pp. 733-749.
34 Bearg, D.W. (1993) *Indoor air quality and HVAC systems,* Lewis Publishers, Chelsea, MI
35 BEIR VI (1998) *Health effects of exposure to radon: BEIR VI Committee on the Biological Effects of Ionizing Radiation (BEIR VI),* National Research Council, National Academy Press.
36 Berman, S.M. (1992) Energy efficiency consequences of scotopic sensitivity, *Journal of the Illuminating Engineering Society* 21(l): 3-14
37 Bluyssen, P.M., de Oliveira Fernandes, E., Groes, L., Clausen, G., Fanger, P.O., Valbjorn, O., Bernhard, C.A., and Roulet, C.A. (1996) European indoor air quality audit project in 56 office buildings, *Indoor Air* 6(4): 221-238.
38 Brightman, H.S., Womble, S.E., Girman, JR., Sieber, W.K., McCarthy, J.F., Buck, R.I., and Spengler, J.D. (1997) Preliminary comparison of questionnaire data from two IAQ studies: occupant and workspace characteristics of randomly selected buildings and complaint buildings, *Proceedings of Healthy Buildings/IAQ 97, vol.* 2, pp. 453-458, Healthy Buildings/IAQ 97 Washington DC
39 Bnmekreef, B.(1992) Damp housing and adult respiratory symptoms, Allergy 47:498-502.
40 Busch, J. (1992) A tale of two populations: thermal comfort in air-conditioned and naturally ventilated offices in Thailand, *Energy and Buildings* 18: 235-249.
41 California EPA (1997) *Health effects of exposure to environmental tobacco smoke, final report September 1997,* California Environmental Protection Agency, Sacramento, CA.
42 Carpenter, S.E. (1996) Energy and IAQ impacts of CO_2-based demand-controlled ventilation, To be published in *ASHRAE Transactions* 102(2): 80-88.
43 Charlesworth, P.S. (1988) *Air exchange rate and air-tightness measurement techniques—an application guide air infiltration* and Ventilation Centre, Coventry, Great Britain.
44 Cohen, T. (1994) Providing constant ventilation in variable air volume systems, *ASHRAE Journal* 36(5): 38-40.
45 Collins, B.L., Fisher, W., Gillette, G. and Marans, R.W. (1990) Second-level post-occupancy evaluation analysis, *Journal of the Illuminating Engineering Society* 19(2): 21-44
46 Committee on Health Effects of Indoor Allergens (1993) *Indoor allergens: assessing and controlling adverse health effects,* Pope, A.M., Patterson, R., and Burge, H., editors, National Academy Press, Washington, D.C.
47 Daisey, J.M., Hodgson, A.T., Fisk, W.J., Mendell, M.J., Ten Brinke, J.A. (1994) "Volatile organic compounds in twelve California office buildings: classes, concentrations, and sources," *Atmospheric Environment* 28(22), pp. 3557-3562.
48 Dales, R.E.; Burnett, R.; and Zwanenburg, H. (1991) Adverse health effects among adults exposed to home dampness and molds, *American Review of Respiratory Disease* 143: 505-509.
49 De Almeida, A.T. and Fisk, W.J. (1997) Sensor based demand controlled ventilation, LBNL-40599, Lawrence Berkeley National Laboratory, Berkeley, CA.
50 De Dear, R. and Brager, G.S. (1998) Developing an adaptive model of thermal comfort and preference, to be published in *ASHRAE Transactions* 104(l).
51 De Dear, R. and Fountain M. (1994) Field experiments on occupant comfort and office

building thermal environments in hot-humid climate, *ASHRAE Transactions* 100(2): 457-475.

52 Dillon, R. and Vischer, J. (1987) *Derivation of the tenant survey assessment method: office building occupant survey data analysis,* Public Works Canada, Ottawa.

53 Division of Respiratory Disease Studies, National Institute for Occupational Safety and Health (1984), Outbreaks of respiratory illness among employees in large office buildings—Tennessee, District of Columbia, *MMWR* 33(36): 506-513.

54 Drees, K.H., Wenger, J.D., and Janu, G. (1992) Ventilation air flow measurement for ASHRAE Standard 62-1989, *ASHRAE Journal* 34(10): 40-45.

55 ECA (1989) Report No. 4, *Sick building syndrome—a practical guide, European Collaborative Action Indoor Air Quality and its Impact on Man, COST* Project 613, Office for Official Publications of the European Communities, Luxembourg.

56 ECA (1992) *Report no. M: guidelines for ventilation requirements in buildings,* FUR 14441 EN, European collaborative action, indoor air quality & its impact on man, Office for Official Publications of the European Communities, Luxembourg.

57 ECA (1997) *Report 19, total volatile organic compounds (TVOC) in indoor air quality investigations,* European collaborative action, indoor air quality & its impact on man, Office for Official Publications of the European Communities, Luxembourg.

58 Eklund, N.H. and Boyce, P.R. (1996) The development of a reliable, valid, and simple office lighting survey, *Journal of the Illuminating Engineering Society* 25(2): 2540.

59 Elixmann, J.H., Schata, M., and Jorde W. 1990. "Fungi in filters of air conditioning systems cause the building related illness." *Proceedings of the Fifth International Conference on Indoor Air Quality and Climate,* Toronto, vol. 1, pp. 193-202. Published by the International Conference on Indoor Air Quality and Climate, Inc., 2344 Haddington Crescent, Ottawa, Ontario, KIH 8J4.

60 Emmerich, S.J. and Persily, A.K. (1997) Literature review on CO_2-based demand controlled ventilation, *ASHRAE Transactions* 103(2): 229-243.

61 EPA (1992) *Respiratory health effects of passive smoking: lung cancer and other disorders,* EPA/600/6-90/006F, US Environmental Protection Agency, Washington, DC.

62 EPA (1994) *A standardized EPA protocol for characterizing indoor air quality in large office buildings,* Office of Research and Development and Office of Air and Radiation, U.S. Environmental Protection Agency

63 EPA (1996) *Air quality criteria for particulate matter volume II of III,* EPA/600/P95/001bF, US Environmental Protection Agency.

64 EPA/NIOSH (1991) *Building air quality: a guide for building owners and facility managers, US* Environmental Protection Agency and National Institute for Occupational safety and Health, EPA/400/1-91/033, Superintendent of Documents, PO Box 37195, Pittsburgh, PA 15250

65 Fang, L.; Clausen, G.; and Fanger, P.O. (1997) Impact of temperature and humidity on acceptability of indoor air quality during immediate and longer whole-body exposures, *Proceedings of Healthy Buildings/*IAQ 97, vol. 2: 231-236, Healthy Buildings/IAQ 97 Washington DC

66 Fanger, P.O. (1970): *Thermal comfort analysis and applications in environmental engineering.* McGraw-Hill, New York.

67 Faulkner, D.; Fisk, W.J.; Sullivan, D.P.; and Thomas, J.M. Jr. (1998) Characterizing building ventilation with the pollutant concentration index: results from field studies, To be published in the *proceedings of IAQ and Energy 98,* October 24-27, New Orleans, ASHRAE, Atlanta

68 Federspiel, C.C. (1996) On-demand ventilation control: anew approach to demand-controlled ventilation. *Proceedings of Indoor Air 96, vol.* 3, pp. 935-940. SEEC Ishibashi, Inc., Japan.

69 Federspiel, C.C. (1998) Statistical analysis of unsolicited thermal sensation complaints in commercial buildings, *ASHRAE Transactions* 104(1): 912-923

70 Fisk, W.J. and deAlmeida, A.T.(1998) Sensor based demand controlled ventilation: a review, LBNL-41698, Lawrence Berkeley National Laboratory, Berkeley, CA, *to be published in Energy and Buildings.*

71 Fisk, W.J. and Rosenfeld, A.H. (1997) Estimates of improved productivity and health from better indoor environments, *InDoor Air* 7(3): 158-172.

72 Fisk, W.J. and Rosenfeld, A.H. *(1998)* Potential nationwide improvements in productivity and health from better indoor environments, *Proceedings of the ACEEE 1998 Summer Study on Energy Efficiency in Buildings,* Energy Efficiency in a Competitive Environment, August *23-28,* Asilomar, CA, pp. *8.85-9.97.*

73 Fisk, W.J., Faulkner, D., Sullivan, D., and Bauman, F *(1997).* Air change effectiveness and pollutant removal efficiency during adverse mixing conditions, *InDoor Air 7: 55-63*

74 Fisk, W.J., Mendell, M.J., Daisey, J.M., Faulkner, D., Hodgson, A.T., Nematollahi, M., and Macher, J. *(1993)* "Phase 1 of the California Healthy Building Study: a summary" *Indoor Air 3(4): 246-254.*

75 Fountain, M.E. and Huizenga, C. *(1996):* "A thermal comfort prediction tool," *ASHRAE Journal 38(9): 3942.*

76 Gammage, R.B. and Berven, B.A., eds. *(1997) Indoor air and human health,* second edition. Lewis Publishers, Boca Raton, FL.

77 Givoni, B. *(1997) Passive and low energy cooling of buildings,* John Wiley & Sons, New York

78 Hanley, J.T., Ensor, D.S., Smith, D.D., and Sparks, L.E. *(1994)* Fractional aerosol filtration efficiency of in-duct ventilation air cleaners, I*ndoor Air 4(3): 169-178.*

79 Heiselberg, P. *(1994)* Draught risk from cold vertical surfaces, *Building and Environment 29: 297-301.*

80 Heiselberg, P., Overby, H., and Bjorn, E. *(1995)* Energy-efficient measures to avoid downdraft from large glazed facades, *ASHRAE Transactions 101 (2): 1127-1135.*

81 Hodgson, A.T. *(1995)* A review and limited comparison of methods for measuring total volatile organic compounds in indoor air, *Indoor Air 5(4): 247-257.*

82 Hygge, S. and Loftberg, H. *(1998)* User evaluation of visual comfort in some buildings of the daylight Europe project, *Proceedings of the 4th European Conference on Energy Efficient Lighting, vol. 2, pp. 69-74,* DEE Congress Service, Coppenhagen

83 IES *(1993) Lighting handbook, reference & application,* M.S. Rea Ed., Illuminating Engineering Society of North America, New York

84 International Energy Agency *(1990) Demand-controlled ventilating systems: state Of the art review,* (editor W. Raatschen) Published by the Swedish Council for Building Research, Stockholm.

85 International Energy Agency *(1992), Demand controlled ventilation systems—source book,* Energy Conservation in Buildings and Community Systems Programme, Swedish Council for Building Research, Stockholm.

86 ISIAQ *(1996)* ISIAQ guidelines: Task Force II—general principles for the investigation of IAQ complaints (DRAFT -May *31,1996),* International Society for Indoor Air Quality and Climate, Ottawa (http://www.isiaq.org)

87 *ISO (1994) ISO 7730:1994 Moderate thermal environments—determination of the PMV and PPD indices and specification of the conditions for thermal comfort,* International Organization for Standardization

88 Jaakkola, J.K.; Hemonen, O.P.; and Seppanen, O. *(1991)* Mechanical ventilation in office buildings and the sick building syndrome. An experimental and epidemiological study, *Indoor Air 2: 111-121.*

89 Janu, G.; Wenger, J., and Nesler, C. *(1995)* Outdoor air flow control for VAV systems, *ASHRAE Journal 37(4): 62-68.*

90 Katzev, R. *(1992)* The impact of energy-efficient office lighting strategies on employee satisfaction and productivity, Environment *and Behavior 24(6): 759-778.*

91 Koenigsberger, Ingersoll, Mayhew, Szokolay (1973) *Manual of tropical housing and building: part 1, climatic design,* 320 pp., Longman Group Limited, London and New York.

92 Lagus, P. and Persily, A.K. (1985) A review of tracer gas techniques for measuring airflows in buildings, *ASHRAE Transactions* 91(2b): 1075-1087.

93 Martikamen, P.J., Asikainer, A., Nevabiner, A., Jantunen, M., Pasanen, P., and Kalliokski, P. (1990) "Microbial growth on ventilation filter materials" *Proceedings of the Fifth International Conference on Indoor Air Quality and Climate,* Toronto, vol. 3, pp. 203-206. Published by the International Conference on Indoor Air Quality and Climate, Inc., 2344 Haddington Crescent, Ottawa, Ontario, KIH 8J4.

94 Mendell, M.J. (1993) Non-specific health symptoms in office workers: a review and summary of the epidemiologic literature, *Indoor Air* 3(4): 227-236.

95 Menzies, D.; Tamblyn, R.; Farant, J.; Hanley, J.; Nunes, F.; and Tamblyn, R. (1993) The effect of varying levels of outdoor-air supply on the symptoms of sick building syndrome, *The New England Journal of Medicine* 328(12): 821-827.

96 Ministers (1970) *Law for maintenance of sanitation in buildings,* Ministers of Justice, Health and Welfare, Labor, and Construction, Japan

97 Molhave, L., Bach, B., and Pedersen, O.F. (1986) Human reactions to low concentrations of volatile organic compounds, *Environment International* 12: 167175.

98 Molhave, L., Lui, Z., Jorgensen, A.H., Pedersen, O.F., and Kjaergaard, S.K. (1993) Sensory and physiological effects on humans of combined exposures to air temperatures and volatile organic compounds, *Indoor Air* 3(3), pp. 155-169.

99 Morey, P.R. and Williams, C.M. (1991) "Is porous insulation inside an HVAC system compatible with a healthy building?" *Proceedings of IAQ '91: Healthy Buildings, pp.* 128-135. American Society of Heating, Refrigerating, and Air Conditioning Engineers, Atlanta.

100 Mudarri, D.; Hall, J.D.; and Werling, E. (1996) Energy cost and IAQ performance of ventilation systems and controls, *Proceedings of IAQ 96, pp.* 151-160, American Society of Heating, Refrigerating, and Air Conditioning Engineers, Atlanta.

101 Nagda, N. and Harper, J., (eds.),1989, *Design and protocol for monitoring indoor air quality,* ASTM STP 1002, ASTM, West Conshohocken, PA.

102 Nathanson, T. (1995) *Indoor air quality in office buildings: a technical guide,* A report of the Federal-Provincial Advisory Committee on Environmental Health, Health Canada, available from Communications Branch, Health Canada, Tunney's Pasture, Ottawa, Ontario, Canada KIAOK9

103 National Laboratories (1997) Survey and discussion of models applicable to the transport and fate thrust area of the Department of Energy Chemical and Biological Nonproliferation Program, available from the National Technical Information Service, US Department of Commerce, Springfield VA.

104 Nazaroff, WW., Gadgil, A.J., and Weschler, C.J. (1993) Critique of the use of deposition velocity in modeling indoor air quality, In *American Society of Testing and Materials Standard Technical Publication* 1205, ASTM, Philadelphia.

105 Nelson, N.A.; Kaufman, J.D.; Burt, J.; and Karr, C. (1995) Health symptoms and the work environment in four nonproblem United States office buildings, *Scand. J. Work Environ. Health* 21(l): 51-59

106 NKB (1991) *Indoor climate-air quality: NKB publication no. 61e,* Nordic Committee on Building Regulations, Espoo, Finland (ISBN 951-47-5322-4)

107 NORDTEST (1982) *Standard NT BUILD 232, buildings: rate of ventilation in different parts of a building,* Edition 2, NORDTEST, Espoo, Finland

108 NORDTEST (1988) *Standard NT VVS 019, buildings—ventilation air: local mean age,* Edition 2, NORDTEST, Espoo, Finland

109 NRC (198 1) *Indoor pollutants,* Committee on Indoor Pollutants, National Research Council, National Academy of Sciences, National Academy Press.

110 Owen, MK, DS Ensor, and LE Sparks (1992) Airborne particle sizes and sources found in indoor air, *Atmospheric Environment* 26A(I 2): 2149-2162

111 Persily, A. and Dols, W.S. (1990) The relation of CO_2 concentration to office building ventilation, *ASTM Special Technical Publication 1067-1990, pp.* 77-91, American Society for Testing and Materials, West Conshohocken, PA.

112 Persily, A., (1993) "Ventilation, carbon dioxide and ASHRAE Standard 62-1989," *ASHRAE Journal,* 3 5 (7): 40-44.

113 Preller, L.; Zweers, T.; Brunekreef, B.; and Boleij, J.S.M. (1990) Sick leave due to work-related complaints among workers in the Netherlands, *Proceedings of the Fifth International Conference on Indoor Air Quality and Climate, vol. 1:* 227-230, International Conference on IAQ and Climate, Ottawa.

114 Raterty, P.J. (1993) *The industrial hygienist's guide to indoor air quality investigations,* Stock # 144-EQ-93, American Industrial Hygiene Association, Fairfax, VA.

115 Schiller, G., E. Arens, F. Bauman, C. Benton, M. Fountain, and T. Doherty. (1988). "A field study of thermal environments and comfort in office buildings." *ASHRAE Transactions, Vol.* 94(2): 280-308.

116 Sepparren, O., Fisk, W.J., Eto, J., and Grimsrud, D.T. (1989) "Comparison of conventional mixing and displacement air conditioning and ventilating systems in U.S. commercial buildings." *ASHRAE Transactions* 95 (2): 1028-1040.

117 Sieber, W.K., Petersen, M.R., Staynor, L.T., Malkin, R., Mendell, M.J., Wallingford, K.M., Wilcox, T.G., Crandall, M.S., and Reed, L. (1996) Associations between environmental factors and health conditions:, *Proceedings of Indoor Air 96, vol.* 2, pp. 901-906. SEEC Ishibashi, Inc., Japan.

118 Skov, P., Valbjorn, O., Pedersen, B. V. and DISG (1989) "Influence of the indoor climate on the sick building syndrome in an office environment," *Scand J Work. Environ. Health* 16, pp. 1-9.

119 SMACNA (1993) *HVAC systems—testing, adjusting, and balancing,* Sheet Metal and Air Conditioning Contractors National Association, Chantilly, VA

120 SMACNA (1995) *IAQ guidelines for occupied buildings under construction,* Sheet Metal and Air Conditioning Contractors National Association, Chantilly, VA

121 Smedje, G.; Norback, D.; Wessen, B.; and Edling, C. (1996) Asthma among school employees in relation to the school environment, *Proceedings of Indoor Air 96,* vol. 1:611-616, Seec Ishibashi, Inc. Japan

122 Solberg, D.; Dougan, D., and Damiano, L. (1990) Measurement for the control of fresh air intake, *ASHRAE Journal* 32(l): 46-5 1.

123 Spengler, J., Neas, L., Nakai, S., Dockery, D., Speizer, F., Ware, J., and Raizenne, M. (1993) "Respiratory symptoms and housing characteristics," *Proceedings of Indoor Air '93,* The 6th International Conference on Indoor Air Quality and Climate, vol. 1: 165-170., Published by *Indoor Air* '93, Helsinki.

124 Stenberg, B., Hansson, Mild, K.H., Sandstrom, M., Sundell, J., and Wall, S. (1993) A prevalence study of sick building syndrome (SBS) and facial skin symptoms in office workers, *Indoor Air* 3 (2): 71-8 1.

125 Sundell, J. (1994) On the association between building characteristics, some indoor environmental exposures, some allergic manifestations and subjective symptom reports *Indoor Air Supplement 2.*

126 Ten Brinke, J. (1995) Development of new VOC exposure metrics and their relationship to sick building syndrome symptoms, Lawrence Berkeley National Laboratory Report, LBL-37652, Berkeley, CA.
127 U.S. Environmental Protection Agency (1996) Air quality criteria for particulate matter, volume II of III, EPA/600/P-95/00lbF.
128 Utterson, E. and Sauer, H.J. Jr. (1998) Outside air ventilation control and monitoring, *ASHRAE Journal* 40(l): 31-35.
129 Vedal, S. (1985) Epidemiological studies of childhood illness and pulmonary function associated with gas stove use, Chapter 23 in *Indoor Air and Human Health,* Proceedings of the Seventh Life Sciences Symposium, October 29-31, Knoxville, TN, Lewis Publishers, Chelsea, MI.
130 Veitch, J.A. [ed.] (1994) Full-spectrum lighting effects on performance, mood, and health, IRC-IR-659, pp. 53-111, National Research Council of Canada, Institute for Research in Construction, Ottawa, http://www.nrc.ca/irc/1ight
131 Veitch, J.A. and Newsham, G.R. (1998) Determinant of lighting quality 1: state of the science, *Journal of the Illuminating Engineering Society* 27(l): 92-106.
132 Veitch, J.A. and Newshain, G.R. (1998) Lighting quality and energy-efficiency effects on task performance, mood, health, satisfaction, and comfort, *Journal of the Illuminating Engineering Society.* 27(l),107-129.
133 Victor Olgyay, V. (1963) *Design with climate; bioclimatic approach to architectural regionalism.* 236 pp., Princeton University Press, Princeton, NJ.
134 Ward, G. and Rubinstein, F.M. (1988) Anew technique for computer simulation of illuminated spaces, Journal *of the Illuminating Engineering Society* 17(l).
135 Watson, D. and Labs, K. (1983) *Climatic design; energy-efficient building principles and practices,* 280 pp. McGraw-Hill Book Company, New York.
136 Weekes, D.M. and Gammage, R.B. (1990) The practitioner's approach to indoor air quality investigations, *Proceedings of the Indoor Air Quality International* Symposium, Stock # 145-EQ-90, American Industrial Hygiene Association, Fairfax, VA.
137 Weschler, C.J. and Shields, H.C. (1997) Is the importance of ventilation partially explained by indoor chemistry, *Proceedings of Healthy Buildings/IAQ 97, vol. 1,* pp. 293-298, Healthy Buildings/IAQ 97 Washington DC
138 Wilkins, A.J., Nimmo-Smith, L., Slater, A.I., and Bedocs, L. (1988) "Fluorescent lighting, headaches and eyestrain. " *Proceedings of the CIBSE National Lighting Conference,* Cambridge (UK), pp. 188-196.
139 World Health Organization (1987) Air quality guidelines for Europe, WHO Regional Publications, European Series No. 23, Copenhagen.
140 Wyon, D.P. (1992) "Sick buildings and the experimental approach," *Environmental Technology* 13: 313-322.
141 Wyon, D.P. (1996) Indoor environment effects on productivity, *Proceedings of IAQ 96, pp.* 5-15, American Society of Heating, Refrigerating, and Air Conditioning Engineers, Atlanta.
142 Yuan, X.; Chen, Q.; and Glicksman, L.R. (1998) A critical review of displacement ventilation, to be published in *ASHRAE Transactions 104(1).*

Appendix D

Resources/Links

Editor's Note: The following resources and links may be useful:

The Association of Energy Engineers
www.AEECenter.org

National Association of Energy Service Companies:
http://www.naesco.org/

IPMVP.org

Federal Energy Management Protocol—
www.eren.doe.gov/FEMP/Finance.html

The Alliance to Save Energy's Financing Energy Efficiency Links

http://www.ase.org/content/article/detail/1346 or just go to www.ase.org

Index

Printed in the United States
by Baker & Taylor Publisher Services